Plant–Animal Communication

H. Martin Schaefer and Graeme D. Ruxton

OXFORD

OXFORD

UNIVERSITY PRESS

Great Clarendon Street, Oxford OX2 6DP

Oxford University Press is a department of the University of Oxford.
It furthers the University's objective of excellence in research, scholarship,
and education by publishing worldwide in

Oxford   New York

Auckland   Cape Town   Dar es Salaam   Hong Kong   Karachi
Kuala Lumpur   Madrid   Melbourne   Mexico City   Nairobi
New Delhi   Shanghai   Taipei   Toronto

With offices in

Argentina   Austria   Brazil   Chile   Czech Republic   France   Greece
Guatemala   Hungary   Italy   Japan   Poland   Portugal   Singapore
South Korea   Switzerland   Thailand   Turkey   Ukraine   Vietnam

Oxford is a registered trade mark of Oxford University Press
in the UK and in certain other countries

Published in the United States
by Oxford University Press Inc., New York

British Library Cataloguing in Publication Data

Data available

Library of Congress Cataloging in Publication Data

Library of Congress Control Number: 2010942950

Typeset by SPI Publisher Services, Pondicherry, India
Printed in Great Britain
on acid-free paper by
CPI Antony Rowe, Chippenham, Wiltshire

ISBN 978-0-19-956360-9 (Hbk.)
      978-0-19-956359-3 (Pbk.)

1 3 5 7 9 10 8 6 4 2

To Veronika and Hazel

Preface

Communication is ubiquitous, and a key feature of life on every organizational level, from intracellular signalling to communication between individuals of the same and of different species. It has most likely evolved independently many, many times between vastly different partners. Although communication is central for the organization of life, communication theory is a concept that has been predominately developed in fields other than biology: namely economics, mathematics, and social sciences. Within biology, evolutionary biology has been the discipline that has contributed most to the conceptual growth of communication theory.

Given that communication often represents the first interaction among organisms, it can structure these interactions and thereby contribute to the organization of biodiversity. It is therefore perhaps surprising that communication theory has been less influential in ecology. In an elegant editorial to a special feature on flower volatiles in *Functional Ecology*, Robert Raguso (2009) laid out why he felt that sensory ecology is now deemed less important in mediating the ecology of pollination than it was one or two decades ago. The main reason for this perspective is that pollinator-driven selection may be overridden by that of herbivores or other selective agents, and that these selective pressures constrain the adaptations of floral phenotypes to communication. Is writing a book on plant–animal communication in 2011 then a day late and a dollar short?

We feel that communication between plants and animals is currently a particularly exciting topic. Plant–animal communication is fascinating because it provides a particularly compelling example of the hidden complexity of seemingly simple questions. If we ask how and why communication evolves, the answers are often far from obvious because evolution works economically on a limited number of traits. The same traits may be adapted for multiple purposes, both within a species and between species. The traits that plants use to communicate with animals provide wonderful examples for the economic nature of evolution. Anthocyanins, for example, are versatile pigments that serve various functions in plants. In fruits and flowers the adaptive significance of anthocyanins is traditionally attributed to the attraction of pollinators and seed dispersers, but these pigments are also involved in defence against abiotic factors and pathogens (Strauss and Whitall 2006, Schaefer et al. 2008a). In vegetative parts, the adaptive significance of anthocyanins is generally attributed to their function as antioxidants directly reducing oxidative stress (Gould et al. 2002); as sunscreens protecting photolabile defensive compounds (Page et a. 1999) and the photosynthetic apparatus against excess light, and thereby reducing the risk of free radical production under unfavourable conditions (Gould and Lee 2002, Gould 2004); to increase cold hardiness (Chalker-Scott 1999); and as defence against herbivorous insects (Archetti 2000, Hamilton and Brown 2001, Hagen et al. 2003). Anthocyanins can thus be involved in the amelioration of plants' responses to various environmental stressors, as well as with the attraction of animal mutualists and the defence against various plant antagonists both in vegetative and reproductive organs.

Coming back to our question on how communication evolves, it is not a trivial task to trace the evolution and original function of anthocyanins. From a purely adaptionist perspective, we might assume that anthocyanins evolved for fulfilling the functions that they currently perform in each organ.

This is unlikely, however, as this view neglects developmental, morphological, physiological, and genetic constraints. Gould and Lewontin (1979) summed up this view as the adaptationist programme or the Panglossian paradigm – where natural selection is assumed to be near omnipotent in fashioning the best of all possible worlds. Moreover, assuming that traits such as anthocyanins evolved to fulfil their current function requires multiple independent origins of these pigments even within the same organs where they fulfil various purposes at the same time. Despite a natural human tendency to want to believe in simple and often beautifully elegant answers for the evolution of complex traits, we are faced with a more difficult but also a more rewarding task: disentangling cause and effect throughout the evolution of these and many other plant traits. Thus, if we ask whether anthocyanins in fruits primarily fulfil a signalling role (i.e., they evolved for attracting mutualists) or whether anthocyanins in leaves evolved as *signals* to deter antagonists, we have to evaluate the fitness effects of one function (say, signalling to animals) relative to the fitness effects of other functions (*photoprotection*, anti-oxidation). Though challenging, this pluralistic approach, encompassing different fields in biology, is more rewarding and gives a more realistic view on the evolution of complex traits than a purely adaptionist perspective.

To exemplify this situation, suppose that anthocyanins evolved first to ameliorate plants' stress responses to abiotic factors and are later used to communicate to animals. In this case anthocyanins are an *exaptation* that evolved for one reason and that has been co-opted throughout evolutionary history to serve a different function, signalling. Similarly, anthocyanins might predominantly function in stress response against an abiotic factor while, at the same time, communicating to animals. In this case, anthocyanins might best be described as a *cue*, not a signal. As we discuss in Chapter 1, the term 'signalling' can usefully be reserved for situations where a trait predominantly functions to communicate to animals.

However, the example of anthocyanins highlights that – in order to understand how plants communicate to animals – we need to evaluate the significance of communication *relative* to the significance of other, non-communicative functions. Only such an integrative approach will allow us to examine questions as to whether (and which) plant traits communicate to animals and benefit both parties (and are thus signals) or whether animals often attend to cues provided by plants. Focussing on the distinction between signalling and cues is a rewarding research avenue for students of communication. However, we feel that the study of plant–animal communication is even more broadly interesting, because it elucidates a central problem: the evolution of complex traits by multiple, often varying or even opposing, selective pressures.

We therefore view the question of how communication arises in the context of multiple selective pressures as an intellectually stimulating question that has not lost its appeal by the insight that non-sensory interactions are important and can often be predominant. On the contrary, as we will lay out in the book, information sometimes arises in plant–animal interactions without being selected for. It is quite likely that 'unintentional' information is more reliable than 'intentional' signalling; a situation that seems counterintuitive at first. We therefore advocate a broad approach to the study of communication.

Based on the above, we conclude that the study of plant communication is not only an ecological and evolutionary endeavour, but one that also involves plant physiology, development, and biochemistry as well as biophysics and considerable knowledge of animal behaviour, their sensory physiology and *perception*. The integrative nature of plant communication originates from two facts. First, the study of communication is always truly integrative, primarily because communication is *per se* a topic concerned with the actions of different partners. Second, the integrative nature of communication is even more apparent when studying how disparate partners, such as plants and animals, interact with distinct physiological processes and pathways involved in the communication system from stimuli production to stimuli perception. The study of communication and signalling has been fraught with problems involving the definition of terms. You might reasonably be concerned that we might not define 'communication' or 'signal' in the same way that you would, or in the way that other texts use these terms. Hence we will begin Chapter 1 by explaining how we define these terms, and thus how they will be used throughout the book.

We should be clear from the start about what this book covers, and what it does not. We adopt a very wide definition of communication, but we are interested in communication (involving stimulation of the sensory systems of animals) rather than the broader subject of 'interactions' between plants and animals. For example, we are interested in questions of how squirrels detect nuts in their local environment, and how they can use their senses to gain information on the relative qualities of different nuts as food. We will not cover how squirrels digest nuts, or how the availability of nuts structures squirrel populations. We are interested in the evolutionary ecology of how traits of plants differentially stimulate animal sensory systems. We touch only very briefly on plant responses to neighbouring plants, which may sometimes usefully be described as plant–plant communication. We similarly do not consider in full the many plastic responses that plants can show to damage by herbivores, some of which could be described as involving within-plant communication. However, we do consider a broad range of ecological processes that involve flexible animal behaviour in response to excitation of their senses by plant traits. These include pollination and seed dispersal by animals, trophic exploitation of plants by animals, and (less usually) trophic exploitation of animals by plants. Of these processes, the most intensely studied has been pollination. Because there have been several excellent, recent books on pollination we have sought to avoid simple repetition of these works. Thus, we do not attempt a comprehensive review of the evolutionary ecology of pollination, but rather take a more comparative approach and look for general points of similarity and difference between how communication influences pollination and how it influences the other plant–animal interactions that we consider here, particularly seed dispersal. Another area where we admit to failing to provide detailed treatment is in belowground influences of plants on subterranean animals and mycorrhiza. We likewise do not cover how plants interact with viruses and bacteria even though these may sometimes influence traits that interact with the senses of animals. The reason here is the opposite of the situation regarding pollination, below-ground plant–animal communication is very poorly known.

Although we are primarily interested in the evolutionary ecology of plant–animal communication, we must sometimes briefly consider selection pressures on plant traits that may restrict or dominate selection pressures related to communication to animals. This will involve brief consideration of plant biochemistry and physiology, but these fields are very much not the focus of this book.

The diversity of fields involved in communication does not require complete mastery in each. Instead, depending on which step in communication is under study, different aspects might become more important than others. To facilitate an encompassing view on plant–animal communication, we first use communication theory to explore whether (and, if so, how) communication structures the interactions between plants and animals. To attain an evolutionary perspective, we will elucidate how interactions among multiple species affect evolutionary dynamics in plant–animal interactions in general, and those of plant–animal communication in particular, in Chapter 1. We then present basic information on plant biochemistry and the sensory world of animals that commonly interact with plants in Chapter 2. In subsequent chapters, we will be mainly concerned with two types of questions. The proximate question: *how does communication between plants and animals work?* and the ultimate question: *did the sensory interactions between plants and animals shape plant traits?* In other words, we ask whether communication between plants and animals is primarily adaptive or whether it is a by-product of other selective pressures. We address both questions by discussing how plants and animals interact in various communication systems: ranging from plants eating animals (carnivory, Chapter 10), to the more common phenomenon of animals eating plants (herbivory, Chapter 9), to animals eating plant rewards (pollination and seed dispersal, in exchange for dispersing their genes (Chapters 3–6)). Chapter 7 discusses the various selective pressures shaping non-green-leaf colouration (including communication), whereas Chapter 8 considers camouflage, *aposematism*, and *mimicry* in the plant kingdom.

As mentioned above, pollination is only the focus of a single chapter because it has already been effectively reviewed from several perspectives in recent years (e.g., Chittka and Thompson 2001, Harder and Barrett 2006, Raguso 2008, Rausher 2008). The reason for our disproportionate focus on communication in herbivory, seed dispersal, and carnivory is that we view our book not as providing definitive and all-encompassing answers to the proximate and ultimate questions we pose. Instead the chapters aim to delineate the unknown and propose research activities that might help reduce the gaps in current knowledge. Also, we agree with Raguso (2009) that our understanding of pollination biology will be enhanced by increased emphasis on other plant–animal interactions. A key theme that emerges in this book is the shaping of plant traits by simultaneous selection pressures from both mutualists and antagonists. In order to develop our arguments, we must be very clear about what we mean by particular key concepts and terms that we use when laying out our arguments. Defining these key concepts is the first task of our introductory chapter.

Acknowledgements

Reviewing book chapters cannot be high on the agenda of any scientist; we are therefore particularly grateful to the following colleagues and friends for feedback on draft chapters and discussions that helped clarify ideas that are included in this book: Aaron Ellison, Kevin Gould, Andrew Higginson, Richard Karban, Doug Levey, Simcha Lev-Yadun, Silvia Lomáscolo, Rodrigo Medel, Daniel Osorio, Helen Ougham, Stefan Porembski, Robert Raguso, Hannah Rowland, Diego Santiago-Alarcon, Thomas Schmitt, Mike Speed, Sharon Strauss, Tom Sherratt, Alfredo Valido, Dave Wilkinson, and Douglas Yu. We apologise to anyone who we have accidentally omitted from this list, and we assure all those on the list that in no sense are they responsible for any errors and omissions that may have made it into the final text – all such mistakes we will blame on each other!

We are extremely grateful to Liz Denton who drew all the figures. Her care, attention to detail, clear thinking, and organizational skills are surpassed only by her patience and good nature.

We are very grateful to Helen Eaton at OUP for her support and advice during the preproduction stage, and to Muhammad Ridwaan and Abhirami Ravikumar for help in turning our submitted manuscript into a book. Lastly, but most importantly, we thank our wives for suffering the consequences of the time investment required to write this book, but more generally for their unflagging support and forbearance.

Contents

CHAPTER 1

Communication and the evolution of plant–animal interactions

'…we shall have reason to believe that the agency of insects has been most powerful in developing the hues of blossoms; while the [colours of] fruits, as we shall see, are rather due to the selective actions of birds and mammals.'

Allen (1879)

'…the evidence that fruits in general have been modified both in edibility and attractiveness in relation to the animals which feed upon them, is by no means so clear as in the case of flowers.'

Wallace (1879)

1.1 Communication

Communication is a ubiquitous characteristic of life. It occurs on every organizational level, from communication of cell organelles within cells to communication between organisms of the same or of different species. The ubiquity of communication demonstrates its key role in the organization of life. Although communication is central for the functioning of biological systems and although it is omnipresent in our everyday lives, a biological definition of communication has proved surprisingly difficult. In its colloquial sense, communication is seen as the transfer of information from a sender to a receiver where the receiver subsequently uses that information when deciding how to respond. This is the definition most behavioural ecologists adhere to. However, philosophers have long stressed that information is difficult to define in the context of communication. Technically speaking, information reduces the uncertainty of the receiver about

alternate states of the sender (Shannon and Weaver 1949). A trait or behaviour that evolved to communicate to others can obviously reduce the uncertainty of the receiver. Although the technical definition by Shannon and Weaver is precise, it is rarely practical. When attending to communication the uncertainty of the receivers is reduced in countless ways, many of which are not of interest to the receiver (Scott-Phillips 2010). For example, upon seeing ripe fruits, a frugivorous animal perceives the information that it might be interested in (fruits are present), but also others which it may not be interested in (the plant presenting fruits is alive, has a specific shape and so on). One challenge in defining information in the context of communication is thus to differentiate between relevant and irrelevant information for the receiver. Relevant information will only be a subset of the available information.

Narrowing the focus to relevant (or functional) information does not solve the dilemma that the term 'information' poses. The information a receiver deems relevant may change depending on its particular interests, motivation, experience, and physiological state. For example, female bower birds make age-specific decisions when assessing male display traits (Coleman et al. 2004); they thus likely differ in the information they extract and in how they value this information. Differences in the information that individuals of different species extract from a given trait are likely to be even more pronounced than the differences that conspecifics extract. To illustrate this point, let us consider saguaro fruits from the giant cactus (*Carnegiea gigantea*), which represent a key resource in the dry environment of the Sonoran desert in southern Arizona. Plants are expected to evolve traits that should enhance seed dispersal. Thus, the information conveyed by the fruits can be 'I am good to eat'. Two dove species consume saguaro fruits but for different reasons: one acquired both nutrients and water from the pulp and the seeds, while the other only acquired nutrients from the seeds (Wolf et al. 2002). The information both doves extract from seeing fruits thus differs according to the nutritional rewards they are interested in. That is, one dove perceives 'there is a good source of nutrients and water' whereas the other perceives 'there is a good source of nutrients'. Furthermore, both dove species are seed predators that are detrimental to the fitness of the cactus, which is why the plant is certainly not selected to communicate to them 'I am good to eat'. This example illustrates that information is not an inherent property of the communicative trait of the sender, but arises from the interpretation of traits by the receiver.

A logical challenge to defining communication in terms of the specific information that it conveys is thus that the same communicative trait can convey very different information to distinct receivers. This is a particular problem in the communication among multiple species that we are interested in. Moreover, we can usually not assess the interpretation of an animal. Some researchers have thus emphasized that definitions based upon information are often not very pragmatic for studying communication and should be abandoned (Rendall et al. 2009, Scott-Phillips 2010, but see Carazo and Font 2010, Seyfarth et al. 2010, Scarantino 2010). Consequently, there are both practical and logical challenges associated with focusing on identifying the information that the receiver gains from the communication.

The principle reason for us not to base our definition of communication upon information is that in evolutionary terms it is not the information that a receiver has (or lacks) that matters, but its response to a communicative trait. The fitness of an individual will be determined by its actions, and not by changes in the informational state and its underlying cognitive processes. Only if changes in the information state lead to changes in behaviour will they be evolutionary relevant. Thus, even if we do not know which information is extracted and processed

during communication, we can measure the efficacy of cues and signals from repeated observations of the reaction of the same and/or different receiving individuals.

To clarify this argument, we use the example of pollination. What matters for the evolution of the interaction between a plant and its pollinator is how the pollinator reacts to the stimulation of its senses by different trait values of the plant. The trait values of concern can be morphological and chemical; interacting with the visual, olfactory, acoustical, and gustatory receptors of pollinators. For certain, there are interesting questions about the nature of the black box that is the decision-making apparatus of the pollinator, but we feel that these are questions for the comparative psychologist or perhaps for the philosopher. The evolutionary ecologist can focus on the same thing that evolution does, the link between stimuli and response. The mechanisms that underlie a particular stimuli-responses relationship of pollinators need not concern us, as long as we can characterize the relationship accurately.

In order to circumvent the challenges laid out above and in order to account at the same time for the great diversity of plant–animal sensory interactions, we adopt a very broad definition of communication.

*There is **communication** between one individual (the sender) and another individual (the receiver) if trait values of the sender stimulate the sensory systems of the receiver in such a way as to cause a change in the behaviour of the receiver (compared to a situation where the trait values of the sender where different).*

Note that this broad definition does not require a communicative intent from the sender. Although the word 'information' does not appear in this definition, it is implicit: the change in the behaviour of the receiver is driven by a change in its informational state. The important issue is that we are very interested in describing the detail of changes in behaviour and how they relate to evolution, but we do not attempt to be specific about the exact nature of the information held by the receiver before and after the communication occurs, only that it changes and that that change produces a change in behaviour that we can define fully. Thus, we make no judgement on the emotional state or level of consciousness of the receiver.

Notice the key role of the sensory systems in communication. If a bee is dissuaded from landing on a flower because of the way it smells, then this is communication; but if the bee is impeded from accessing the flower's nectar by the physical structure of the flower, then this is not communication. Notice also that this definition makes no assumptions as to whether the sender or the receiver benefits from the communication. It thus differs from definitions of communication that have defined communication as an act of manipulating others (Dawkins and Krebs 1978, Rendall et al. 2009). However, it is useful to define different subsets of communication according to costs and benefits.

1.2 Signals vs. cues

The simplest form of communication is inadvertent communication, where traits of an organism can be used by receivers for their own benefit. These traits are called cues. Cues are necessarily informative, otherwise receivers would not respond to them (Maynard Smith and Harper 1995). If inadvertent communication is selectively neutral to the sender, we would not expect it to evolve towards increasingly stimulating the receiver. If inadvertent communication is detrimental to the fitness of the sender, we would expect selection for sender traits that reduce the effectiveness of communication. In such situations, we must explain why cues continue to exist. For example, many biting insects find mammalian hosts by cuing into the carbon dioxide that mammals emit as a by-product of respiration. Here, carbon dioxide is a cue that imposes a cost on the sender. The reason that the cue still exists despite its communicative function imposing a cost on the sender, is that there are counter-balancing factors unrelated to communication that maintain selection pressure for carbon dioxide production. In the extreme case, if the mammal did not respire, it would die. In the less extreme case, if it has a lower respiration rate then likely the communication based cost would decline but there would be even greater costs associated with reduced metabolism

that have a stronger influence on selection on carbon dioxide production.

If inadvertent communication benefits the sender, the cue evolves to become a signal unless constraints prevent changes that would increase the effectiveness of communication. A long-standing hypothesis is indeed that signals evolve from cues as precursors through a process of ritualization that gradually emphasizes traits and behaviours that increase the effectiveness of communication (Tinbergen 1952, Steiger et al. 2011). In our definition of signals we closely follow Maynard Smith and Harper (1995).

> We define **signalling** as a subset of communication where the change in behaviour of the receiver that is consequent to detection of particular trait values of the sender benefits both the sender and the receiver (on average) and where the trait values evolved for that effect. In this case, the specific values of the traits of the sender (as they are detected and perceived by the receiver) is a **signal**, which is an act or structure that alters the behaviour of another organism, that evolved and/or is maintained through selection (at least in part) because of that effect, and which is effective because the receiver's response has also evolved.

There are several aspects of these definitions that merit clarification. Firstly we see signals as a subset of a wider group of acts of communication, that we call cues. The key differences between signals and cues are (i) only signals evolved to communicate and (ii) the benefits derived from communication differ. Signals benefit both parties (although benefits are often not symmetrically distributed among senders and receivers), whereas cues usually only benefit the receiver. This is because if cues also benefit the sender, signalling is likely to evolve. For example, the carbon dioxide that attracts mosquitoes to their mammalian hosts is a cue, not a signal, because it did not evolve for that attractive function and because it only benefits the receiver.

By 'on average' we mean that the reaction towards stimulation by signals needs not to benefit both parties on every occasion, but must do when averaged over a number of essentially identical occasions. For example, floral scent and *colours* are signals if pollinators use and select them to locate and identify flowers in a way that allows effective foraging by the pollinator and effective gene transport for the plant. A given floral scent is a signal even if it sometimes is not associated with a food reward (because floral rewards have been recently depleted by other foragers), as long as the scent is associated with a reward sufficiently frequently that the pollinator benefits on average from responding as it does to the signal. Consequently, a signalling system between flowers and pollinators can be destabilized by permanently non-rewarding flowers if they have similar scents or colours to rewarding flowers and are more common than these. This is because now the pollinator would not benefit on average from responding to floral signals. The issue about 'as they are detected and perceived by the receiver' is that the signal is determined by the inherent trait values of the sender but also by the ability of the receiver to sense and evaluate these trait values. To take a trivial example, the UV reflectance of a fruit is not particularly contrasting to humans and other primates that cannot perceive UV, but it is usually contrasting to animals perceiving UV.

The distinction between cues and signals is obvious when examining the selection pressure upon the communicative trait. Because a signal benefits the sender as well as the receiver, we would expect that the form of the signal has been influenced by selection to improve the efficiency of the signal. That is, there is selection for trait values that improve the ability of the receiver to detect and correctly evaluate the signal.

In general, communication with a particular type of receiver will not be the only selection pressure influencing the communicative trait of senders. Trait values will be influenced by communication with other types of receivers and by selection pressures that have nothing to do with communication. This leads to one final useful distinction that we would like to make between signals and cues. For a signal, the trait values of the sender will be influenced substantially by their function in communication (relative to other selection pressures on those traits), and that selection should change the traits in ways that enhance the efficacy of communication. Efficacy is measured as the probability that a received signal will change the behaviour of the receiver and thereby increase the fitness of the sender (Endler 2000). One

important way to increase efficacy is to increase the detectability of a signal, that is the spatial range over which it can be perceived and the ease of identifying and perceiving it. In comparison, for a cue it is possible that the trait values are entirely set by other selection pressures, and are unaffected by selection imposed by their effect on communication. This can be the case, but need not be. However, if the communicative function of the cue does impose significant selection pressure on the trait values of the sender, this selection should act to decrease the effectiveness of communication.

Signals are also often distinguished from indices, which are a particular type of cue. An *index* is a trait that is informative because it is necessarily related to a given quality (Maynard-Smith and Harper 1995). An animal's body size is an index if it is inevitably connected to fighting ability, and thus informative for putative opponents, but size generally does not evolve substantially in response to selection pressure to convey information to opponents (Maynard-Smith and Harper 1995). Most importantly, indices are honest by definition, they are unfakable. A small animal simply can only appear to a limited extent to be bigger than it actually is by raising hairs and adopting specific postures. Likely examples of indices are also found in plants. Let us come back to anthocyanins as plant pigments – the example we laid out in the preface. Many fruits are coloured by anthocyanins, which are at the same time important dietary antioxidants for fruit consumers. If fruit colour is necessarily linked to the anthocyanin contents of a fruit, and a bird is selecting fruits owing to the antioxidant properties of anthocyanins, fruit colour is an index rather than a signal of anthocyanin contents. However, a plant with low anthocyanin contents might be selected to produce other pigments that mask the anthocyanin colouration and hence reduce the bird's ability to utilize colour as an index. Depending on fitness benefits of the sender, there can thus be selection for communication-related traits that either enhance or reduce the ability of the receiver to accurately evaluate indices. Selection for masking indices is possibly not very common however, since birds might reasonably interpret an unwillingness to be honest about anthocyanin contents as an admission that these are poor. CO_2 emission can be an index in plants because it is necessarily associated with respiration rates. This association may make floral CO_2 emission best seen as an index, rather than a signal, of nectar production in *Datura wrightii* where CO_2 emissions peak in the first 30–60 minutes of floral opening and coincide with peak nectar availability (Guerenstein et al. 2004, Goyret et al. 2008). A common topic throughout this book will be to consider the evidence that plants' sensory traits evolved as signals, as cues, or indices in plant–animal interactions. Finally, our very broad definition of communication includes *deceptive* communication which differs from communication based upon cues, signals and indices (see Box 1.1).

Now that we have defined communication and made a distinction between communication based upon cues and communication based upon signals, we will turn more specifically to introduce plant–animal communication (Section 1.3). Plant–animal communication is tied into the larger context of *coevolution* between plants and animals as disparate partners. In order to provide the backbone for the evolution of communication among plants and animals, we will continue this chapter with an examination of the evolutionary foundations that underlie the interactions between plants and animals (Section 1.4). To illustrate how the evolutionary concepts apply to the book, we will use communicative traits as examples of the relative

Box 1.1 Deceptive communication

Our very broad definition of communication involves situations where the sender but not the receiver benefits (on average) from the behaviour of the receiver in response to sensory stimulation caused by specific traits of the sender. Since the receiver does not benefit from this communication (and indeed pays a cost), the communication is neither a signal, a cue nor an index. Rather we describe this as deceptive communication.

One such situation is *crypsis* (often called camouflage). We consider this communication because if the trait values of the cryptic sender that stimulate the sense organs of the receiver were different, then its likelihood of being detected by the receiver would be different and the receiver's subsequent behaviour would be different (on average). If we take the example of camouflaged prey, then if the predator detects it the predator attacks and consumes the prey; but if the predator fails to detect camouflaged prey then the prey survives. Thus we consider that there is communication between a camouflaged individual and an individual that (i) approaches near enough that it could detect the camouflaged individual if the sender had different traits that interact with the senses of the receiver and (ii) would respond to the camouflaged individual if it detected it. Readers should note, however, that most other published definitions of communication are less broad than ours and generally exclude crypsis.

Another situation where the sender benefits from communication and the receiver experiences a cost is where the traits of the sender make it more likely that the receiver will misidentify or misclassify the sender. An example of this would be those orchids that achieve pollination by duping male bees to copulate with them. We consider this again to be deceptive communication. The orchid benefits and the bee pays a cost (at least an opportunity cost in time wasted) because traits of the orchid stimulate the senses of the bee in such a way as to increase the likelihood that the male bee misidentifies the orchid's flower as a female bee.

For all deceptive communication, if the receiver pays a cost because of its response to the communication, then we must explain why the receiver has not evolved to ignore the communication or respond differently to it. We discuss this issue of the evolutionary maintenance of such communication in Chapter 8.

importance of generalization vs. specialization in interactions (Section 1.5) and how generalization leads to diffuse co-evolution among multiple interacting species (Section 1.6). In Section 1.7 we discuss correlated evolution, before aiming to synthesize the implications of the topics in the previous sections for the evolution of communication (Section 1.8). In the final sections we quickly turn to the genetic architecture of communicative traits and other architectural constraints (Section 1.9), before describing some of the most pressing issues in the final perspective (Section 1.10).

1.3 Plant–animal communication

The diversity of plant colours and odours probably fascinated humans long before the history of writing and long before the start of modern science. It seems intuitively convincing that immobile plants can increase their fitness by attracting (and thereby communicating with) the animals they depend on for dispersal of pollen and seeds. Pollinators and seed dispersers in turn can forage more effectively – i.e., find nutritional flowers and fruits more quickly – if plant colours and odours are matched to their sensory abilities. This occurs obviously if sensory traits, such as odour and colour, are fine-tuned to the sensory abilities of the animals so that they stand out from the environment. The adaptive framework of signalling is even more appealing because of our own inherent biases. It is a long-standing hypothesis that colour vision, the sense that we humans rely upon most for evaluating our distant environment, evolved as an adaptation to locate colourful fruits against a predominantly green background (Allen 1879; see Box 4.1). As such, assuming co-evolutionary dynamics to occur – and to result in fine-tuned communication between plants providing energy rewards and the sensory systems of mutualistic animals that rely upon these rewards – seems an entirely plausible idea.

Studying communication between plants and animals is much more fascinating than the dry description of possible co-evolutionary processes between these partners indicates. This is because plant–animal communication may significantly contribute to the development of communication theory. Communication theory is mainly an evolu-

tionary concept that is interested in explaining the dynamics between senders and receivers leading to the specific design and content of signals. So far, communication theory has mainly been advanced by studies examining communication in the context of social interactions and mate choice (Otte 1974, Maynard Smith and Harper 1995, Hauser 1997, Bradbury and Vehrencamp 1998, Searcy and Nowicki 2005). It is a very useful concept to apply to plant–animal interactions because this theory has generated a number of important concepts such as sensory exploitation as a model for signal evolution and the advantages of multi-modal signalling that may well pertain also to the communication between plants and animals (see Section 8.4.1 for exploitation in flowers and Box 11.1 for multi-modal communication). While many of these concepts can probably be transferred to communication among species, various ecological factors in the interactions among multiple species likely play a more important role in their communication.

Theoretically, communication is commonly treated as a two-agent sender–receiver game resulting in specific signalling equilibria that either result in *honest information* or not. However, in the real world, communication usually takes place among multiple agents because it can be eavesdropped by competitors (such as conspecifics competing for mates) and heterospecifics such as predators. For example, it is well known that predators can eavesdrop on intraspecific communication and that the design of epigamic signals, which indicate the communicative intent of potential mates, represents a shifting balance between sexual selection and predator avoidance (Endler 1980). Because communication between plants and animals commonly involves many receivers and senders, it is a superb example of communication taking place between multiple players.

Studying plant–animal communication can advance communication theory because it can be a good model to understand the evolution of complex communication networks. If such networks among multiple players are highly adapted, they provide a functional understanding for complex tasks such as group cohesion and teamwork (Skyrms 2009). Communication among plants and animals differs from such signalling networks in that communicative traits in plants interact at the same time with a multitude of species that vary distinctly in their sensory and cognitive abilities and that consequently also vary distinctly in their selective pressures upon plants. It is this aspect that differentiates plant–animal communication most from communication among animals. Moreover, in contrast to signalling networks that lead to teamwork, the conflict of interests is likely greater among the many animals responding to plant communicative traits. This characteristic of plant–animal interactions has important ramification for the evolutionary dynamics of communication.

Before we discuss the evolutionary dynamics, we would like to point out some specific characteristics of plant–animal communication. Plant–animal communication is suitable for studying communication because at least visual and gustatory traits used by plants to communicate are comparatively simple and invariant in space and time (compared to animal signals; Schaefer et al. 2004). Moreover, the correlation between communicative traits in plants (the conditional stimuli to receivers) and the nutritional quality of plant rewards (as the unconditional stimuli) is also relatively simple to measure. It is simple to measure because animals' interest in plant rewards is so invariant that we can develop strong predictions on how they should select plant traits. Although many animals have strong and consistent preferences for higher nutritional contents in their food, there is variability among animals in the nutrients they select. For example, European starlings (*Sturnus vulgaris*) differ from many other birds in their lack of sucrase which is why they cannot digest sucrose (Martinez del Rio and Stevens 1989). When fed with sucrose, starlings develop a conditioned aversion to it. Thus, although nutritional quality of plants is much easier to measure than, say, good genes or compatible genes in animals, the selective pressures of species may differ. The challenging task is thus to evaluate how they translate into the evolution of plant communicative traits.

We will now discuss evolutionary dynamics in plant–animal communication using the handicap principle as an example, because it is arguably the best-studied model of how communication can work. Given that the interests of organisms communicating with each other often conflict, the mecha-

nisms preventing cheating in communication and therefore safeguarding the evolutionary stability of *honest signalling* are hotly debated (Zahavi 1975, Grafen 1990, Tibbetts and Dale 2004, Dawkins and Guilford 1991, Lachmann et al. 2001, Cotton et al. 2004, Getty 2006; see Benitez-Vieyra et al. 2010 for an example on pollination). In other words, how do organisms communicate honestly despite the temptation to deceive one another when interests conflict? Note that this discussion is centred on signals because the evolutionary stability of reliable cues does not pose a problem because cues did not evolve for the purpose of communication. The handicap principle has emerged as the prevailing paradigm explaining the evolutionary stability of honest signalling. It predicts that signals are reliable and informative because they represent handicaps that are costly to produce or to maintain (Zahavi 1975, Grafen 1990, Tibbetts and Dale 2004). Signal costs allow receivers to assess individual quality, as they demonstrate the ability to signal and survive in spite of the handicap. Epitomized by the peacock's tail, the handicap principle became a core concept of behavioural ecology and sexual selection theory. The metaphor of signals as handicaps remains controversial, though. First, handicaps are usually thought of in terms of the absolute costs they cause, but the critical components are the marginal fitness costs of increasing signal size at a given value for a given individual. Hence, costly signals are viewed as investments instead of handicaps, and higher quality signallers are better at converting advertising into fitness (Getty 2006). In both views, costs are central to ensure the evolutionary stability of honest information transfer.

Nature does not abound with examples of handicaps, as many communication systems seem to work without very costly signals. In the communication between plants and their mutualists, for example, there are few demonstrations of cost-enforced honest signals (Edwards and Yu 2007). Given that senders and receivers rarely share the same interests, a strategy of cheating can likely invade most reliable plant–animal signalling systems that are cost-free, unless receivers can verify the indicated quality of a signal immediately (Lachmann et al. 2001). Analysing the importance of verification for the evolutionary stability of communication has not been high on the agenda of many researchers because it is rarely possible in the communication system of mate choice from which most concepts have been derived. For example, the good genes of a potential mate are at best only apparent in the offspring long after mate choice has taken place. Verification, as we will see, can occur more easily (than during mate choice) in many other communication systems between plants and animals. For example, fruit stimuli can be verified when frugivores consume fruits, which enables frugivores to evaluate the match between the appearance of fruits and their nutritional contents via taste or post-ingestive feedbacks. Although verification is likely to be important, it is currently unknown whether the verification of signals alone affects the design and evolutionary stability of plant–animal communication and thereby explains the lack of handicap signalling in plant–animal communication.

The rarity of cost-enforced signals in plant–animal communication can partly be explained by the ease of verification and partly because selection by mutualists is diluted into the patterns of community-driven selection and may thereby be not strong enough to lead to the establishment of cost-enforced honest signalling. To illustrate this point, imagine that a pollinator selects for cost-enforced signals in the flowers it pollinates. This selective pressure is diluted into those exerted by other, possibly more opportunistic pollinators or even antagonists such as nectar robbers that may select flowers differently. The dilution of species-specific selective pressures is not specific to communication but a common theme in the interactions between plants and animals (Herrera and Pellmyr 2002). We therefore now turn to the co-evolutionary framework of plant–animal interactions.

1.4   The co-evolutionary framework of plant–animal interactions

In the eighteenth century, Kölreuter (1761) and Sprengel (1793) were apparently the first to interpret floral traits from the perspective of pollinators that interact with these flowers. In particular, Sprengel hypothesized that the distinct colours of a flower guided pollinators to the reward provided by the flower. This adaptive framework on plant reproductive traits has guided research activities ever since,

and developed into what can be termed conventional wisdom (Rausher 2008). The evolutionary ecology of herbivory started a rise to prominence in the 1960s (e.g., Ehrlich and Raven 1964) and seminal papers on frugivory followed in the 1970s (Snow 1971, McKey 1975, Howe and Estabrook 1977). But, because frugivory does not have the long history of pollination biology and lacks the agricultural importance of herbivory, the field has developed more slowly. In the 1980s, frugivory was propelled into the mainstream of evolutionary ecology through a collective focus on a central paradigm stating that co-evolution between fruit-eating vertebrates and fruiting plants would lead to suites of morphological and chemical characters defining 'specialized' and 'generalized' seed dispersal syndromes (McKey 1975, Howe and Estabrook 1977) similar to the pollination syndromes that had been postulated earlier (e.g., Faegri and van der Pijl 1966, Grant and Grant 1968). Syndromes are sets of flower or fruit traits hypothesized to evolve under the selective pressure of the animals they are interacting with. The syndrome hypothesis rests upon the assumption that distinct animal groups are associated with consistent differences in phenotypic selection on plants. Communicative traits such as colours form an integral part in defining syndromes and are functionally linked to differences in the sensory abilities of animals (Lomáscolo and Schaefer 2010, Lomáscolo et al. 2010). The syndrome hypothesis remains contentious, mainly because plants vary strongly in their degree of specialization towards particular animal mutualists and because phylogenetic constraints are important in many plant traits (for example morphology, Jordano 1995a) and may restrict the evolutionary potential for adaptations towards specific agents (see Section 1.8). Nevertheless, most research activities have assumed that distinct selection pressures of animals can contribute to the diversity of flowers and fruits in terms of morphology, biochemistry, colouration, and odours.

Communication can be an important driver for the evolution of syndromes because animal classes differ strongly in their sensory biology, in the way they see colours, smell odours, and taste the quality of their food. These differences in the sensory abilities of animals can be important. For example, pollinators may be very important selective agents because differential visitation in response to certain plant traits (e.g., floral scent and floral colour) can represent important pre-zygotic filters promoting specialization in plant-pollinator interactions. Compared to floral colours, there are fewer studies on odours, but those that have been published often report that pollinators react to odour (as they do to colour) in a consistent and predictable way that is compatible with a possible role in signalling. For example, *Helicoverpa* moths fly upwind to floral odours, using them as long-distance signal or cue (Cunningham et al. 2004). Consistent with the adaptive framework on plant-pollinator signalling, behaviourally active scent components in the sexually deceptive orchid genus *Ophrys* that mimics female bees and tricks male bees into trying to copulate with the flower (and thereby pollinating it) showed less geographic variation than non-active scent components (Mant et al. 2005). Similarly, the monoterpene lilac aldehyde is a common component of floral scent. In *Silene latifolia* it is also the key component involved in attracting a noctuid moth as a pollinator. Interestingly, the geographic variability in isomers of lilac aldehydes is much less than that of the total flower bouquet, which makes these compounds a reliable cue for identifying flowers and a possible target of stabilizing selection (Dötterl et al. 2005).

The studies cited above support the adaptive framework on plant–animal communication. They are congruent with numerous studies showing that colour traits in fruits, flowers, and vegetative tissue are not selectively neutral. However, there are no studies showing that communication to animals represents the selective pressure that led to the evolution of such colour and odour traits. The lack of such studies does not imply that these selective pressures do not occur, rather it testifies to the difficulty of unequivocally establishing cause and effect in the evolution of plant traits. This is partly due to the usual difficulties of reconstructing selective pressures and the evolution of traits based upon the contemporary distribution and function of traits. If animals drive the evolution of certain colour and odour traits, three basic assumptions need to be fulfilled: (i) that animals discriminate among colours or odours, (ii) that this discrimination affects plant fitness, and (iii) that this selective pressure is stronger than other selective pressures acting upon these traits.

Otherwise colours and odours are better considered as cues and their evolution towards increasing the effectiveness of communication is constrained. Many studies have supported the first assumption, some remarkable studies also the second, but we know of no studies that also supported the last assumption. This is partly because evolution is all too often treated as a univariate problem that can be solved by examining the selective pressures of single species or single guilds of species. However, inferences from such studies have to be treated with some caution when applied to interactions among multiple species (see Strauss and Whittall 2006). Consequently, the lack of studies supporting the third assumption likely reflects the difficulty of this task – which requires truly integrative research covering different disciplines and measuring the relative selective pressures of distinct agents upon traits – rather than the irrelevance of selective pressures of animals upon communication.

Because the patterns of selection are influenced by the community of animals interacting with a plant trait (Stinchcombe and Rausher 2001), we now examine how often plant–animal interactions are specific and pairwise or generalistic and involving many different species.

1.5 Generalization vs. specialization

The extent of specialization in plant–animal interactions – commonly analysed as networks – differs. Pollination networks are more specialized than seed dispersal networks and obligate plant–ant networks are more specialized than facultative plant–ant networks mediated by nectar (Blüthgen et al. 2007). There is also considerable variability in the degree of specialization within a given type of network. For example, the degree of specialization has been a controversial topic in pollination systems. In temperate areas pollination is often not very specialized, because the majority of flowers interact with a large number of species (Waser et al. 1996). There is a higher degree of specialization in tropical and subtropical vegetation communities that remain relatively stable over time (Waser et al. 1996, Johnson and Steiner 2000). Moreover, flowers can be specialized because the multiple species they interact with can be grouped into a few – and often only one –

functional groups that exert similar selective pressures (Fenster et al. 2004).

The strategies of specialization and generalization can have advantages both in pollination and seed dispersal (Fleming et al. 1993). If plants are pollinated by specialists, they can benefit from a high specificity of gene transfer because a low proportion of pollen is lost to heterospecifics. Conversely, if plants are pollinated by generalists, they can benefit from a high redundancy of pollination services from different generalistic pollinators and thereby minimize the risk that suitable pollinators are lacking. It is less clear why animals specialize on particular plants. In some cases, they may do, as suggested above, because certain plants are easier to find or easier to exploit (because of their morphology). Flower morphology can influence the degree of specialization. Open, actinomorphic flowers are pollinated by a broad taxonomic diversity of insects, whereas zygomorphic, tubular flowers restrict access to flower rewards and are significantly more specialized (Fenster et al. 2004; for a discussion of floral symmetry see Section 6.4.2). A higher degree of specialization in pollination can also be related to the fact that pollinators need 'to pay upon delivery of nutritional returns', whereas the plant 'pays in advance' in seed dispersal without any control of successful dispersal events, which are rare anyway. If flower morphology restricts the movements of animals, animals can learn to associate the profitability of flowers with their sensory aspects. Contingent on the profitability, this type of learning can lead to avoidance and preferences for certain flower stimuli. These preferences can in turn increase foraging success of animals and are reliant on communication as we have defined it.

Specialization is often achieved by very specific communication. For example, figs include approximately 850 species that are pollinated by species-specific aganoid fig wasps breeding inside the figs. Figs emit species-specific blends of volatiles that differ quantitatively and qualitatively from the odours of other figs and that attract the specific aganoid wasp (Hossaert-Mc Key et al. 2010). Hence, variation in odour profiles leads to pre-mating isolation among sympatric figs. A similar case of specialization is found in the orchid bee pollination syndrome that consists of approximately 700 spe-

cies of orchids that are exclusively pollinated by male orchid bees that collect volatile substances from flowers (and also from fungi) and store them in hind-leg pockets. Males later emit these substances as pheromone analogues at mating sites and near females. Interestingly, closely related but genetically differentiated morphotypes of the orchid bee *Euglossa viridissima* show differential sensitivity to floral volatiles. Given that these volatiles are probably used in species recognition, olfaction-driven differences in perfume perception and collection supports a non-gradual diversification of the insects' pheromone communication system (Eltz et al. 2008). The non-gradual shift is explicable by the discrete nature of odours (Section 2.7).

The two examples in the paragraph above involve cases of extreme specialization. Even in less specialized pollination systems, pollinators commonly differ in their spectral sensitivities (Section 2.5). These differences may explain why floral colour contributes to structuring plant-pollinator networks into distinct modules (Olesen et al. 2007a). Differences in spectral sensitivities can translate into differential visitation rates of pollinators because certain colours are more contrasting – and therefore easier to detect – for one type of pollinator compared to another. Similarly, the lack of suitable chemoreceptors may make one animal species insensitive to certain components of floral scent, while another species having the appropriate chemoreceptor may pick up the scent from afar. Consequently, specialization may be explicable by differences in the sensory biology of animals which – in theory and at least partly – may contribute to the extraordinary diversity of plant reproductive traits. This conjecture links communication to a controversial hypothesis dating back to Darwin (1876) on the extraordinary diversification of the angiosperms (see Box 1.2).

Given that pollination networks are more specialized than most other plant–animal networks, it is not surprising that the most spectacular examples of co-evolutionary processes pertain to pollination systems. Most famously, Darwin predicted the existence of a particularly long-tongued pollinator that would be able to pollinate the orchid *Angraecum sesquipedale* that had a spur of up to 30 cm in length. Indeed, a few decades after Darwin's death the

Box 1.2 The diversification of angiosperms

The diversification of angiosperms has intrigued evolutionary biologists for a long time, and it has been attributed at least partly to selective pressures exerted by pollinators and seed dispersers (Darwin 1876). Biotic dispersal of pollen and seeds can increase speciation rates and reduce the probability of extinction owing to a higher probability of long-distance dispersal and directed dispersal to favourable sites, such as forest fragments (see Montoya et al. 2008). For example, groups such as orchids that have highly specialized animal-based pollination systems are particularly species-rich. Indeed, it has been proposed that the extraordinary diversity of orchids is an evolutionary consequence of the commonness of pollination by deceit (Cozzolino and Widmer 2005). Roughly a third of all orchid species are rewardless and most achieve pollination by either mimicking other rewarding species or by exploiting the sensory biases of pollinators to respond to large and contrasting floral displays (Naug and Arathi 2007). The most widely accepted hypothesis proposes that animal-mediated pollination could enhance speciation rates by facilitating the development of pre-zygotic isolation (Bawa 1995). Support for the link between animal dispersal and speciation rates come from broad-scale comparisons between lineages with animal pollination and those with abiotic pollination (e.g., Jesson 2007). Yet, it is difficult to establish cause and effect, since high species diversity may also drive specialization owing to increased character displacement in sympatric species to avoid clogging by heterospecific pollen (Armbruster and Muchhala 2009). Well-marked differences in the sensory biology of mutualistic animals suggest that communication to the animals that disperse the genes of plants can contribute to the diversification of plant traits.

moth *Xanthopan morganii* was discovered that has a proboscis measuring 30 to 35 cm in length. Similar, equally compelling, and more recent examples include that of a long-tongued fly (*Prosoeca ganglbaueri*) and its primary floral food plant (*Zaluzianskya microsiphon*) in South Africa. The dimensions of the fly's proboscis and the corolla length of the flower covary across geographical regions, and the match between tube length of flowers and tongue length of flies was found to affect plant fitness (Anderson and Johnson 2008). This example suggests that co-evolution is a compelling explanation for the geographical covariation in flower depth and fly proboscis length.

Similar but less well-known examples of adaptations through co-evolution occur in communication. For example, the widespread shift in floral colour – occurring in at least 77 plant families – once pollination has occurred (Weiss 1991). This floral change is thought to direct pollen arrival efficiently to yet unpollinated flowers, while enabling pollinators to forage more efficiently on flowers still containing nutritional rewards. Interestingly, floral colour change can involve the whole flower or only parts of it, and has evolved

independently many times, presumably owing to selective pressures of pollinators (Weiss 1995).

Adaptations resulting from co-evolutionary processes are more contentious in other plant–animal networks (see Section 3.9 for the famous examples of the dodo and the tambalocoque tree). Still, there are consistent differences in sensory traits such as fruit type, shape, colour, and odour between plants relying on different animals as dispersers (Lomáscolo et al. 2010) and between plants relying on animals for dispersal and those relying on abiotic agents for dispersal. Such differences are consistent with a certain degree of adaptation to communication to seed dispersers (see Chapters 3–5). Similarly, there are numerous studies supporting a scenario of antagonistic co-evolution of escalating defences between plants and their herbivores that lead to the bewildering diversity of plant secondary defences. The continuous selective effects of herbivore attack and plant defence are thought to be largely responsible for the incremental elaboration, proliferation, and intricacy of plant secondary compounds and insect detoxification mechanisms (Becerra 2007). This view is supported by the positive correlation between the co-evolutionary specialization in plant–herbivore interactions and the dissimilarity in plant defensive chemistry at the community level (Becerra 2007).

While the extraordinary examples of adaptations to pollinators provide particularly compelling evidence for co-evolutionary mechanisms between plants and animals, such obvious examples are comparatively rare. They are probably rare because the interactions between plants and animals commonly involve many species in tritrophic interactions. For example, seed set in sweet clover (*Melilotus officinalis*) was reduced by 50 per cent within 200 m distance to breeding colonies of cliff swallows (*Hirundo pyrrhonota*) that consume insects and thereby apparently reduce pollinator densities (Meehan et al. 2005). Similarly, in a factorial predator-exclusion experiment, the duration of flower visits by a Satyrid butterfly and subsequent seed output were 2–4 times greater if lizard predators were excluded, while the exclusion of a flycatcher species had no effect (Muñoz and Arroyo 2004). Indirect effects can also occur because browsing or trampling by herbivores may affect the absolute and relative population density of plants. The population densities of plants may in turn affect pollen deposition patterns, although such effects were only found in one out of ten animal-pollinated species in the temperate forest of the southern Andes (Vázquez and Simberloff 2004). These studies show that the abundance of other species can dramatically change the interaction strength among two or more species.

Interactions among multiple species have several important implications for the evolution of plant–animal communication. First, the strength and consistency of selective pressures of animals upon communicative traits generally vary according to variation in the relative abundance and species composition of animals that interact with a plant trait (Strauss and Irwin 2004). Second, the composition of animal communities that plants interact with is often unstable in space and time, thereby reducing the likelihood of consistent selective pressures upon plant traits (Herrera 1985, 1996). That is, in most systems, selection will not be pairwise between one plant species and one animal species but depend on the temporal dynamics of the composition of the animal community interacting with a given plant

population (including mutualists and antagonists). This pattern of evolution is termed 'diffuse' co-evolution (Janzen 1980).

1.6 Diffuse co-evolution

Diffuse co-evolution between plants responding to the selective pressures of mutualists and antagonists is an important concept because it can explain why resources that are made to be eaten like nectar and fruits can taste bitter to the extent that they are rarely consumed. The composition of nectar likely reflects the distinct selective pressures of pollinators and nectar robbers that do not pollinate flowers. Various secondary compounds in nectar may act as chemical defence against microbes and insect nectar robbers (Adler 2000). The situation in fleshy fruits is similar. Biochemical defences in fruits invariably lead to the conflict that fruits need to be deterrent to non-seed-dispersing organisms while maintaining palatability to the legitimate seed dispersers (Section 5.5). A first important conclusion to be drawn from diffuse competition is that while pollinators and seed dispersers may often be the most apparent animals interacting with plants, their selective pressure on plant traits is likely to be modulated, or even abolished, by the effects of initially less apparent but often more common agents of selection.

Diffuse co-evolution implies that the selective pressures of different species are not independent of each other. This interdependence is found, for example, in the selection for resistance against herbivores in the ivyleaf morning glory (*Ipomoea hederacea*). Significant negative genetic correlations exist between resistance to deer and to generalist insect herbivores (Stinchcombe and Rausher 2001). In addition, the resistance loci under selection differ according to the composition of the local herbivore community. Finally, selection for deer resistance depends on the presence of insects. In the absence of insects, deer resistance is effectively neutral. Thus, the sign and strength of selection can be influenced by the composition of the local community of animals and plants. This is important to note because animal communities can fluctuate widely in species composition and relative abundance of species over time and space. These fluctuations are likely to be pronounced because the relative abundances of very distinct groups such as pollinators and various antagonists (including herbivores, nectar thieves, parasites, fungi, and microbes) that interact with plants are unlikely to remain constant over time. The corollary of this conclusion is that, in general, fluctuations in the species composition of communities, as well as gene flow among populations with distinct animal communities, will retard the adaptation towards specific agents and reduce the degree of specialization of plant species.

Owing to the multiple species generally involved in plant–animal interactions, we need to account for (that is, sum up) the many selective pressures acting upon plant communicative traits. The multiple species vary continuously in their impacts on plant fitness from negative to positive and from direct to indirect effects, complicating the overall assessment of the net effects (Strauss and Irwin 2004, Strauss and Whittall 2006). If the selective pressures align, they can accelerate the evolution of plant traits. Interactions among multiple species can involve direct or indirect competition among species. If mutualists and antagonists select the same communicative trait similarly (e.g., if seed-dispersing toucans and seed-predating parrots both locate fruits visually), their selective pressures upon the visual traits of fruits will be opposed. In general such opposing selection pressures often result in fluctuating selection, because the relative abundances of animals commonly vary over time, resulting in variable relative interaction strength of each animal species with a given plant (Irwin 2006). This is important because fluctuating selection may retard or even prevent the fixation of alleles that are favourable for a given animal. To evaluate the net selective pressure acting upon plant traits, it is thus essential to quantify the relative selective pressure of each species or functional group interacting directly with a trait – information that is all too often not available.

It is generally difficult to measure selective pressures of species upon traits. It is more easily achievable with direct interactions among distinct animal species, because these are easier to observe and measure than indirect interactions among species. A good example for direct interactions is the flower polymorphism of *Clarkia xantiana*. Here, the predominant pollinator, a bee, exerts positive frequency-dependent selection, while less common or less dominant bee species exert negative frequency-dependent selection, preferring the rarer phenotype to avoid competition with the predominant pollinator (Eckart et al. 2006). Direct competition also occurs with herbivores that eat flowers, reducing the number of flowers or fruits available for pollination and seed dispersal, respectively. Direct interference can also occur because predators of pollinators like crab spiders catch pollinators and alter flower visitation patterns by bees in laboratory experiments (Ings and Chittka 2009). The interactions between herbivory and pollination can also be more subtle than the direct competition for resources. Indirect effects may occur because foliar damage may alter flower morphology, and decrease pollen production and quality (Strauss 1997).

While the evaluation of net selective pressures acting directly upon a trait is already a formidable task, it is often much more difficult to uncover to what extent selective pressures of one species or functional group may depend upon indirect effects imposed by animals that interact with different plant traits. Yet, these interactions also occur; for example, because of systemic responses to herbivory that include altered allocation to root/shoot biomass (including allocation to reproductive organs) and the induction of structural and chemical defences in reaction to damage by herbivores that can also affect pollinators (Mothershead and Marquis 2000). Such effects can be subtle; florivory, the consumption of flowers, may affect gender expression in flowers and the mating system of plants (McCall and Irwin 2006). As yet, florivory is a relatively under-studied process whose evolutionary significance is not well understood. Plants may use different mechanisms to adapt to florivory: using resistance (by defending themselves) and tolerance (by regrowing quickly). Thus, a species' net selective pressure may depend upon direct and indirect effects that we need to keep in mind when considering the conventional wisdom that plants, traits may be adaptations to specific animal groups. Taken together the conclusion of theoretical and experimental studies is that one needs caution to infer the net selective pressure on plant traits from studies on one or few species, and one or few populations. As a corollary, the community context of species interacting with plants is important for understanding the selective pressures upon plant traits (Strauss and Irwin 2004).

The paragraphs above are intended to alert the reader to the complexity of species networks that typically characterize the interactions among plants and animals. Yet, it is important to emphasize that the complexity of interactions as well as temporal and spatial heterogeneity of interactions do not prevent adaptations and co-evolutionary processes per se, they rather enrich their dynamics. Thompson (2005a) elaborated this view elegantly in his book *The Geographic Mosaic of Coevolution*. Since many species are phylogenetically conservative in their interspecific interactions, this conservatism can hold interspecific relationships together for extended time periods. This holds even though genetically differentiated populations may specialize on interactions with different species, resulting in interactions that may differ within and among communities. Thus conservatism in interactions (see Fenster et al. 2004 for compelling examples involving pollination) coupled with genetic differentiation among populations and a spatially variable community context present the malleable raw material for co-evolution among species to occur. Studies on intraspecific adaptation to local (biotic and abiotic) selection regimes are thus important for understanding the precursors to diversification at the species level. Recent advances in population-genomic approaches indeed reveal adaptive floral divergence in morphometric traits, even in populations that are strongly connected by gene flow (Herrera and Bazaga 2008). In general, the rate of immigration constrains local adaptation, which then only occurs if local selection is stronger than the constraining role of gene flow (Postma and Noordwijk 2005). Yet, the consistent differences in the biotic and abiotic environment envisioned by the geographic mosaic of co-evolution theory may result in natural selection overriding the effects of gene flow between populations. The geographic mosaic of co-evolution is a particularly useful concept for studying communication because there is considerable spatial variation in many communicative traits. For example, capsaicin, the compound that produces pungency in wild chillies, is likely involved in

gustatory communication to mammals, which destroy the seeds of *Capsicum* species (Tewksbury and Nabhan 2001). *Capsicum chacoense* is a naturally polymorphic species for the production of capsaicinoids, displaying pronounced geographical variation (see Section 5.5.2).

1.7 Correlated evolution

Although syndromes are usually interpreted as plant adaptations to specific types of animals (e.g., pollinators), diffuse co-evolution implies that a large array of distinct and taxonomically diverse selective agents (from microbes to insects and grazing mammals) will shape the evolution of floral traits. Correlated evolution ensues if the selective agents attend to different cues, and if these cues are genetically correlated among each other. Such correlations can occur between different traits of a floral display, or if plants' systemic response to a specific selective agents (e.g., herbivores) alters their interaction with other selective agents (e.g., pollinators). The latter scenario originates from plants' inability to escape predators in space resulting in their need to mediate the interactions with antagonists mainly through plant biochemistry, which will often also affect mutualists.

There is mounting evidence for correlated evolution among the processes of herbivory, pollination, and seed dispersal, which are traditionally considered separately in studies. In a clever experiment, Herrera et al. (2002) showed that the sign and magnitude of selective pressures by herbivores depended on the interaction with pollinators and vice versa. In different populations of the buttercup *Helleborus foetidus* herbivory had only detrimental population effects in the presence of pollinators, whereas pollinators had only a positive effect on the number of seedlings produced if herbivores were absent. The most likely explanation for this complex interaction is that herbivores strongly prefer plants with many open flowers or developing fruits, thereby reducing the number of seeds produced to levels that are equivalent to those where no pollinators are present. This scenario leads to correlated selection resulting in the joint evolution of traits that mediate the interactions with mutualists and antagonists.

The term 'correlated evolution' entails important implications for how communication among plants and animals will evolve. It suggests that communication by cues will probably be common because ubiquitous selective agents like microbes and fungi can influence the evolution of plant traits even though they do not communicate with these traits. For example, anthocyanins, the ubiquitous plant pigments, are fungicides (Schaefer et al. 2008a). Since the action on fungi is direct, and there is no communication involved between plants and fungi, anthocyanins are cues for animals if they are mainly selected by fungi. Likewise, capsaicin not only communicates to mammals as seed predators, but also protects fruits against a *Fusarium* fungus that kills the seeds. The spatial variation in capsaicin production is linked to variation in the damage of chilli seeds caused by the fungus (Tewksbury et al. 2008a). Capsaicin further modulated retention times of seeds in the guts of legitimate avian seed dispersers, having a constipative effect (Tewksbury et al. 2008b). Importantly, while longer seed retention may increase the potential for long-range dispersal, it also had negative effects upon germination probabilities in pungent seeds because a larger proportion of these seeds were damaged in the gut. These complex effects of capsaicin show that in *Capsicum* fungi defence, gustatory communication to seed predators, seed passage, and germination are not independent traits, but expected to evolve in a correlated fashion.

Correlated evolution may be particularly important for pigments, which fulfil (like many other secondary plant compounds) multifarious ecological and physiological functions (Section 2.6). These multiple functions greatly restrict the evolutionary potential for species-specific adaptations. Correlated evolution of communicative traits used in defence and attraction is illustrated by the balancing selection of herbivores and pollinators that result in the flower colour polymorphism of the wild radish (*Raphanus sativus*). In that species, pollinators and herbivores both prefer the anthocyanin-recessive yellow morph. Herbivores perform differentially on both colour morphs and may thus counter the selection exerted by pollinators (Irwin et al. 2003). It remains unknown in this example whether herbivores use floral colour to choose their hosts (an instance of communication) or whether their preferences are caused by traits correlated to floral colour; a situation that has been found in fruit colour polymorphisms (Whitney and Stanton 2004; and see Section 4.6).

Correlated evolution is probably common, shaping biochemical defence in plants. In general, predation selects for higher defences, which often lowers the attractiveness of flowers and fruits for mutualists. This has been well documented in flowers of native tobacco (*Nicotiana attenuata*). Nicotine-deficient flowers were more often attacked by nectar robbers and herbivores. Nicotine contents in the nectar also modulated the visitation pattern of hummingbirds and hawkmoths with higher contents enforcing lower nectar intake, leading to higher visitation and outcrossing rates (Kessler et al. 2006). Similarly, manipulating the uptake of alkaloids in a hemiparasitic plant, Adler (2000) documented that the levels of alkaloids not only (predictably) decreased herbivory but also increased visitation by pollinators and lifetime seed production. It was not known which cues pollinators used to assess damage by herbivores in this species, but since damage often alters the volatile profile of plants and may result in induced resistance, pollinators may have used volatiles as a cue to avoid damaged plants. This study suggests that pollinators eavesdrop on the interaction between plants and herbivores, which provides a functional explanation for why processes such as herbivory and pollination are interrelated.

Correlated evolution is an important concept in plant–animal communication, it is difficult to analyse experimentally, and it implies that we need to assess the adaptive significance of any trait in its wider physiological and ecological context. This broader perspective on the evolution of plant traits is particularly important for biochemical traits, which are the building blocks for visual, olfactory, and gustatory communication. Excitingly, the diverse selective pressures exerted by different groups may act both as constraints and as pre-adaptations in the evolution of plant traits. Constraints can be imposed by multiple selective pressures because these can restrict the evolutionary potential to respond independently to selection by distinct animals. At the same time, adaptations to one group of animals (e.g., herbivores) can function as pre-adaptations to another group (e.g., pollinators), which we will discuss in the next section before considering the genetic and morphological architecture of plants.

1.8 Evolutionary significance of complex interactions for communication

The interplay between constraints, pre-adaptations and natural selection in shaping evolutionary patterns is generally not well resolved, although this is often considered to be one of the major questions in evolutionary biology (Arnold 1992, West-Eberhard 2003). Direct and indirect interactions among species yield important insights into the evolution of sensory aspects of plant–animal communication. A fundamental challenge is to differentiate between traits that are selected for their communicative function, that is signals, and those that are not, but which still can be informative (that is, cues). This distinction is often not easy to draw. For example, UV reflectance in flowers has often been attributed to selection by pollinators that are sensitive in the UV range of the spectrum. UV reflectance can be imparted by flavonoids and by de-aromatized iso-prenylated phloroglucinols. The latter are found not only in the facial surface of flowers but also in the abaxial surface, the ovary wall, and the anthers. The secondary de-aromatized isoprenylated phloroglucinols can be toxic to herbivores (Gronquist et al. 2001). These secondary compounds may thus serve a dual role in attraction of pollinators and defence against herbivores (see Section 4.6 for examples for fruit colour). If selection for their defensive properties overrides selection for communication, these traits are best regarded as cues (Maynard Smith and Harper 1995). The UV-reflecting compounds can still attract and guide pollinators to the flower, but they may not have evolved for that purpose. Often, although evidence is consistent with a signalling role, it will be impossible to unambiguously assign a signalling function to a given trait. In these circumstances it would be more prudent to use the term cue. The distinction between signals and cues is important because it enables us to trace the evolutionary dynamics of plant–animal communication in a wider context.

The evolutionary dynamics of cues and signals can be more complex than the simple distinction between them suggests. The primary function of a given trait may change over evolutionary time according to changes in the strength of selective pressures which may be either associated with fluctuations in the composition of animal communities or to changes in correlated plant traits. For example, a signal may evolve from a trait that was originally selected for functions other than communication. In this case, the signal is an exaptation, a derived function from a pre-existing trait. Perhaps the most convincing example so far of the evolutionary dynamics of communication and of biochemical exaptations is the resin secretion of *Dalechampia* vines. The resin is excreted by blossom glands and is collected by pollinating bees for use in nest construction. This resin reward system appears to have originated as a taxonomically widespread defence system in plants where they exude terpenoid resins in response to wounding, such as many conifers do (Section 2.10.1). Foliar and floral resins are chemically similar; both deter feeding and leaf-cutting. Thus, the most parsimonious explanation is that the resin excretion was secondarily adopted as a reward, when resin-collecting bees began visiting flowers incidentally to steal resins (Armbruster et al. 1997). Pollination of *Dalechampia* vines relies primarily upon resin-collecting bees, which choose among individuals based upon bract size, a trait that is phenotypically correlated to the amount of resin offered by the plant (Armbruster et al. 2005). The consistent phenotypic correlation between bract size and gland size has a strong genetic basis. Thus, bees select directly for larger bract size as the most visible advertisement of rewards and thereby indirectly for larger resin glands (Armbruster et al. 2005). When the resin defence of flowers was lost by conversion into a reward system, *Dalechampia* species evolved a series of further defences. These include the nocturnal closure of involucral bracts that protect stamina and pistillate, as well as the deployment of sharp, detaching trichromes that defend developing seeds (Armbruster 1997). Interestingly, phylogenetic analyses show that bracts probably evolved first as a floral advertisement system before they assumed a defensive function as an additional exaptation. This clearly illustrates the changing nature of the balance between cues and signals in plants' simultaneous communication to herbivores and pollinators. These studies further show how a broad historical perspective on plant–animal interactions allows for a better understanding of how adaptations to herbivory and pollination may mutually influence each other as pre-adaptations and thereby significantly alter the evolutionary trajectories of communication between plants and herbivores and communication between plants and mutualists.

1.9 Constraints on communication

1.9.1 Genetic architecture

A current limitation to understanding of the form and strength of multiple selection pressures is that the genetic variation for communicative traits in flowers, fruits, and leaves is often unknown in natural populations. This may come as a surprise given that commercially grown fruits and flowers are often selectively bred for their sensory properties and therefore offer ideal model systems to analyse the underlying genetic variation for these traits, as well as the correlations among them. Widespread polymorphisms in floral and fruit colours document substantial genetic variation for floral colour (Rausher 2008). In some, possibly exceptional, cases floral colour changes can be controlled by a single gene. For example, by introgressing different flower colour alleles into near isogenic lines of two *Mimulus* species, Bradshaw and Schemske (2003) showed that a shift in floral colour resulted in a concomitant change in pollinator spectrum (hummingbirds vs. bees) that could, in theory, be initiated by a single mutation. This and a similar study on flower colour in *Petunia* (Hoballah et al. 2007) are extremely interesting examples, because the repeated evolution of pollination by hummingbirds from ancestral insect-pollinated species is a recurrent theme in the North American flora. These studies are important in the context of our book because they illustrate that effective communication to animals depends on their sensory biology, and that changes in communication underlie the shift in pollinator spectrum.

Although the genetics underlying communicative traits will probably be more complex in many species than those determining floral colour in *Mimulus* and *Petunia*, colour often seems to be an evolutionary flexible trait that shows no strong phylogenetic signal in fruits and flowers (Lomáscolo and Schaefer 2010, McEwen and Vamosi 2010). Although it is difficult to evaluate the phylogenetic distribution of floral scent compounds, the available evidence suggests that floral scents evolve relatively quickly, since scent is of little value for reconstructing phylogenies at high taxonomic levels (Knudsen et al. 2006).

Evolutionary biologists have long recognized that the genetic structure of traits, that is, the genetic covariation among them, may significantly – and in addition to non-genetic factors such as epigenetics – influence the evolutionary trajectory of species. There are several distinct mechanisms that might cause non-random patterns of correlation among traits and more complex processes of phenotypic evolution: single genes that generate several phenotypes (*pleiotropy*), interactions among genes (epistasis), and physical linkage of genes. Although the genetics underlying the biosynthesis of anthocyanins, one of the main classes of pigments in plants, has became a model system in plant molecular biology, evidence for pleiotropic effects upon plant colours is scarce. Arguably the best evidence is the study by Schemske and Bierzychudek (2007), which found that spatial differentiation in the blue and white colour morphs of the annual desert herb *Linanthus parryae* is explicable by local differences in the regimes of natural selection (see Section 6.7). Similar to the *Mimulus* example above, the colour polymorphism in this species is caused by a single gene. Previously, the floral colour polymorphism of this species has been regarded as the prime example of random genetic drift, but the study by Schemske and Bierzychudek showed that allozyme markers did not differ between the colour morphs, and that each morph fared best in transplant experiments in their local habitats even though the sole pollinator, a Melyrid beetle, does apparently not differentiate between colour morphs. Rather, the authors suggest that the association between floral colour and environments is the result of pleiotropy. Pleiotropic effects of floral colour have been attributed to plant vigour, drought resistance, herbivore deterrence (Irwin et al. 2003), plant biochemistry, and response to edaphic stress. Pleiotropic effects apparently also oppose the selective advantage of increased selfing in the rare white coloured morph of *Ipomoea pupurea* flowers because this morph suffers from reduced survival from germination to flowering (Coberly and Rausher 2008). Such pleiotropic effects seem puzzling at first glance because we perceive colour as a trait unrelated to survival probabilities or soil type. Yet, pleiotropic effects are understandable in light of the biochemistry of pigment synthesis. Floral colour polymorphisms (particularly white – blue, blue – red) are often associated with the inactivation of regulatory sequences in the anthocyanin pathway. Such inactivation may also block the synthesis of beneficial secondary compounds that are produced by the same pathway (see Section 2.6.3) and thereby reduce overall plant fitness (Rausher 2006).

Genetic correlations are well known for many morphometric traits, particularly in flowers. Detailed studies on the selective pressures of bumblebees as pollinators and ants as predators of flowers of the Alpine skypilot (*Polemonium viscosum*) showed that plants' response to escape predators through morphological adaptations are constrained both by selection imposed by pollinators and by strong (r = 0.7) genetic correlations among different morphological traits (Galen and Cuba 2001). This conflicting selection pressure explains why adaptation towards predation occurs only in populations with high predator pressure, because it comes at the costs of reduced pollination rate. This study elucidates the widespread role of genetic constraints, but it is also a good example of the geographic mosaic of co-evolution.

The genetics of floral colouration offer a particularly convincing example of how adaptive change may constrain the potential direction of future evolutionary change. The ancestral floral colour of the genus *Ipomoea* is blue and associated with bee pollination. In one region a transition to red-flowered species that are pollinated by hummingbirds occurred. Floral colour change is caused by switching the flux from one branch in the biosynthesis of anthocyanins to another. Notably, this switch, which

resulted in the inactivation of one pathway, resulted in subsequent degeneration of further enzymes in that pathway; thereby supporting Dollo's law that character elimination is irreversible. This degeneration makes it unlikely that red-flowered species may evolve back to the ancestral blue colour because it would require the restoration of two or three independent steps (Zufall and Rausher 2004).

1.9.2 Architectural constraints

Flowers, fruits, and leaves all are complex functional structures that are characterized by certain developmental, genetic, and physiological architectures. As such, functional constraints as well as non-random genetic correlations among traits might cause complex processes of phenotypic evolution (e.g., Pigliucci 2001, West-Eberhard 2003, Hansen 2006, Wagner et al. 2008). In particular, these correlations may greatly constrain the plasticity of a trait to respond to a certain selective pressure. While many of these correlations are probably adaptive for plants, they are not necessarily adaptive from the perspective of mutualistic animals because they may constrain the evolutionary potential to respond to selection by mutualists. The concept of phenotypic integration has mostly been applied to morphological traits, particularly in flowers (see Section 6.4.3), but also to dry fruits (Pigliucci et al. 1991) and to avian plumage colours (Bleiweiss 2008). Studying integration, Frey (2007) found that floral colour evolution in the eastern spring beauty *Claytonia virginica* (Portulacaceae) may proceed relatively unconstrained from associations with other traits (see Section 4.3.2 for a similar example in fruits). It would be extremely interesting to expand studies on phenotypic integration to include other ecologically relevant traits (for example, we know of no study analysing the integration of odour related traits) that are involved in communication to mutualists and antagonists.

1.10 Perspective

The common thread through this chapter is that the interactions between multiple species of plants and animals shape the evolution of plant–animal communication. Selection upon communicative traits thus depends strongly upon the community context of both plants and animals and can best be envisioned through the geographic mosaic of co-evolutionary theory. The important corollary is that the evolutionary trajectories of communication will differ among species but also among populations of the same species. In some populations selection unrelated to communication can prevent the evolution of signalling, whereas in other populations of the same species signalling may evolve. Furthermore, adaptations to selective pressures unrelated to communication can constrain communication, and also channel communication in specific directions owing to pre-adaptations and exaptations.

Currently, many fundamental issues of communication theory remain unresolved in the communication between plants and animals. Most obviously perhaps, we lack understanding of the covariance between sensory traits and plant qualities in most systems. Do the colours and odours of flowers and fruits function only as positional cues that indicate their presence? If this is the case, we expect evolution for increased effectiveness to increase the range over which the stimuli can be perceived by animals. Alternatively, are cues correlated with the presence of specific nutritional rewards such as micro- or macro-nutrients? If this is the case, we can expect evolution for increased effectiveness to increase the association between stimulus and nutrients and to also increase the range over which the stimuli can be perceived by animals. Similarly, do induced volatiles in plants simply indicate herbivorous attack or indicate more specifically the type of herbivore present on the plant? Owing to these uncertainties we can only speculate on the mechanisms that may underlie the evolutionary stability of plant communication systems. Clearly, the evolutionary stability of signals will be very different from that of cues.

We have argued that the many selective pressures of animals and microbes upon plant traits will vary according to their relative abundance and according to their relative effects on plant fitness. This fluctuating selection upon plant communicative traits results in a dynamic evolution of communication because cues can evolve into signals and signals can evolve into cues.

Attaining a broader perspective on the evolution of plant–animal communication is still an ambitious project because there are many gaps in our current knowledge. Some of these concern the genetic (both underlying genetic variability as well as genetic covariation among traits) and developmental structure of plant traits, others concern the multiple functions of many traits that simultaneously interact with different selective agents (both biotic and abiotic). Only if these underlying mechanisms – that again may act as constraints and as a driving force throughout evolutionary history – are better known, can we truly understand local adaptation, a cornerstone of the geographic mosaic of co-evolution.

As so often, our current knowledge on plant–animal communication is imbalanced. The sensory ecology of communication among plants and animals is much better known in pollination than it is in seed dispersal, herbivory, and carnivorous plants. This is partly attributable to the relatively higher degree of specialization seen in pollination systems. Given the interdependence among at least some of these communication systems, it is even more urgent to work on as yet neglected fields. Imbalance is also apparent when comparing visual to olfactory communication. As yet, visual communication is much better known, partly because it is so more obvious to our own senses and partly because visual reception and the initial stages of neural perception are easier to model (a topic that we explain in the next chapter). The perception of cues and signals, starting from the neuronal coding to the extraction of information, is still largely unresolved, preventing progress on evaluating how cognition shapes communication and its ecology (Chittka and Brockmann 2005), as well as preventing progress on defining information. A major challenge will thus be to study how the information obtained from different sensory modes is integrated in the brain of the animals responding to (plant) cues and signals.

In summary, we have stressed the importance of geographically variable co-evolution as a framework to guide research on communication. How fluctuating selection shapes communication is possibly the most valuable aspect that the study of plant–animal communication can contribute to the development of communication theory. Within this book, we will explore how distinct species within and across functional groups interact with plant traits, and how such sensory interactions may influence the evolution of plant–animal communication.

In particular, we will ask whether communicative traits evolved as signals, that is, *in order to* communicate to animals, or whether they are cues that are primarily constrained because they interact (without having a communicative function) directly with other animals or are constrained by indirect interactions (i.e., by correlated selection) or are influenced by non-communicative (sometimes abiotic) factors. Throughout the book it will be apparent that the communicative traits of plants can be informative even without being selected for that function, partly owing to shared biochemistry between the stimulus and the plant reward and partly as an effect of the interactions among multiple species.

CHAPTER 2

Animal sensory ecology and plant biochemistry

2.1 Introduction

In 1974 Thomas Nagel titled a famous essay on consciousness: 'What is it like to be a bat?' Nagel used the example of bats as a life form with a range of activities and a sensory apparatus so very different from ours in order to expose the subjective character of consciousness. In essence, Nagel argued that we cannot imagine what it is like to be a bat because our imagination is tied to our own sensory and mental experience, which is likely to be very different from that of a bat.

Although we do not focus on consciousness in this book, the key issue is that most animals perceive the world around them quite differently from the way we do. It is important to remember this almost trivial point throughout the book, because we can only understand the evolution and ecology of plant–animal interaction if we attempt to analyse plant traits as they are sensed and perceived by the animal interacting with them. Perceptual differences among animals exist in all sensory modes. We humans are mainly visual animals, but we can obviously be strongly influenced by smell and sound. Our spectral sensitivities to light reflected from objects around us are different from those of most other animal groups. Birds and primates can discriminate flowers and fruits from their background by their colours over a range of tens of metres (Schaefer et al. 2006), whereas insects can use the chromatic information of flowers only on a range of centimetres (Spaethe et al. 2001; at least for common flowers whose diameter do not exceed a few centimetres). To illustrate the extent of the differences in the ability to use colour from afar, imagine standing a few metres away from a flower-

ing plant (see Plate 1). For non-green flowers, our eyes allow us to perceive a colour contrast between that flower and its leafy background. The perception of colour differences in birds likely exceeds that of our visual system, which is why they are likely to perceive the differences between a flower and its background better than we do. Yet, for bees the flowery patch must be 26 cm in diameter in order to be detectable from a distance of 1 m by its colour (Spaethe et al. 2001). Similarly, many animals are sensitive to a larger range of acoustical stimuli, and many animals perceive smell better and over larger distances than we do. For example, moths can perceive very low quantities of plant volatiles over ten or more kilometres; whereas humans perceive plant smells usually only within a range of a few to tens of metres. It is thus paramount to account for the sensory abilities of animals if we are to understand how selection imposed by animals acts on plant traits that have a communicative function. In this chapter, we give a short introduction to the sensory systems of those animals that are commonly involved in plant–animal communication.

In trying to link the sensory ecology of animals to their selective pressures on plant traits, Thomas Nagel's cautious note (1974) about interferences still applies. To get around it, most researchers work with models of the sensory space of the focal animal. Yet even these models have their limitations. For example, and although the predictions of models on bird vision are often in agreement with behavioural data in the field (e.g., Cazetta et al. 2009), these models are based only upon very few experiments (Maier and Bowmaker 1993) on colour discrimination in the lab. This does not mean that the models are wrong, but that more validation (particularly involving ecologically realistic tasks) would be welcome.

Because the senses constitute the first interception of external stimuli, it is important to understand the way animals sense their environment. There are several stages involved in communication from the production of a stimulus to the decision-making of the receiver. Endler and Basolo (1998) identify nine stages of communication (Box 2.1). This compilation shows that selection on communicative traits may not only be caused by the sensory abilities of animals, but also by the neuronal coding and higher brain processes of animals. While considerable progress has been made in modelling the sensory world of animals, the way their cognitive abilities may shape communicative traits remains in most cases a black box (Chittka and Brockmann 2005 for a

---

Box 2.1 The nine stages of communication (modified after Endler and Basolo 1998)

1) **Stimulus production and emission:** Cue production depends on the internal state of the sender and its physical and psychological state. For example, the number of flowers produced can depend on the availability of resources such as water and light. The emission properties of the stimulus largely depend on its biochemistry.

2) **Stimulus transmission:** Stimulus transmission depends on the environmental conditions. Visual stimuli can be degraded by ambient light, by the prevalence of fog (e g., in cloud forests), whereas the transmission of olfactory traits is influenced by the prevailing wind conditions as well as the biochemistry of chemical volatiles that can influence their propensity to bind with other molecules.

3) **Stimulus reception:** The senses of an animal represent the first contact with a stimulus. The signal-to-noise ratio will affect stimulus collection; that is, the likelihood that the sensory organ produces an output that differs from the one corresponding to background noise.

4) **Transduction of the stimulus:** The structure of the receptors and the degree and rate of physiological adaptation to previous stimuli affects the collection and transduction of the stimulus.

5) **Stimulus coding:** In the first neurological steps after the reception of a stimulus, these are coded. The coding influences which properties can be extracted from the stimulus. Coding mechanisms can also lead to biases towards new stimuli, such as peak shift phenomena (see Section 6.2).

6) **Perception:** The mechanisms of processing stimuli can affect the attention towards them. For example, stimuli can be processed by the same or by different brain regions, which can influence the attention given to them.

7) **Recognition and classification:** Pattern recognition and classification of stimuli can bias receivers' responses to stimuli. For example, bees quickly discriminate visual stimuli that sport radial stripes and dark centres; features that occur in their nest holes and their floral resources (Biesmeijer et al. 2005).

8) **Cognition:** The assessment of a stimulus requires higher brain functions to extract information from it. The design of a stimulus and its overlap with confounding stimuli influence the information that can be extracted.

9) **Decision-making:** Receivers decide among alternative states (eat vs. reject). Decisions can depend upon memory functions, the availability of alternatives, and the internal motivation. For example, bees choose among floral colours depending on their experience with these colours 24 hours previously. The retrieval of colour memories in the honey bee thus follows a circadian rhythm that seem to match the circadian rhythm of nectar and pollen production in flowers (see e.g., Prabhu and Cheng 2008). Bees thus possess a *Zeitgedächtnis* (time memory) allowing them to adaptively fine-tune their foraging behaviour to the peak availability of nectar associated with visual stimuli.

review). This is (in part) why in Chapter 1 we define aspects of communication without reference to information and to how the receiver represents its environment cognitively. Here, we lay out the fundamentals that underlie visual, olfactory, and gustatory communication, focusing on the sensory underpinnings. We do not include temperature-sensing because there is currently no evidence that pollinators seeking heat rewards perceive heat remotely (Section 6.2). Animals not only rely on communication to locate plants and their rewards but can also find plants by use of their spatial memory. Here we are primarily concerned with the sensory underpinnings of communication; memory functions are briefly alluded to in Box 2.2. In each sensory mode, we first discuss the sensory world of the animals before switching to the side of the sender to discuss the biochemistry of the production of cues and signals in plants. Consideration of the biosynthetic pathways involved in the production of visual and olfactory traits is important for understanding the possible information of these traits. We start with vision because of its limited complexity compared to the perception and production of scent. We then (Section 2.7) consider olfaction and scent before turning to taste and plant biochemistry (Section 2.11). We finally (Section 2.12) discuss acoustics.

2.2 Chromatic and achromatic vision

Most animals are able to extract information from visual traits using two different aspects, *achromatic* information on variation in brightness or the intensity of reflected light from a given surface, and *chromatic* information on colour. Variation in brightness is perceived on the scale of white to black and often referred to as luminance if analysed according to the sensory abilities of animals. We humans perceive saturation and hue as chromatic aspects of visual information. Saturation describes the colours' similarity to a neutral grey; colours that contain no or very little grey are deeply saturated, whereas a

---

Box 2.2 Memory functions

Throughout the book we have made the admittedly simplifying assumption that animals strongly rely on sensory cues to locate food resources. While this seems intuitive, animals can rely solely on their spatio-temporal memory of the locations of known food resources and use any visual or olfactory cues only for identification of the food resource at close distance. In particular, territorial animals that remain in a given area for long periods can rely strongly on their spatial memory to track food resources. Landmark learning, for example, is believed to be important in many trap-lining hummingbirds. Also, birds such as toucans often patrol their home ranges and check on the ripeness of valuable fruit resources. They often fly directly to the fruiting trees from long distances even if the tree is not visible to them suggesting that they rely primarily on spatial memory (HMS pers. obs.). We predict that cue-based foraging will be particularly important in animals that are nomadic, as many frugivorous birds are but also fruit- and nectar-feeding bats, for example and which track fruit resources over considerable distances outside of their well-known home ranges (Fleming 1992, Moegenburg and Levey 2003, Telleria et al. 2008). As yet, the relative reliance of memory vs. cue-based foraging is unknown in most foragers and represents an exciting future research topic.

It is not surprising that there are strong interactions between cue-based foraging and memory functions because memory is an important property for learning. Again, circadian shifts in flower colour preferences in honey bees according to the circadian rhythm of the retrieval of colour memories are a good example. This circadian rhythm apparently tracks the circadian rhythm of nectar and pollen production in differently coloured flowers (Prabhu and Cheng 2008). This is a particularly interesting example of how cognitive abilities shape the way that receivers respond to plant communicative traits. We expect that such interactions are commonplace but as yet remain mostly uncovered.

grey with a little tint of colour has a low saturation (see Kelber et al. 2003 for a thorough review). In contrast, hue defines colour differences that are related to human colour categories such as red, yellow, and green. While there is no evidence that animals perceive hue and saturation as humans do, there is some evidence that bumblebees have a perceptual dimension of saturation (Lunau et al. 1996), and experiments on chickens suggest that they can categorize colours in a way similar to humans (Jones et al. 2001). These authors suggest that chickens like humans generalize across colours; that is, they treat stimuli that can be differentiated as equivalent.

The saying 'at night all cats are grey' illustrates the fact that colour perception requires relatively high light intensity, at least in most vertebrates and honey bees. This is because the noise in each photoreceptor is increased when the light intensity is reduced because fewer light quanta actually reach the photoreceptor. The noise originating in the cones sets a limit for the usefulness of colour vision to discriminate objects (Vorobyev and Osorio 1998). There are exceptions to the rule that colour vision requires relatively high light intensity; nocturnal hawkmoths, geckos, and the obligatory nocturnal Indian carpenter bee *Xylocopa tranquebarica* can all use colour vision even in starlight (Somanathan et al. 2008). In general, however, we might expect that that visually guided diurnal foragers select for colour traits (see Section 2.3 for discussion of background variability), whereas nocturnal foragers are more likely to select for visual plant traits that are conspicuous at night because their brightness stands out from the surrounding vegetation. While this conjecture appears plausible, the available evidence is limited given that many bat-pollinated flowers are not white but green or brownish-green (von Helversen and Winter 2003).

The first neuronal steps involved in the processing of colour stimuli may either sum photoreceptor signals (in achromatic vision), or compare them by some type of inhibitory interaction (in chromatic vision), to give the ratio or difference of receptor signals (Kelber et al. 2003). In the honey bee, which has become a model system for studying colour reception and perception, at least seven types of colour opponent neurons have been found in the optic lobes, although it is currently not clear how they contribute to colour perception (Chittka and Wells 2004). Colour opponency mechanisms are assumed to operate in most animals, but how they weigh the outputs of different photoreceptors is generally unknown. Achromatic and chromatic vision differ in the speed of neuronal processing. Typically, processing of achromatic stimuli exceeds that of chromatic stimuli. For example, in bumblebees the green receptor responsible for achromatic vision has an impulse response of 7.9 ± 1.1 ms, whereas those of the blue and UV receptors are 12.3 ± 1.8 ms (Skorupski and Chittka 2010).

Differences in neuronal processing entail that achromatic and chromatic vision are used for different tasks. The quicker processing of achromatic stimuli makes achromatic vision ideal for motion detection. Achromatic vision is also thought to be particularly relevant for another important task in vision, the identification of objects, particularly small objects. For humans, chickens, and pollinators such as bees and bumblebees, the discrimination of small objects is mostly colour blind (Livingstone and Hubel 1988, Giurfa et al. 1997, Osorio et al. 1999, Spaethe et al. 2001). Object identification is often based upon segregating images based upon high-contrast borders into different items. Edge detection is mediated primarily by achromatic contrasts because achromatic variation on the edges of objects typically exceeds, and is therefore more reliable than, variation in chromatic contrasts (Osorio et al. 1999). Chromatic stimuli reliably indicate the surface of objects and are therefore often more useful than achromatic stimuli for obtaining information about objects. This is particularly so because the intensity of light reflected from an object varies depending on whether it is directly illuminated by sunlight or whether it is blocked from direct illumination. As such, the luminance characteristics of objects can vary dramatically according to light conditions, while the apparent colour of an object changes to a far lesser extent with intensity, owing to the phenomenon of colour constancy (reviewed in Kelber et al. 2003).

2.3 Visual detection

When animals forage on plants, they need to detect them by discriminating the target plants (or their reproductive organs) from the background of the surroundings (which often consists of vegetation). Detection can thus be viewed as a deceptively simple task where the receiver decides whether a signal is present or not. This decision will depend on the signal-to-noise ratio. With increasing difference between the target and the background, correct discrimination of the target should follow an asymptotic curve approaching 100 per cent correct responses (Bradbury and Vehrencamp 1998). Importantly, vegetation which makes up an important proportion of the background of natural scenes varies, dramatically (that is, over 3,000-fold) in brightness (Sumner and Mollon 2000a, Regan et al. 2001). The strong variation in brightness is caused by the mosaic of sunlit spots and shadows created by leaves that intercept sunlight. Vegetation does not vary strongly in colour because it is mainly green. Thus, against a background of vegetation with pronounced achromatic variation, we can predict that chromatic stimuli are more salient and as such are more important for detecting objects than achromatic stimuli. This is simply because the signal-to-noise ratio is relatively high in chromatic vision but low in achromatic vision because any achromatic variation among senders (or among fruits and flower and their respective backgrounds) is likely to be masked by the strong achromatic variation among foliage in the background.

There are surprisingly few tests of the idea that colour (chromatic) differences between target and background are more important for object detection than achromatic differences against the background of foliage. This is likely because it is not easy to measure the distance of detection in free-ranging animals. Experimental evidence from large aviaries and from measuring the detection probability of differently coloured fruits support the hypothesis that chromatic contrasts between fruits and their background of foliage are more important than achromatic contrasts (Schaefer et al. 2006, Cazetta et al. 2009). If the background was relatively invariant, birds apparently also used achromatic contrasts to detect fruits (Schaefer et al. 2006). We encourage experiments testing this relationship in animals with distinct visual systems to assess the generality of these results.

Based on the above, we can predict that a stronger signal-to-noise ratio of communicative traits will lead to a higher detectability (see Plate 2 for another phenological strategy to achieve conspicuousness). Thus, traits that plants use to attract pollinators, seed dispersers, prey (in the case of carnivorous plants), and the predators of herbivores should be characterized by a high signal-to-noise ratio. However, the relationship between signal-to-noise ratio and plant fitness is not expected to be linear. This is because the detectability of a given stimuli cannot be enhanced above a certain threshold, simply because all stimuli exceeding that threshold will always be detected and correctly classified. While this theory is well accepted, the thresholds of detectability are poorly understood in natural scenes where the viewing distance and angle as well as the background noise often vary widely. We know of only one study (Cazetta et al. 2009) that demonstrated the theoretically predicted asymptotic relationship between detection and chromatic contrasts (Figure 2.1). Analysing the proportion of artificial fruits that birds detected, the authors found that fruit detection cannot be enhanced above a certain threshold. Given the pronounced differences between the sensory worlds of many animals, it is likely that the thresholds for detection (as well as

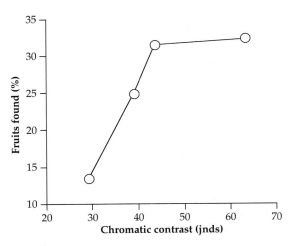

Figure 2.1 The number of artificial fruits that frugivorous birds detected is a sigmoid function of their colour contrasts to the overall background. This relationship demonstrates that the conspicuousness of visual targets cannot be increased above a threshold of approx. 44 jnds (just noticeable differences to birds). Redrawn from Cazetta et al. (2009).

the inclination of the slope prior to the threshold) vary among taxa. Analysing this variation would be important to evaluate the relative selective pressures that different receivers may exert on the conspicuousness of communicative traits in plants.

If the signal-to-noise ratio is strong, animals can employ a pre-attentive search mode where they scan their visual field in parallel for potential targets (Treisman and Gelade 1980). However, if objects are very similar to the background, pre-attentive visual search is complicated, and viewers scrutinize potential targets serially, categorizing each of them one at a time and identifying them as real targets or so-called distractors. It is likely that both types of search modes commonly occur when animals forage for plant rewards. Often, the colours of fruits and flowers may be sufficiently distinct from that of leaves so that animals can use a pre-attentive search mode. However, if the colours of differently rewarding flowers are similar, bumblebees sacrifice foraging speed and made slower foraging decisions (Dyer and Chittka 2004).

2.4 Ambient light

Visual traits – with the exception of biolumines-cence – are dependent on illumination by an external light source, mostly the sun, but also the moon in the case of flowers that are pollinated by nocturnal pollinators such as moths and bats. While the spectral composition of sunlight is essentially constant, the spectral composition of the prevailing ambient light varies across habitats depending on the amount of vegetation between the sun and the focal trait (Figure 2.2). Vegetation filters the incoming light because light is transmitted through the leaves, and it is reflected at various angles off the leaves. Endler (1993a) defined five main light habitats that are present in terrestrial habitats on sunny days. These are open areas (including large gaps in the forest), small gaps in the forest, woodland shade, forest shade, and a light habitat specific to dusk and dawn. Open areas are characterized by white light, the typical spectrum of sunlight. Small gaps are characterized by a relatively high proportion of reddish light, woodland shade by an overrepresentation of blue light (because the sun is blocked through vegetation and most light enters indirectly from the

Figure 2.2 Visual traits with the exception of bioluminescence are dependent on quality and quantity of ambient light. Light quanta (S) are emitted from light sources (the sun during the day, the moon in the night) and can be scattered and absorbed by particles (P). The ambient light illuminating the target and the background is the sum of direct illumination from the light source (S), and light reflected from and through vegetation (A). The reflectance spectrum (R) reflected from the target depends thus on illumination by S and A. The reflectance spectrum passes through the air, where it is scattered and partially absorbed by particles (P) thereby forming the transmittance spectrum T. The quanta (Q) that are transmitted to the eye of the receiver are those that are processed by its sensory system.

blue sky), and forest shade by green light because much of the incoming light is filtered through the vegetation. The dusk light habitat is characterized by purplish light because illumination is indirect since the position of the sun is below the horizon. On cloudy days all light habitats converge on the light habitat of open areas because all incoming light is filtered by the clouds.

The qualitative differences in the prevailing wavelengths among light habitats imply that colours are transmitted differently in each light habitat. For example, the transmission of light reflected off a red fruit would be enhanced in small gaps because there is much reddish light available in this habitat. In contrast, a red fruit is less well transmitted within the forest because the prevalence of green light results in a relatively higher transmission of green wavelengths (compared to red ones). Consequently, the resultant spectrum of a red fruit that reaches the eye of the receiver is more saturated (see below) in a small gap than it is inside the forest. Theoretically, these qualitative differences could lead to adaptations in visual traits so that the reflectance of a fruit is optimally transmitted in a given light habitat. For example, plumage colours of birds displaying in certain light habitats within forests are suggested to be adapted to these habitats so that their colours are transmitted optimally in that light habitat (Endler and Thery 1996, Heindl and Winkler 2003). As yet, there is no evidence that plant colours are adapted to communicate in a specific light habitat. For example, we have not found that the diversity of fruit reflectances differs among light habitats (HMS unpubl. data). However, adaptations could also be more subtle. For example, it is feasible that plants growing in locations with variable light environments (e.g., pioneer species or those confined to forest edges) have a higher prevalence of colours that are less likely to be degraded in a specific light habitat. These colours would be either very saturated or achromatic (being white or black). We do not know of any study that has investigated the saturation of colours in distinct light habitats, and would welcome more studies on this subject.

Depending on whether illumination is direct or indirect (e.g., owing to cloud cover), there are pronounced changes in light intensity. Variation in the chromatic composition of ambient light is smaller than variation in the achromatic composition. This, in concert with the pronounced achromatic variability in the background of natural scenes (Section 2.3), makes chromatic contrasts reliable cues to detect objects under changing light conditions (Kelber 2005). The changes in light intensity have little effect on colour perception owing to the phenomenon of colour constancy. Colour constancy means that our perception of a banana is similarly yellow independent of whether we view it in full sunlight or in the shade created by foliage. Colour constancy is achieved if the relative excitation of the photoreceptors (their quantum catches) from a given target (the banana) are related proportionally to the quantum catches of its background (the foliage). In this case, quantum catches change with changes in illumination but the rescaled vector resulting from the ratio of quantum catches between target and background changes substantially less

(Kelber et al. 2003). Chromatic adaptation of the photoreceptor cells, lateral interaction of neurons in the retina or higher up in the visual pathway, and cognitive processes contribute to colour constancy and lead to a relatively stable perception of colours independent of illuminating light (Hurlbert 1998, Komatsu 1998). Colour constancy is best known in humans but has been demonstrated for numerous other animals (reviewed by Neumeyer 1998).

Many animals can perceive the polarization of light. It is known that plant surfaces can affect the plane of polarized light, but we have not found evidence that polarization of light from plant surfaces affects the foraging behaviour of animals.

2.5 Visual sensitivities

In order to extract colour information, an animal needs at least two distinct receptors that differ in their spectral sensitivities because their differential output, when viewing a stimulus, is a prerequisite to distinguish between colours that do not differ in brightness. Obviously the sole presence of pigments with different wavelength sensitivities is not sufficient for colour vision. To discriminate among colours and not simply show wavelength-specific behaviour such as phototaxis, a subsequent neural pathway has to compare the output of the different photoreceptors (Pichaud et al. 1999). The relative input of each receptor type then defines colour. A corollary is that colour is not an intrinsic character of any object, but colour only appears in the eye of the beholder. We admit that we have used colour in the colloquial sense of being a characteristic feature of plant traits throughout this book, mainly for trying to enhance readability by not using the technical term of reflectance.

Colour vision is variable among animals and can be characterized according to the number of distinct receptors that are found in their retinas. Monochromats have only one receptor type (and can thus not perceive colour), *dichromats* which include many mammals have two receptor types, primates and many insects are *trichromats* with three receptor types, while most fishes yet known, most birds and lizards, and some insects like the Pieridae butterflies are *tetrachromats* (Figure 2.3). The colour vision of mantis shrimps (stomatopods)

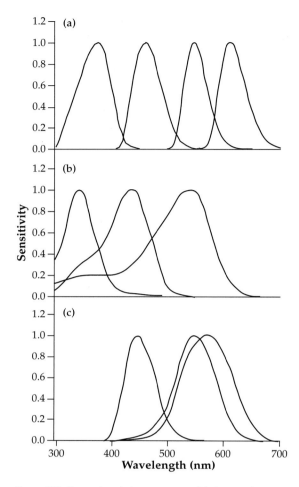

Figure 2.3 The number of photoreceptors and their spectral sensitivities differ among animal groups. Birds are tetrachromatic possessing four cone types with distinct sensitivities (a), whereas bees (b) and humans (c) are trichromatic possessing three different cone types. The spectral sensitivities of bees and humans differ in that bees have a UV-sensitive cone (with peak sensitivity below 400 nm), whereas humans have a cone sensitive to red light with peak sensitivity at 570 nm.

is exceptional in that they have a poorly under-stood 12-channel colour visual system (Marshall et al. 2007). Animals differ not only in the number of photoreceptors but also in their sensitivities to par-ticular wavelengths. Primates are not sensitive to ultraviolet light, while most insects, birds, reptiles, and fishes are. Conversely, many insects are not particularly sensitive to colour variation in the red

part of the spectrum, while primates, birds, and fish are sensitive in this area.

The colour visual system of terrestrial taxa tends to be evolutionarily conservative. There is variation within phylogenetic lineages; for example, the vis-ual systems of insects vary from dichromats to tet-rachromats, and both trichromats and tetrachromats are found within the Hymenoptera (Briscoe and Chittka 2001). However, there is little evidence for adaptive variation in insect vision that maximizes the ability to discriminate among reflectance stimuli such as those of flowers and fruits. This is partly explicable by the fact that colour vision in many animals did not evolve in a co-evolutionary process between plants and animals. For example, the col-our vision of arthropods probably evolved *circa* 570 million years ago, thereby preceding the evolution of flowers and fruits in angiosperms by roughly 400 million years (Chittka 1997). The available evidence suggests that colour vision is adapted to respond to a broad range of stimuli that involve detection tasks (of food, mates, and predators), communication in social groups as well as general orientation during movement (see Box 4.2 for optimization of primate colour vision).

Evidence of adaptive variation in animals' sen-sory systems is stronger in aquatic environments, where light gradients are particularly pronounced, and ambient light changes consistently with increas-ing water depth. Evidence for adaptive tuning of photoreceptors has been found in several sister spe-cies comparisons in marine and limnetic environ-ments (Boughman 2001, Cummings 2006, Seehausen et al. 2008). In these fish (sticklebacks, surfperch, and cichlids), the perceptual sensitivity to colour varies according to the prevailing ambient light in the water column, and males evolved nuptial col-ours that match the perceptual sensitivities of females. These studies provide the best evidence for sensory drive, where adaptation to the general environment drives the evolution of the sensory system and, consequently, also that of the epigamic signals used for mate choice. It is probably the more pronounced and more consistent changes in ambi-ent light that make aquatic habitats home to clearer examples of sensory drive than terrestrial habitats.

Owing to the pronounced variability in visual sensitivities, it is necessary to quantify the reflectance of targets and backgrounds using a spectrometer. The spectrometer measures the reflectance of surfaces, such as fruit skin and the structures against which fruits are presented, relative to standard white and black references. The resulting reflectance spectra (Figure 2.4) can than be analysed according to the spectral sensitivities of photoreceptors. Additionally, the spectrometer can quantify the irradiance, which characterizes the spectral composition of the ambient light at the place where the target is presented and communication takes place.

The pronounced differences in spectral sensitivities (and those found in other sensory modes as well) can be important for the evolution of plant communicative traits. This is because differences in spectral sensitivities can translate into differential conspicuousness of plant colours. For example, a

red flower will be more conspicuous to a hummingbird that has a photoreceptor sensitive to red light than to bees lacking such a photoreceptor. Such differences in the relative conspicuousness of flowers can lead to ethological isolation if pollinators restrict their visit to flowers that they detect easily. Because pollinators include various groups with distinct spectral sensitivities such as birds, reptiles, and marsupials, as well as many distinct insect groups (e.g., hymenoptera, lepidoptera, coleoptera, diptera, rhynchota), this functional explanation may commonly underlie ethological isolation and the evolution of pollination syndromes.

Owing to the differences also in spatial acuity of colour vision among animal classes, it seems logical that very distinct groups also use different signal components for long-distance detection. However, we know of no study that has formally compared

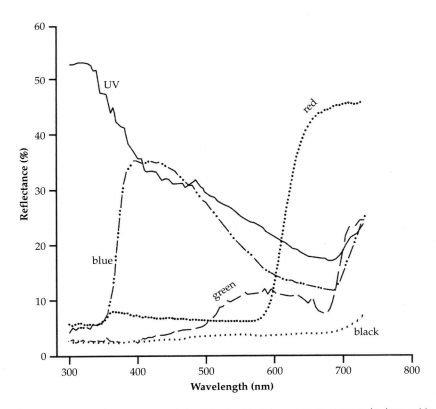

Figure 2.4 Fruit reflectance spectra over the spectral range that is visible to birds. Colour categories correspond to human vision (except for UV) and depend on the spectral area where the peak reflectance occurs.

the distances of detection among distinct pollinators (e.g., insects and vertebrates) and related them to selection upon variation in floral traits. Yet, it is known that bees detect flowers from a distance larger than a few centimetres by their achromatic contrast in the green part of the spectrum. As such, we can hypothesize that bee pollinators select for an increased difference in the intensity of the reflected light of a target flower relative to the background, whereas pollinators such as hummingbirds can feasibly also select for colour differences in other parts of the spectrum.

2.6   Colour production

Most plant colours are produced by a mixture of three common pigment classes: the chlorophylls, carotenoids, and anthocyanins – with betalains replacing anthocyanins in a phylogenetically restricted group of plants. Pigment-based colours are not polarized and diffusing, meaning that they are perceived similarly from varying angles. While pigment-based colours dominate in plants, structural colours also occur in the vegetative and reproductive tissue of some plants. Structural iridescence is the change of hue with variation in the angle of observation. In other words, iridescence occurs if a fixed object changes colour with the movement of the observer. Structural colours are common in animals but relatively rare among plants. Structural iridescence occurs in plants that live in low-light environments and has been proposed to function in photoprotection or optimization of photosynthesis (Glover and Whitney 2010). Iridescence can be produced in two different ways. Leaf iridescence can be caused by multiple layers in the outer cell wall of the epidermis or within the cell (Hebant and Lee 1984, Lee 2007). Iridescence is also found in flowers where finely ridged surfaces act as diffraction gratings (Whitney et al. 2009). Modelling bee vision suggested that bees can perceive the changes in hue resulting from iridescence, and experiments revealed that they did so independent of underlying pigments, UV colouration, or polarization effects (Whitney et al. 2009). Given that iridescence in plants is still poorly known, we lack evidence for its role in communication to animals. It is conceivable that the change of hue that accompanies insects'

movements on a flower may help the insect orient and find nutritional rewards. We would welcome studies investigating the adaptive significance of iridescence in flowers and other plant parts.

2.6.1   Chlorophylls

Chlorophyll is the primary photosynthetic pigment imparting green colouration to plants. It is found in almost all angiosperms throughout all life stages and all individuals. Green colouration is produced because the reflectance of chlorophyll peaks at 555 nm. However, there is considerable variation in the reflectance of leaves, which is partly achieved by variation in the amount of chlorophyll present in the leaves, by variation of accessory pigments (e.g., carotenoids), and partly by variation in the waxy layers that cover the epidermis of plant organs. Chlorophyll is most important in leaves, but it is also the primary pigment of developing fruits and flowers. Green stems, leaves, and some developing reproductive organs are optimized for light harvesting and are characterized by net photosynthetic assimilation using mainly atmospheric CO_2. For example, chlorophyll contributes to the production costs of developing peaches by fixing 15 per cent of the requirements of atmospheric CO_2 (Pavel and DeJong 1993). Beside a net photosynthetic carbon uptake, chlorophylls are important in refixing respiratory CO_2 from reproductive organs (Watson and Caspar 1984, Ashan and Pfanz 2003). Flowers are typically green in the early developmental stages when the relative importance of floral photosynthesis in contributing to the carbon requirements of flower production may be highest and range between 0–63 per cent (Moehs et al. 2001, Ashan and Pfanz 2003). The chlorophyll contents of flowers typically decrease during floral development, although there are flowers from many families that remain green and photosynthetically active when mature (Heaton and Marangoni 1996). Similar to flowers, the chlorophyll contents in unripe fruits are usually high and decrease during ripening (Ashan and Pfanz 2003). However, in mature fruits and flowers the photosynthetic activity of chlorophylls is often secondary. Here chlorophyll plays a different role in refixing respiratory CO_2, particularly from reproductive organs such as climacteric fruits. The degradation of chlorophyll is induced by the plant hormone ethylene (see Box 5.1). Although most fruits are non-green when ripe, there is a distinct set of species with ripe green fruits which are usually large and typically dispersed by bats or large terrestrial mammals (e.g., Korine et al. 2000, Guimarães et al. 2008).

Chlorophyll in developing fruits and flowers is not only advantageous because it helps in reducing the costs of producing plant reproductive organs, but also reduces the conspicuousness of the developing organs against the predominantly green background. Although camouflage is a long-standing hypothesis to explain the chlorophyll contents of developing flowers and fruits, we know of no study that evaluated the selective pressures related to communication (camouflage) relative to those related to plant physiology (fixation of CO_2).

2.6.2   Carotenoids

Carotenoids are lipophylic pigments that impart yellow, orange, and red colours to vegetative and reproductive plant tissue. They belong to the extremely large group of terpenoids (see Section 2.10.2). The occurrence of carotenoids in leaves is not transient; it is usually masked by chlorophylls. Therefore, the yellow and orange colours imparted by carotenoids appear usually only in senescent leaves where chlorophyll is broken down (Section 7.3.1). In vegetative tissue carotenoids mainly function to absorb light for photosynthesis and also in photoprotection. Similar to the masking in leaves, carotenoids in fruits are often masked by anthocyanins which are more concentrated and impart more light per molecule than carotenoids (Sims and Gamon 2002, Schaefer et al. 2008b). During fruit ripening there is no overall trend for changes in carotenoid concentrations, as these might increase or decrease (Minguez-Mosquera and Garrido-Fernández 1989, Fraser et al. 1994, Schaefer et al. 2008b). In flowers, carotenoid concentrations can increase over a 100-fold during colour development (Moehs et al. 2001). The appearance of yellow to red hues is thus associated with carotenoid accumulation rather than, as in senescent leaves, with the unmasking of already present carotenoids due to the degradation of chlorophyll (Wilkinson et al.

2002). This implies that the appearance of yellow or red colour is associated with distinct volatiles in flowers (those related to growth) and in leaves (volatiles related to senescence). In contrast to vegetative tissue, where red carotenoids are scarce, they are commonly found in fruits and flowers (e.g., lycopene).

Carotenoids comprise over 600 known substances that are C_{40} isoprenoids with numerous double bonds and often with cyclic end groups that comprise carotenes (mainly orange) and xanthophylls (mainly yellow), the latter being the oxygenated derivates of carotenes. Variation in carotenoid colour depends on the number of double bonds, the presence of end groups, and conjugation with other molecules (Goodwin and Goad 1970). Carotenoid biosynthesis occurs in the plastids which later differentiate into chromoplasts.

2.6.3   Anthocyanins and flavonoids

Anthocyanins impart blue, red, or violet colour; at high concentrations they may appear purplish-black, as seen in many fruits (e.g., blackberries and cherries). Anthocyanins are the main pigments in plant reproductive organs, but are also present in most other organs from roots to leaves. They are water-soluble substances of the large group of phenols that are the end product of the phenylpropanoid (or shikimic) pathway. Phenolic compounds include a wide variety of substances that possess an aromatic ring with a hydroxyl (OH) substituent. The phenylpropanoid pathway is one of the best-known pathways that produces many biologically active substances such as tannins (employed in defence against herbivores and pathogens, excess light), flavonols and flavones (both employed in defence against pathogens and excess radiation). Because the molecular basis of the anthocyanin biosynthesis is well understood, flower and fruit colour are often the target of genetic engineering (Davies 2009).

Anthocyanins are found in all plant tissues, but they are often transiently distributed, appearing in some stages (e.g., young and senescent leaves) but disappearing in others (e.g., most mature leaves; Gould et al. 2009). They are also environmentally transient, appearing and disappearing with changes in photoperiod and temperature (Chalker-Scott 1999, Gould and Lee 2002, Kong et al. 2003). In fact, anthocyanin synthesis is seen as a biochemical model system for studying mechanisms of photoregulation (Mancinelli 1983). It is precisely the environmental induction of anthocyanins that links to their adaptive significance in the amelioration of abiotic stress reactions (Section 7.3.2).

Anthocyanins occur mainly in or just below the upper epidermis of leaves, although they are also present in the lower epidermis and in the mesochym in some species. More than 500 anthocyanins are currently known, from 6 different major groups. Cyanidin is the most important form, being present in 80 per cent of leaves, 70 per cent of fruits, and 50 per cent of flowers (Kong et al. 2003). Anthocyanins differ according to their hydroxyl groups, the nature and number of sugars attached to the molecule, and the position of the sugar attachment. The colour imparted by anthocyanins differs according to hydroxylation, methylation co-pigmentation, complex-forming with metals, their overall concentration, and the acidity of the vacuoles where they are stored.

As already mentioned in the Preface, anthocyanins are very versatile pigments with multifarious ecological and physiological effects. Anthocyanins increase during fruit ripening, and they may mask carotenoids in ripe fruits, being on average 16 times more concentrated than carotenoids and imparting more colour per molecule (Schaefer et al. 2008b, see Section 4.5.3). Frugivorous birds prefer artificial food with anthocyanins over those without (Catoni et al. 2008), and they have a strong anti-fungal effect in fruits (Schaefer et al. 2008a). Owing to their function as antioxidants, they might be induced by wounding in leaves (Gould et al. 2002) and thus be indirectly associated with herbivore attack (Costa-Arbulú et al. 2001). However, since anthocyanins share the biosynthetic pathway with deterrent phenols, the production of anthocyanins might be a cue indicating to animals that this important pathway for plant defence is currently activated (Fineblum and Rausher 1997, Schaefer and Rolshausen 2006, see Section 7.4.2). There are two possible scenarios following from this conjec-

ture. On one hand, deterrent compounds might be more quickly synthesized from a currently activated pathway because precursors (e.g., for anthocyanins) can be channelled towards the production of deterrent substances. On the other hand, allocation trade-offs of the precursor forms might result in a lower production of deterrent substances. Several trade-offs have been proposed: for example, between lignin (which shares the first precursors with the phenylpropanoid pathway) and anthocyanins. However, in general, the issue of biochemical trade-offs between different functions remains understudied.

A very diverse array of secondary compounds is produced by the phenylpropanoid pathway; these compounds are often referred to as flavonoids. Flavonoids include flavanones which are colourless, whereas other flavonoids, flavones, and flavanols absorb UV light and thus produce a specific colour. Flavonoids can be excreted by leaves and possibly function as protection against desiccation. However, they can also have an anti-microbial function (Cooper-driver and Bhattacharya 1998) and are often involved in plants' stress responses (Winkel-Shirley 2002). Flavonoids often co-occur with anthocyanins. We know of no broad interspecific study that has investigated whether colour induced by flavonoids can be used by animals as a cue or whether it is (usually) masked by anthocyanins or other compounds. The role of flavonoids in plant–animal communication is therefore poorly known. Clearly, owing to their various ecological effects as protection against abiotic and biotic factors, there is the potential that they indicate plant defence and can thus be used as informative cues as animals. Anthocyanins do not apparently co-occur with another group of red to purple and black pigments: the betalains.

2.6.4 Betalains

Betalains impart similar colouration to anthocyanins, but they are restricted to the suborder of Chenopodiniae comprising 13 families (Stafford 1994). Betalains are most commonly found as betanidin; with well-known examples being the colouration of prickly pear fruits (*Opuntia* sp.), red beet

(*Beta vulgaris*), and the colours of pokeweed fruits (*Phytolacca* spp.). However, large colour variation may be found in flowers (e.g., yellow to violet in *Bougainvillea*). Red and violet colour is achieved by different substitution patterns (Stinzig and Carle 2004). Betalains can be deposited as crystals in vacuolar bodies but usually occur in diluted form inside the vacuoles. Betalains may function as osmolytes to maintain physiological processes. The precursor of betalains is tyrosine. The origin from an amino acid could imply that pleiotropic effects occur where variation in betalain colours is correlated with specific physiological or stress-related plant traits. Our idea is speculative and has not been tested yet. It is based upon the well-known pleiotropy of melanin, another amino acid-derived pigment, in animals (see Ducrest et al. 2008).

2.7 Olfaction

Olfaction is complex. Odours consist of volatile organic compounds (VOCs) with a molecular weight below 300 and that are – like pigments – discrete. Unlike pigments, VOCs do not combine to form a continuous, physical trait such as reflectance. Instead, the discrete VOCs that make up blends are sensed differently by many distinct types of odour receptors. From the limited evidence available (Chittka and Brockmann 2005) it seems as though they are also perceived separately – that is mapped to distinct region in the brain of receivers – according to chain length and functional group (e.g., alkanes, alcohols, ketones, aldehydes). Moreover, and in contrast to the limited diversity of plant pigments, there is an extraordinary diversity of plant odours; more than 1,700 VOCs have been identified in the bouquets of flowers alone (Knudsen et al. 2006), with some flowers emitting more than 100 VOCs (Levin et al. 2003). The complexity of odour blends may be further enhanced by the chirality and the position and stereochemistry of double bonds (Dötterl et al. 2007). For example, the two different forms (enantiomers) of the common plant volatile linalool influence oviposition rate (but not feeding decisions) differentially in the nocturnal moth *Manduca sexta* (Reisenman et al. 2010). Females prefer laying eggs on plants emitting (+)-linalool. The different reactions to linalool in distinct contexts suggest different neuronal pathways, with the enantiomere-specific pathway found only in females. Even we humans, with a sense of smell that is considered poor, can smell thousands of odours and distinguish among them. The discrete nature of odours leads to an unparalleled dimensionality of scent production, reception, and perception that has complicated the conceptual growth of that field until a few decades ago, when gas chromatography coupled with mass spectrometry allowed increasing identification of compounds and molecular advances shed light on the diversity of scent perception.

How do animals perceive the great diversity of odours surrounding them? Not surprisingly, there is a greater variation among animals in the number of odour receptors they possess compared to the variation in photoreceptors used to intercept light (Section 2.5). More than 1,000 odour receptor genes were found in mammals, making it an exceptionally diverse gene family, although as many as 50 per cent of the genes are thought to be pseudogenes in humans with a lesser proportion of pseudogenes in other terrestrial taxa (Sosinsky et al. 2000, Touhara and Vosshall 2009). Fish and amphibians have approx. 100 odour receptor genes, whereas adult *Drosophila* possess 60 and honey bees more than 130 odour receptor genes (Firestein 2001, Chittka and Brockmann 2005). The consequence of these differences in the quantity of odour receptors is that vertebrates should have a larger sensory odour space that contains more dimensions than that of insects.

Vertebrates perceive smell with their noses, whereas in insects the odour receptors are located in the antennae. Invertebrate receptors are not homologous to the receptors of vertebrates, and indeed they have very little similarity to each other. Odour receptors differ in a region of hyper-variability that provides the binding sites for different ligands thereby forming the molecular basis for the reception of distinct odours. Similar to the photoreceptors, odour receptors can be classified according to their sensitivity to different odours. In general, there is a wide range of sensitivity with some receptors being relatively specific and others more broadly stimulated (Hallem and Carlson; Hallem et al. 2006; for a review on odour perception and its molecular underpinnings, see

Firestein 2001). For a general understanding of the sensory reception of odours it is sufficient for now to keep in mind that most odours can stimulate different receptors, and most receptors recognize several odours that are chemically related.

The great diversity of odour receptors makes neuronal coding of odour receptor outputs a rather complex affair. As in the coding of photoreceptors, the coding can be inhibitory or additive, but because more receptors can be involved in neuronal coding (compared to vision), the stimuli can be graded on a very fine scale by comparing outputs across receptors. For example, studies on odour reception in fruit flies showed that odours can be discriminated by combining the outputs of broadly tuned odour receptors (Hallem et al. 2006). This coding mechanism represents a generalist odour perception system. Many animals additionally possess a specialist odour perception system where specific substances such as sex pheromones are detected by specific, narrowly tuned receptors. For example, the silk moth (*Bombyx mori*) famously uses only one compound (bombykol) to elicit the full range of sexual behaviour. This compound is received by only two receptors that are insensitive to other odours to which silk moths respond (Nakagawa et al. 2005). Interestingly, in a specialist odour perception system, filtering is not a big issue because the background is not expected to produce the specific substance that the receptors are tuned to.

Neurones collect stimuli from the different odour receptors of a given type. In three species of Heliothine moths four neurone types were identified, three of them responding to inducible plant volatiles, and one to the common plant volatile: geraniol (Stranden et al. 2003). Each of these neurone types had a narrow tuning, responding only to a few structurally related compounds of one of the following groups: monoterpenes, sesquiterpenes, homoterpenes (which are C-11 and C-16 compounds that are induced in response to herbivore damage see Section 2.10.1), and monoterpene alcohols. Finally, the reception of scent is not only influenced by the sensitivity of odour receptors but also by the quantity of specific receptor types that an animal possesses. Some 60,000 odour receptors are distributed along the antennae of honey bees (Galizia and Menzel 2000). Unfortunately, equivalent numbers are not widely known for distinct types of herbivores, pollinators, and seed dispersers.

The olfactory system functions by counting the binding events of molecules to which receptors are sensitive. By integrating binding events over relatively long times, neurons are able to include even low-probability binding events in generating their response (Firestein 2001). Temporal resolution is thus lost (owing to a long integration time) and traded for a higher specificity. The extent to which this notion holds generally for olfactory systems is contentious because honey bees, for example, are exceptionally fast in learning odours (Menzel 1985). They can discriminate faster among odours than among colours and discriminate among strong odours within 0.47s of perceiving them, while they are slower in discriminating lower amounts of odours (Wright and Schiestl 2009).

The remarkable diversity of odour receptors and neurones can have important implications for the evolution of chemical communication. Many of the emitted compounds will be widely smelled by many different animals. However, the variability among odour receptors across animals entails a higher probability than in other sensory modes that some compounds are not received widely. Therefore, private communication channels can evolve which match the sensitivities of the intended receivers but not those of unintended receivers such as predators. We know of no study investigating the hypothesis that private communication channels are more likely to evolve in olfaction than in other sensory modes. Given the diversity in floral colours and scents, flowers are obviously a good study system to address this question in the context of plant–animal communication.

Robert Raguso (2008) compiled a list of potential floral odour candidates that might represent private channel communication. Not surprisingly, these odours are narrowly distributed among angiosperms, and they fulfil specialized roles. Interestingly, these odours have diverse biochemical origins and functions including the use by Euglossini bees and fruit flies as nuptial gifts, the mimicry of female sex pheromones in wasps, and the attraction of bees to oil glands. All of these pollination systems are more specialized than the traditional food-rewarding systems of presenting nectar or pollen rewards, leading

Raguso to pose the question as to whether specialization in food-rewarding system is more likely to be achieved by a combination of sensory and morphological filters rather than a sensory filter only. We know of no study that has quantified this interesting hypothesis.

Olfaction differs from vision in that it can provide information on time since an event has occurred because the decay rates of chemical compounds differ. For example, groups that are attached with single bonds are less stable than those attached with double bonds. Receivers are thus able to assess when a scent mark has been left behind by comparing the decay of different compounds relative to a template of a fresh scent mark. Information on time is probably less important in plant–animal communication than it is in the communication among animals, but it might be an important source of information in the chemical footprints that pollinators leave behind on flowers (Box 6.1).

2.8 Olfactory detectability

As in vision, the detectability of stimuli should depend on the signal-to-noise ratio. However, if olfactory receptors are very specific because they are narrowly tuned, the response can be an almost binary 'all or nothing'. Supporting such a stepwise model, single molecules are sufficient to elicit a neuronal response in silk moths (Kaissling and Priesner 1970). Similarly, *Manduca sexta* moths showed no behavioural change over a 1,000-fold range of concentration of *Datura* bouquets (Riffell et al. 2009; Section 6.6). Often we cannot measure olfactory detectability along a single axis of increasing contrasts because most odours consist of multiple blends. Instead, there are theoretically as many dimensions of detectability in a blend as there are distinct receptor types to detect different odours; the importance of each of these dimensions will vary according to the sensory capacity of a given animal and according to the composition of background odours. Background odour can influence detection by masking target odours, by reducing the sensitivity of odour receptors (to compounds that are commonly found in the background) or by enhancing the sensitivity of odour receptors (to compounds that are found in small quantities in the

background) (see Kelling et al. 2002). It is thus not surprising that the detectability of a volatile depends on its relative commonness in the overall environment.

Large-scale assessments of background odours are lacking because they are technically not easy to perform, because background odours consist of a variety of components that differ quantitatively and qualitatively on a fine spatial scale. There is further temporal variability in background odours as odour production can be rhythmic in plants, and odour composition is strongly influenced by prevailing temperature and wind conditions. As yet, assessments of the quantitative and qualitative differences in background odour in natural settings are missing. Even in the absence of these data we can (tentatively) conclude that selection for the detectability of olfactory traits can be quite distinct from that of visual traits.

Everything else being equal, smaller molecules are more volatile than longer-chained molecules and as such are more likely to be used for long-distance communication (Brennan and Zufall 2006). If substances are used for long-distance signalling that are common in the background or received by multiple receptor types whose output is compared (e.g., if the stimulus consists of a ratio of compounds), a gradual stimulus – response curve such as that of Figure 2.1 for visual traits can be expected. As outlined above, this response curve is likely to be very steep in narrowly tuned receptors. Selection on detectability can thus differ depending on whether a specific substance is used in communication or whether it is based upon a blend.

At first glance, selection for an increase in the emission of compounds seems more likely in a gradual stimulus – response curve because the size of the stimulus is correlated to the response. Yet, since the diffusion of odour molecules is gradual and depends on air movement, an increase in the emission of very specific substances still increases the likelihood that these molecules reach an intended receiver. In general, selection for increased quantities of emitted volatiles can be common in private communication channels because it is not countered by the selective pressures of predators. The enlarged antennae of many male moths testify to the importance of increasing the probability of capturing scent molecules of females. Although there is evidence that animals can select for higher quantities of plant volatiles (e.g., Majetic et al. 2009; Figure 6.11), pollinators do not have generally larger antennae than other insects, and we currently do not know whether such a selection pressure is a common phenomenon in plant–animal communication.

Odours have a low directionality compared to visual traits. Pollinators thus need to track odour gradients in order to arrive at the source. There are strong differences among animal classes (and also among species of the same class) in their ability to perform tracking. Silk moths are not exceptional in their ability to track low quantities of odours (eels and dogs can do the same). However, not all animals can perform such long-distance tracking equally well. For example, stingless *Trigona* bees use scent trails by depositing saliva on leaves to guide nestmates to rewarding food resources (Schorkopf et al. 2007). If stingless bees are unable to use scent marks (e.g., across a water body), they fail to recruit foragers even over distances of a few hundred metres (Lindauer 1975). Stingless bees deposit more scent marks close to the food source (Stefan Jarau pers. comm.) suggesting that these bees rely on repeated encounters with scent marks to encode the distance to the food source rather than on the plume of the food source itself. Comparing silk moths and stingless bees, it is easy to envision that foragers differ in their selective pressure upon plant odour depending on their abilities to spatially track these odours. As yet, the extent of these sensory differences is mostly unknown. However, given that floral visitors, for example, commonly comprise species from very different groups (Coleoptera, Diptera, Hymenoptera, Thysanoptera), it is quite likely that the differences in their sensory abilities lead to differential selection pressures upon blends. Analysing the number of foragers attracted to a food source in response to manipulation of the quantities of different compounds within a blend could shed light on how strongly insects differ in their sensory perception of odour compounds.

2.9 Evolution of plant odours

Odours are likely to be the primary communication channel of higher plants because fragrances are nearly universally employed among the early angiosperms (Pellmyr and Thien 1986). There are several possible scenarios for the evolution of plant odours depending on in which context they are used. Plants use induced volatiles to recruit the enemies of herbivores in response to attacks by herbivores. This communication system may have evolved from communication within a plant because airborne signals are quicker to inform distant parts of the same plant than a systemic response (Heil and Bueno 2007). Alternatively, the lesion of plant tissue induced by herbivore activity leads to the unavoidable emission of volatile compounds. If predators and parasites of herbivores used this unavoidable consequence of feeding activity to locate herbivores, these odour compounds could have been the precursors for an intentional communication system between plants and the enemies of their enemies (see Chapter 9).

There are alternative scenarios for the evolution of floral odours. One scenario for the evolution of flower odours is that they were originally employed in defence against herbivorous insects and only secondarily co-opted for the attraction of pollinators (Pellmyr and Thien 1986). This scenario is not unlikely given that many odour compounds in flowers also occur in vegetative tissue (Figure 2.5) and that odour compounds such as monoterpenes are chemically similar to compounds that can have defensive properties (Section 2.10.1). Furthermore, herbivory evolved among insects prior to pollination. In particular, Pellmyr and Thien (1986) emphasized that the volatile compounds of the resins of gymnosperms, which function to deter herbivores, are chemically very similar to those emitted by flowers. As such, the defensive function is likely to precede the evolution of attraction. This view matches the sequence of events in the evolution of specialized pollination systems such as those of resin-collecting pollinators. Armbruster (1997) showed that floral resins are chemically similar to foliar resins and both types of resins significantly reduced feeding activity. Interestingly, when the resin-rewarding pollination (which is an exaptation of the original defence) was lost in evolution, resins were independently employed again several times for their defensive function (Armbruster et al. 2009; see Section 6.4.1). The relative importance of defensive and attractive functions can thus change

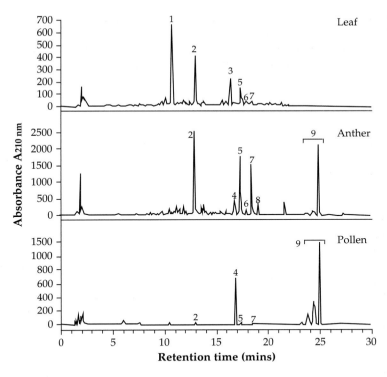

Figure 2.5 Phenolic compound production in leaf, anther and pollen tissue of *Solanum peruvianum*. The HPLC chromatograms show distinct peaks which correspond to specific plant odours. The position of peaks is a function of the molecular weight and chemical structure of the molecule. The emission of some volatiles is correlated in the different tissues. Numbers correspond to: 1, neochlorogenic acid; 2, chlorogenic acid; 3, quercetin glycoside; 4, 7, 8, unidentified flavonoids; 5, rutin; 6, quercetin glycoside; 9, coumaroyl derivatives). Redrawn from Kessler and Halitschke 2009.

over time, creating a particularly dynamic pattern of chemical evolution.

Alternatively, Schiestl (2010) proposed that plants exploited the communication systems among insects by using those compounds that insects were already pre-adapted to respond to. Schiestl showed 87 per cent overlap in the VOCs produced by plants and insects. Such an overlap might occur because these compounds are particularly easy to synthesize or because they are particularly suitable for transmission through the air or because such compounds are adaptive in another context. Examining the overlap between insect and plant volatiles according to the functional group of insects (herbivores and pollinators) revealed a pattern suggesting an adaptive basis for the observed overlap. There was only a correlation in the VOCs between the

angiosperms (that are often pollinated by insects) and pollinators, but not between primarily wind-pollinated gymnosperms and pollinators. Conversely, both gymnosperm and angiosperm volatiles were correlated to those of herbivores. Interestingly, the patterns of correlation varied according to chemical classes, with aromatics being more closely associated with pollinators and monoterpenes more closely associated with herbivores. These differences link to different functions of both compound classes with aromatics being more involved in attraction and monoterpenes being more involved in defence (Rodriguez and Levin 1976, Schiestl 2010).

The conjecture that different classes of odour compounds fulfil different functions represents a functional explanation for the pronounced diversity of odour compounds. Some of these compounds may have evolved primarily for their defensive function, some primarily for the role in attracting mutualists, and some primarily as waste products or unavoidable consequences of metabolic activity (e.g., CO_2 emission). The discrete nature of chemical compounds and the perception thereof entails that the diverse selective pressures acting upon a plant can – possibly more easily than in other sensory modes – lead to the evolution of multiple odour components. In other words, we propose that the trade-off to balance distinct, often opposing selective pressures (such as those commonly associated with defence and attraction) can be mitigated by using multiple components within a blend rather than by resolving the trade-off within a single compound. This is obviously a speculative idea that awaits further testing. Importantly, this idea can also be extended to encompass different selective pressures associated with communication. For example, compounds with a low molecular weight and high volatility are usually less reliably associated with individual quality than larger, less volatile compounds because these can function as defence against microbes or in the protection against abiotic factors. Again, it is conceivable that some odours evolve for long-distance attraction that contain no other information than that of locality, while other less volatile compounds can also indicate aspects of quality.

A corollary of the multiple dimensions of plant odours is that it is necessary to establish first which of the many compounds elicit a behavioural response. This is the first step to later differentiate between cues and signals in order to trace the evolution of olfactory plant communication.

To analyse odours, these are often collected from flowers through a filter that absorbs volatile molecules (headspace analysis). The molecules that accumulated on the filter are then separated by gas chromatography and analysed by mass spectrometry for identification. After odour identification, there are different ways to analyse behavioural reactions. In an electrogram, the sensitivity of the antennae of an insect to known chemical compounds is tested. If there is a positive reaction, the insect is assumed to be able to detect the compound. This method does not measure, however, how the insects respond to volatiles because this involves

the cognitive dimension of decision-making. This is commonly done in olfactometers, where receivers are faced with alternative blends. By measuring whether receivers consistently prefer (and approach) one odour over the other, their discrimination abilities between alternative blends is evaluated. More sophisticated experimental setups measure the response of receivers to blends in wind-tunnels where – similar to olfactometers – different concentrations of blends can be tested.

2.10 Plant odours

This section aims to give a short introduction to odour production; for an encompassing overview readers are referred to the review by Knudsen et al. (2006) on floral scent. The essential theme of this section is that the biochemical pathways involved in odour synthesis differ between odour classes. Consequently, we can expect that the biochemical origin of the compounds will affect the tightness of their association with other plant qualities (nutrients and secondary compounds) that animals consuming plant material can be interested in. If odours share the biosynthetic pathway with other substances, their association to them may be relatively tight and invariant; a situation that we term biochemical pleiotropy. Although *biochemical pleiotropy* between compounds of a shared biosynthesis is a good null model, a positive association between compounds is by no means the only possible outcome. Allocation trade-offs between end products of a pathway can lead to a negative association between the products of a biosynthetic pathway. Given the variability in the possible outcomes, we currently lack a framework to evaluate whether a common biosynthetic pathway usually leads to the production of (informative) cues where odours or pigments indicate the presence of another, less-easily detected compound of the same pathway or whether signalling commonly occurs. Unless proven otherwise, we cautiously assume that a common biosynthetic pathway leads to the production of cues via biochemical pleiotropy.

Consideration of biosynthetic pathways is important because signalling is a more likely explanation if odours are indicative of other substances that are produced by different pathways. This is because this association requires cross-talk between pathways and biochemical pleiotropy is not as likely to arise as if compounds share a pathway.

While consideration of the biosynthetic pathway is important, it does not solely explain how volatiles function and why receivers react to them. The chemical structure of volatiles influence the range they travel through air or water. Subtle changes in chemical structure can also lead to differential reactions of animals. For example, females of *Manduca sexta* oviposit more on *Datura wrightii* plants that emit a specific enantiomer of linalool (Reisenman et al. 2010) compared to plants that emit the other enantiomer. Enantiomers are stereoisomers with identical chemical and physical properties except that they are mirror images of each other like a left and right hand. They occur because a chemical group binds at a different orientation in each enantiomer. Interestingly, foraging behaviour in *Manduca sexta* is not enantioselective, but oviposition is. This is an interesting example illustrating the complexity of odours and the resulting difficulties of predicting the responses of animals.

Odours are typically lipophilic molecules with high vapour pressure; that is, they can pass the membrane freely and evaporate into the atmosphere (if they are not held back by chemical conjugation or barriers to diffusion). Scent production is mainly regulated at the site of emission, which is the upper epidermis cells in flowers, particularly on the petals (Pichersky and Gershenzon 2002). Knudsen and co-authors compiled a list of over 1,700 VOCs, which are grouped into seven major floral compound classes. These classes are also found in the odours of vegetative tissue. Odour molecules are aliphatic (straight chain) or aromatic (ring structure) with varied carbon backbones, and they are distinguished by their diverse functional groups including aldehydes, esters, ketones, alcohols, alkenes, carboxylic acids, amines, imines, thiols, halides, nitriles, sulphides, and ethers.

2.10.1 Chemical classes

Terpenoids
Terpenoids are the largest group of natural products and comprise over 40,000 substances that are assembled in various ways (Yu and Utsumi 2009). They are derived from 5-carbon isoprene (see 2.10.2)

via the enzyme terpene synthases. According to the number of 5-carbon isoprene units, terpenoids can be classified into monoterpenes (C-10) and sesquiterpenes (C-15), diterpenes (C-20), sterols (C-30), and carotenoids (C-40). Monoterpenes and sesquiterpenes constitute the majority of plant odours. For example, six monoterpenes occur in roughly 50 per cent of all flowers investigated thus far (Knudsen et al. 2006). Importantly, terpenoids have many defensive functions. Monoterpenes are involved in the amelioration of stress induced by high temperatures and in reducing damage caused by oxidative stress (Loreto et al. 2004). Terpenoids make also up an important part of resins excreted for defensive functions, which often serve as first line of defence against herbivores (Theis and Lerdau 2003). Resin is a viscous secretion common in coniferous trees consisting of mono-, sesqui-, and diterpenoids. Resin is released upon tissue injury and herbivores can be trapped and killed within resins, as is well known for amber. Moreover, resins can seal the wounds inflicted by herbivores. Other functions of terpenoids include protection against microbes, viruses, and fungi. It is likely that these protective effects explain the commonness of terpenoids in flowers.

The concentrations of terpenoids are typically higher in reproductive tissue than in foliar tissue and in some species more terpenoids are found in young leaves compared to old leaves (Rapparini et al. 2001, Fischbach et al. 2002). Owing to the defensive properties, it is not surprising that terpenoids are often linked to defensive functions in flowers (e.g., Knudsen et al. 2006, Junker and Blüthgen 2008). However, terpenoids can also attract specialized insect pollinators such as orchid bees (Euglossini; Dressler 1983) and undergo functional changes from defence to rewards throughout plant evolution (Armbruster et al. 1997; see Section 6.4.1). These functional changes are explicable because ancestral insects that originally searched for pollen as food on flowers may have inadvertently pollinated them leading to new selective pressures to increase the attraction of pollinators (Pellmyr and Thien 1986). In this scenario the defensive function has been co-opted to an attractive function. Another scenario is that herbivorous insects can adapt to defences and subsequently use defensive compounds in host recognition. If the defensive compounds are still beneficial in deterring other herbivores, they will be retained by the plant without a function change.

Monoterpenes are also employed in attracting parasitoid wasps that lay their eggs into herbivores (Theis and Lerdau 2003). It is easy to envision that this function is an exaptation of the defensive function (see Section 9.4.1).

Interestingly, some odour compounds of this group are not derived directly from isoprenoid but from the cleavage of carotenoids (Pichersky et al. 2006). This biochemical origin can, again, result in either a consistent positive or a consistent negative relationship between carotenoids and monoterpenes or diterpenes. Positive covariation would be explicable if carotenoid cleavage occurs only when carotenoids are plentiful. In this case, biochemical pleiotropy would lead to multi-modal stimulation of animals because these are confronted simultaneously with stimuli in distinct sensory modes (vision and olfaction). There is strong evidence that multi-modal stimulation results in better discrimination and faster learning abilities among pollinators (Gumbert and Kunze 2001). These two effects explain why multi-modal stimulation is generally seen as being highly adaptive (Bro-Jørgensen 2010). Our hypothesis that multi-modal stimulation can also arise through biochemical pleiotropy remains speculative because it has not been tested so far.

Indirect evidence supports the concept of biochemical pleiotropy. For example, anthocyanins protect photolabile phytoalexins that are synthesized and accumulated in plants after exposure to micro-organisms and that possibly serve as plant antibiotics. Anthocyanins are therefore associated with the presence of phytoalexins. Likewise, terpenoids are employed in various plant defences, which could lead to a consistent association between these volatile compounds and anthocyanins, even though they originate from different biosynthetic pathways. We believe that biochemical pleiotropy occurs more widely (see also the next class of odour compounds).

Phenylpropanoids and benzenoids

These compounds have a benzene ring in their skeleton. They are both derived from the phenylpropanoid pathway that is also responsible for the synthesis of anthocyanins and for the synthesis of many deterrent secondary compounds. It thus seems likely that the emission of phenylpropanoids and benzenoids can indicate the activity of the phenylpropanoid pathway. Consequently, phenylpropanoids and benzenoids can be related in a similar way to plant defences as anthocyanins – which are visual indicators of that pathway (Section 7.4.2). Likewise, it is possible that phenylpropanoids and benzenoids are linked with the expression of anthocyanins through biochemical pleiotropy. This conjecture awaits analytical confirmation. Some substances of this group are among the commonest volatiles found in flowers. For example, benzaldehyde is found in 50 per cent of plants (Knudsen et al. 2006).

Other odour classes

There are four other major odour classes. First, C-5 branched compounds that are characterized by an isopentane carbon skeleton probably derived from the amino acids: valine, leucine and isoleucine (Knudsen et al. 2006). They are divided into saturated and unsaturated compounds. Second, aliphatics, which have a straight chain, are believed to be derived from fatty acids through oxidation. They are compounds without rings and include a diverse range of compounds: alkanes, alkenes, acids, ketones, alcohols, esters, ethers. The third group comprises sulphur-containing compounds, which are found primarily in plants interacting with two distinct pollinator groups, carrion flies and bats. The sulphur compounds include dimethyl disulfide, dimethyl trisulfide, and dimethyl tetrasulfide. Given that these compounds are emitted by unrelated bat-pollinated plants, and that they are mostly absent in plants pollinated by other animal groups, and that dimethyl disulfide is a strong attractant for free-ranging bats, convergent evolution of bat-pollination odour syndrome has been proposed in flowers (von Helversen and Winter 2003). Similarly, the presence of sulphur compounds but not other volatiles, flower morphology, or flower colour distinguish *Eucomis* spp. pollinated by flies from those that are pollinated by wasps (Shuttleworth and Johnson 2010). Importantly, this study showed that experimental addition of sulphur compounds to wasp-pollinated species led to an ecological shift towards fly pollination thus apparently contributing to ethological isolation in these species. There is currently no evidence that sulphur compounds form a private communication channel; although humans, for example, are often deterred by them.

2.10.2 Volatiles from vegetative tissues

Plants commonly smell. Most of us will know the typical smell of grass that has been recently cut or of the herbs used in Italian cuisine. The overall diversity of volatiles released from vegetative tissue is less than that of floral scents and with the exception of induced volatiles, the evolutionary ecology of emissions from vegetative tissue is not well known. Here we will quickly discuss some of the major compounds.

Isoprene is a common volatile from vegetative tissue that is thought to increase photosynthesis at high temperatures either through quenching reactive oxygen or by stabilising the photosynthetic apparatus (Pichersky and Gershenzon 2002). It is possible that the emission of isoprene is correlated with photosynthetic activity and is thereby potentially informative about general aspects of plant quality. Again, this idea is speculative and awaits experimental confirmation. Importantly, terpenoids and C-5 branched compounds are derived from the isoprenoid pathway. Whether their production is linked in a predictable way to photosynthesis and is thus potentially informative is unknown.

Ethylene is an important plant hormone regulating growth that is emitted by plants and can be informative about growth-related processes to herbivorous and possibly also to frugivorous animals (see Section 5.7.1). For example, ethylene production peaks in climacteric fruits during the final stages of ripening. Finally, volatiles such as methyl-methionine are thought to function as general defence against herbivores in addition to induced volatiles.

Induced volatiles
Induced volatiles can have various functions (see Chapter 9). They can deter herbivores, reduce the probability of egg-laying, and they can attract the enemies of herbivores, be it their predators or parasites. Although odours can travel large distances, induced volatiles have thus far proven to be only effective over small ranges of 10–15 cm (Heil and Karban 2010). In some plants, there is evidence for a threshold, whereas in others the effect of induced volatiles to elevate defences in other plants diminishes linearly with distance (Heil and Karban 2010).

As in the scent of flowers, induced volatiles usually consist of a blend of odours with terpenoids – particularly monoterpenes and sesquiterpenes – constituting a major fraction of these blends owing to their defensive properties. Aliphatics and aldehydes like hexanal and (Z)-3-hexenal are also involved in the resistance against herbivores, the latter for example in potatoes (Pichersky and Gershenzon 2002).

2.11 Taste

The variety of plant compounds that can interact with taste receptors of animals roughly parallels the diversity of odours. However, the dimensionality of taste is much less than that of odour perception. Humans and probably many mammals have four primary taste modalities (sweet, salty, bitter, and sour), with umami, a savoury taste associated to specific amino acids being a fifth modality (Yarmolinsky et al. 2009). Although taste has evolved quite independently in insects, they detect the same food qualities. Sweet and umami promote feeding behaviour, whereas bitter and sour taste reduce food intake. Animals often trade off the intake of attractants, such as sugar, against those of mild feeder repellents, such as tannins (Remis and Kerr 2002). The effect of salty taste varies and can either be attractive or repellent depending on concentration and nutritional state of the animal. Taste preferences are usually innate but can – to a certain extent – be modulated by experience and physiological state of the animal. Innate regulation of taste preferences is useful because taste mainly serves to discriminate between foods that are beneficial to ingest and those which are not. This is why taste receptors are located in the mouths and in insects additionally in their legs and wings. When insects walk on food resources, their gustatory cells can thus inform them about food quality. Stimulation of the leg taste neuron with attractants such as sugar induces the insect to extend the proboscis and start feeding (Yarmolinsky et al. 2009).

Taste receptors are related to odour receptors and function roughly similarly. They show remarkable sequence divergence in mammals, which is thought to be related to different foraging habits (Yarmolinsky et al. 2009). For example, cats have lost their receptivity to sweet taste. The binding affinity of sugar receptors is relatively low, whereas that of bitter receptors is much higher in order to avoid lethal mistakes in food consumption (Yarmolinsky et al. 2009). The bitter receptors are very diverse, which is assumed to reflect the diversity of bitter substances they respond to.

Interestingly, in herbivorous insects the number of taste receptors can change within an individual's lifetime in response to the chemical complexity of its food (Rogers and Simpson 1997). Plants can use flavours to attract seed dispersers and pollinators which can identify nutritionally rewarding food resources by taste and visit them repeatedly. Likewise, bitter flavours can by used by plants as repellents, for example, in the diverse secondary compounds that are astringent. Before we start to sum up how taste functions, it is important to clarify that animals can respond to variation in nutritional contents without relying on their gustatory senses. Obviously, only those compounds that are intercepted by the sensory system can have communicative functions. Compounds such as toxins that directly act upon a receiver without being intercepted by the senses are by definition not used in communication but in direct defence. As a corollary, if insects sequester toxins from their plant diet by absorbing them in their guts and incorporating them into their own bodies as defences, this acquired insect defence is not a form of communication. Insects can potentially use gustatory cues to identify the plants they sequester toxins from, but this idea is not well studied. Hence, unlike vision or olfaction, the first necessary step in the study of gustatory communication is to establish that the interaction is actually mediated by taste and not by alternative mechanisms such as direct defence or post-ingestive feedbacks (see below).

2.11.1 Nutrients

Carbohydrates are the most common plant rewards found in flowers and fruits. There is an inverse relationship between the contents of hydrophobic lipids in fruits and the contents of hydrophilic carbohydrates (see Section 5.3.2). Protein contents in fruits and flower rewards are generally relatively low. Unlike fruit and flower rewards, vegetative tissue consists mainly of protein and carbohydrates, but these nutrients are heavily diluted in a bulk of indigestible material.

There is abundant evidence that animals react to nutritional contents of plant food. As expected, animals from herbivores to seed dispersers usually prefer more nutritious food over less nutritious food, and nutrients such as sugars and amino acids are phagostimulants that induce feeding in insects (Bernays and Simpson 1982). Although taste is surely important for discrimination, we should not assume that all of these food preferences are mediated by taste. For example, ruminants are not believed to taste the nutrients in their plant food directly but rather to rely on post-ingestive feedbacks to achieve a balanced diet (Provenza 1995). Post-ingestive feedbacks can be based upon measuring the concentrations of amino acids and carbohydrates in the blood and associating food with its consequences for general well-being. These feedbacks involve the digestive system, they can be informative to an animal, but they do not rely on the sensory system and consequently do not represent a communication system. However, the situation can be complex; animals (including humans) associate the post-ingestive consequences of consuming a specific food with its sensory properties (that is taste). Learning could thus enable consumers to use sensory cues for identifying rewarding food resources even if the reward status is only assessed by post-ingestive feedbacks. For example, humans exhibit flavour–nutrient learning enabling them to achieve a higher caloric intake (Brunstrom and Mitchell 2007). The extent to which non-human animals use flavour–nutrient learning is less clear in general and in particular to what extent flavour–nutrient learning relies on post-ingestive feedbacks.

Very distinct animals such as herbivorous *Schistocerca* locusts and frugivorous birds can also regulate the intake of specific nutrients (Simpson and Raubenheimer 2000b, Schaefer et al. 2003a). This is not an easy feat: since each food possesses an invariant ratio of nutrients, it is necessary to switch among foods in order to achieve certain nutrient intake goals. Animals are thus often forced to over-eat a certain nutrient (relative to their ideal intake) within a food and to eat less than their ideal intake of another nutrient. If *Schistocerca* locusts or frugivorous tanagers are forced to prioritize among nutrients, they regulate protein intake over sugar intake, which might be driven by the fact that protein is a relative scarce nutrient in plants and that protein deficiencies are deemed to be more detrimental to animal survival than deficiencies in carbohydrates which can be balanced by lipid intake. The extent to which nutrient regulation relies primarily upon gustatory communication or upon post-ingestive feedbacks is not well known, even for locusts (Simpson and Raubenheimer 1999). As a general rule, decisions mediated by the senses occur more rapidly than those relying on post-ingestive feedbacks.

We may ask why animals have taste receptors if they could rely on post-ingestive feedbacks. There are several obvious reasons. First, identification of reward status by post-ingestive feedbacks is necessarily slower than by taste. As such the former are not ideal to identify harmful or deterrent substances because an animal may already have ingested a lot of these substances. Moreover, post-ingestive feedbacks are often not very accurate, particularly if the food item is not frequently consumed and if the time lag between post-ingestive consequences and consumption increases (Yearsley et al. 2006). Importantly, post-ingestive feedbacks and the gustatory system are interdependent systems. The electrophysiological responses from mouthpart taste hairs to either sugars or protein depend on their previous exposure to these nutrients such that receptors responded most vigorously to the nutrients that were absent from what the locusts fed on previously (Figure 2.6; Simpson and Raubenheimer 1999). Taste and post-ingestive feedbacks are thus two interrelated systems that provide information on nutritional quality at different speeds and possibly with distinct accuracy.

Communication by taste relies on gustatory taste receptors that are located in the mouth of verte-brates and in the labial or antennal glands of insects. Herbivorous insects can use these gustatory receptors to sense the sugar content of ingested food directly (Simpson and Raubenheimer 2000b). However, as in odour receptors, there seems to be high variability among animals in the extent and sensitivity of gustatory receptors. For example, honey bees have only a handful of putative sugar receptors and are missing receptors sensitive to bitter taste (Touhara and Vosshall 2009). This is not related to their diet because nectar often contains defensive compounds (Adler 2000), which can reduce the time both pollinators and nectar robbers probe nectar and stay on the flowers (Adler and Irwin 2005). Because alkaloids affected the quality but not the quantity of visits in this study, the author proposed that insects perceive the alkaloids via taste. No negative effects of nectar alkaloids on female reproduction were found.

The sugar receptors mediate foraging behaviour in honey bees because a hungry individual extends its proboscis reflexively when the sucrose receptors on the mouth parts are stimulated. If an olfactory stimulus precedes the sucrose stimulation, the honey bee will associate the olfactory stimulus as the conditioned stimulus with sucrose as the unconditioned stimulus of sucrose (Menzel 1999). This form of associative learning is similar to that found in the classical conditioning experiments on mammals. In general, learning abilities obviously depend on the memory properties of an animal, the time lapse between the conditioned and unconditioned stimuli (short if mediated by gustatory responses, long if mediated by post-ingestive feedbacks), and the frequency of encountering a particular food.

2.11.2 Secondary compounds

The term 'secondary compounds' encompass all metabolic products without a known function in the primary metabolism of plants. In other words, all substances are included under this umbrella term that are not known to be involved in the vital functions of growth, development, maintenance, and reproduction of plants. This definition is obviously problematic as it is based upon a 'negative' classification system. Furthermore, it has become clear that many secondary compounds play

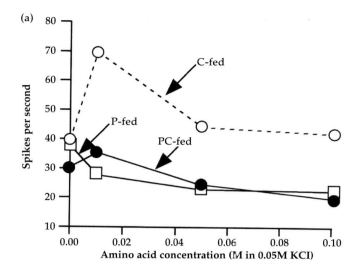

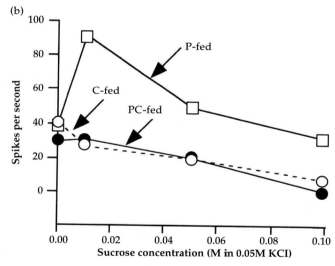

Figure 2.6 Gustatory responses of the migratory locust (*Locusta migratoria*) depend not only on the nutrients in the food they are eating but also on the nutritional content of the food they were eating before. Data show the electrophysiological responses from mouthpart taste hairs to stimulation with a range of amino acids and carbohydrate solutions. The C food contains digestible carbohydrates but no protein; the P food contains protein but no digestible carbohydrates; whereas the PC food contains both nutrient groups. Receptors respond more strongly to nutrients that were lacking in the pre-treatment food, that is more strongly to amino acids if fed with carbohydrates previously (a) and more strongly to carbohydrates if fed with amino acids previously (b). The different responses are mediated by levels of amino acids and sugars in the blood. Redrawn from Simpson and Raubenheimer 1999.

important roles in mediating plants' interactions with the biotic and abiotic environment. Plants synthesize an enormous variety of secondary compounds, with more than 200,000 substances identified up to now and many new substances being identified each year. Many of these substances are thought to be defence-related, although some of the pigments and volatiles discussed above obviously function in attracting plant mutualists such as pollinators, seed dispersers, and the enemies of herbivores. Owing to the chemical and functional diversity of plant secondary compounds, there will not be a single, all-encompassing hypothesis that explains the overall distribution of secondary compounds across different plant taxa and across different organs within plant taxa.

There are several alternative hypotheses to explain the presence and diversity of secondary compounds. The arms race hypothesis is concerned with plant–animal interactions. It predicts 1) that secondary compounds defend plants against their antagonists and 2) that antagonists overcome these defences over evolutionary timescales. The co-evolutionary arms race hypothesis dates back to Ehrlich and Raven (1964) and is characterized by three steps. First, plants producing biologically active compounds may gain fitness benefits by reducing the fitness of their enemies. Second, some enemies in turn adapt to these substances by developing resistances. This step is facilitated by two factors, the significantly shorter generation times of herbivores and pathogens (compared to most plants) and the higher population density of antagonists. Third, plants develop new secondary compounds to fend off the antagonists that developed resistance to the previously evolved substances. A given plant still retains the original secondary compound because not all of its enemies will have evolved to overcome in. Thus, within-plant diversity of secondary compounds is driven (at least in part) by the ecological diversity of the enemies a plant may face.

Interestingly, the evolutionary arms race hypothesis could also be extended to plant–plant interactions because the genetic variation in the concentration of sinigrin, an allelopathic secondary compound, mediates the interactions between *Brassica nigra* and its competitor species (Lankau and Strauss 2007). Genotypes with high sinigrin concentrations were better invaders of heterospecific communities but less successful invaders of conspecific communities than genotypes with low sinigrin concentrations. Thus, within-species diversity in secondary compounds is apparently linked to competition within and between species in *B. nigra*.

In support of the arms race hypothesis, many secondary compounds do have defensive functions (Firn and Jones 2003), and specialized insects do

adapt to some of the defences of their hosts (Ratzka et al. 2002, Wittstock et al. 2004). Moreover, there is at least some evidence that positive selection by herbivores is driving diversification in plant secondary compounds in well-studied species (Benderoth et al. 2006). The diversity of secondary compounds between species may then partly be explained by the distinct herbivores they face and very likely also by chance processes that lead to the evolution of alternative defences.

Alternative hypotheses include the carbon sink hypothesis that posits that secondary compounds act as carbon sinks to absorb excess photosynthetic carbon. This hypothesis does not easily explain the immense diversity of secondary compounds. The resource availability hypothesis explains that secondary compounds mainly depend on overall resource availability, the carbohydrate/protein balance, and plant growth rate (Coley et al. 1985). According to this hypothesis, environmental variation in resource availability contributes to observed variation in secondary compounds. Finally, many secondary compounds protect against abiotic factors such as excess light and high or low temperatures. Given that pronounced variability in many abiotic factors may severely impact plant fitness, abiotic factors might contribute to explaining the diversity of secondary compounds.

For most secondary compounds the mode of action on animals still needs to be explored. What is clear is that some of them are sensed by gustatory receptors (and consequently involved in communication), some are directly deterrent (and thus function as an unconditioned stimulus that could be related to a conditioned visual, olfactory, or gustatory stimulus), a few may be sensed by post-ingestive feedbacks and some are not sensed at all. Most defensive compounds that taste bitter or astringent are likely to be perceived by gustatory receptors in most animals. However, as in other sensory modes, there are pronounced differences among animals in their taste perception. For example, capsaicin, the compound that makes chilli hot, binds to a receptor of the vanilloid receptor subtype 1 in mammals, whereas birds are not deterred by it (see Section 5.5.1). Similarly, amygdalin, a cyanogenic glucoside and common fruit compound, can be toxic to mammals, but frugivorous birds are not repelled by it

(Struempf et al. 1999). As yet, information on which secondary compounds interact with the gustatory receptors of non-verbal animals is almost completely lacking. Consequently, the role of taste in the communication between plants and animals is currently difficult to evaluate.

2.12 Acoustic communication

Acoustic communication is limited to special plant adaptations to pollination and seed dispersal by bats. Neotropical bats use echolocation as a sonar system to orient in space during flights and also to find prey. It goes without saying that the form of acoustic conspicuousness differs from conspicuousness in the other senses. Conspicuous acoustical features are those that clearly stand out from background clutter. Thus one way of being conspicuous is to present the flowers and fruits on stalks away from the vegetation. Classical examples of this type of presentation are the elongated fruits of *Piper* species. Spatial distance to the background is not the only feature of acoustic conspicuousness, which also depends on the frequency of echoes reflected back to the bat. The background clutter is produced by leaves that are positioned at various angles to the receiver and therefore reflect echolocation calls in various directions; few of them will be reflected back to the receiver. Consequently, the shape of a fruit or flower is the second important determinant of acoustical conspicuousness because the morphology of a fruit or flower will determine the way in which echoes are reflected. Given that the shape of fruit is typically different from that of leaves and given that fruit shape is constrained by physical laws because most fruits consist primarily of water (which is most economically contained in a spherical form), specific adaptations in fruit shape are probably not widespread, although fruits of *Piper* species are a likely candidate for such adaptations. Clearly, one of the most obvious adaptations towards acoustic conspicuousness is the mirror-like structure of bat-pollinated *Mucuna holtonii* flowers that strongly reflects echolocation calls back to the sender (von Helversen and von Helversen 1999; see Section 6.4 and Figure 6.3).

We might speculate briefly on why acoustic communication is so little used by plants. It may in part be a constraint of physiology, since animal–animal acoustic communication requires movement: either of one body part on another or of air through or past a body part. However, this may not be the whole reason, since one could imagine for example winged seeds being selected to whistle as they move through the air, or flower stalks selected to make a noise as they sway in the wind. Perhaps in part the explanation can be found in timescales: many animal calls (e.g., alarm calls) function to trigger instantaneous reaction in the receiver, and plants' need to communicate with animals takes place over more leisurely timescales. Another attraction of sound to animals is that it can allow signalling over long distances in complex environments such as forests. However, volatile emissions allow plants to communicate over longer distances and (as already discussed) plants are much more likely to be pre-adapted to the development of olfactory signalling to animals than to acoustic signalling. Certainly, fish can react to the noise of fruit falling into the water, as discussed in Chapter 3, but this noise is a necessary by-product of fruits' contact with water, and we consider it unlikely that fruits are specifically adapted to produce more sound in order to attract fish. Similarly, the noise of fruits falling to the ground can also attract foragers to plants which have a large fruit crop, but we know of no evidence that fruit traits are selected to enhance this.

2.13 Summary and perspective

The many different topics of this chapter make a short summary necessary. In this chapter, we have discussed the sensory modes separately, but we deem the effects of multi-modal stimulation of the receiver as a particularly important future research area (Box 11.1). We have seen that vision and olfaction differ in some important ways, namely that reflectance is a continuous variable that is received by a few distinct receptor types, whereas 1–2 orders of magnitude more receptor types are involved in olfaction. These differences entail that conspicuousness can vary between these two sensory modes because only in olfaction can it be both gradual (when contingent on a ratio of compounds) and stepwise (when contingent on few molecules of a specific compound). Because animals differ in their sensory perception, it is crucial to analyse plant traits according to the senses of the animals interacting with a plant. In both sensory systems information can arise through biochemical pleiotropy which we predict to be a common mechanism providing reliable information in plant–animal communication. The information content of plant traits also provides starting points for the evolution of plant–animal communication that is based upon signalling. This is because if plants incidentally provide information, animals can start reacting to it and exert a selective pressure upon the plant that may or may not come to dominate the original selective pressure leading to information as a by-product. Importantly, in visual and olfactory communication there is variation in the biochemical pathways producing the plant traits that stimulate the sensory receptors of animals. Consequently, there is the potential that within both sensory modes communicative traits are reliably associated with information and that this kind of information differs between biochemical pathways.

Animals' responses to the plant material they ingest can be mediated by gustatory communication, by post-ingestive feedbacks and by direct effects that are not mediated by the senses (e.g., in toxins). The extent to which animals rely on gustatory communication is not well resolved. Pursuing this question represents an important research avenue in plant–animal communication. Vision is by far the best-studied sensory system. Encouragingly, current experiments suggest that an understanding of sensory stimulation can translate into predictions on how receivers select visual traits. This may mean that our current lack of understanding of the cognitive mechanisms by which sensory stimulation is converted into decision-making is less of a weakness than it could have been. However, before we can become more confident about this, we need to attain a functional understanding of sensory physiology in more ecologically relevant situations.

Having now overviewed the sensory modalities in which plant–animal communication takes place, we now move to the next chapter, where we overview one of the key ecological situations in which such communication occurs: dispersal of plant seeds by animals.

CHAPTER 3

Animals as seed dispersers

3.1 Introduction: the advantages of seed dispersal

Dispersal is a fundamental life-history process. Most animals are mobile for at least one stage of their development; in contrast, plants are fundamentally sessile with their only chance of dispersal being as seeds. Seeds are not self-powered and almost exclusively rely on some external agent to provide transport. As well as abiotic agents such as air and water currents, plants make extensive use of the self-powered movement of animals.

Dispersal of seeds away from the parent may have a number of selective advantages. Firstly it may reduce the exposure of the seed to predators or pathogens that are attracted to, or supported by, the parent. Further, it reduces the potential for competition between parent and offspring and among offspring. Even in the absence of these effects there may still be selection for dispersal if this reduces the risk of the parent and offspring simultaneously experiencing similar adverse conditions. That is, dispersal may be selected by the need for spreading of risk in a spatio-temporally variable environment where localized and unpredictable disasters can occur. These concepts have been encapsulated in the Janzen–Connell hypothesis for the high species diversity in many plant communities (introduced independently by Daniel Janzen (1970) and Joseph Connell (1971)). This hypothesis suggests that diversity is maintained by two mechanisms: (i) mortality of seeds and seedlings increases as their density increases, and (ii) survival of seeds or seedlings increases with increasing distance from the parent. Although it is a highly simplified description, the Janzen–Connell framework continues to

be a powerful theory in plant community ecology (e.g., Harms et al. 2000).

Another key concept in the dispersal of plants is the dispersal kernel. This is a plot of either the probability of a seed being dispersed or of dispersal and successful establishment, as a function of distance from the parent plant (see Cousens et al. 2008a for discussion of the conceptual difficulties arising from the usage of these two definitions). At small distances the two types of kernel look very different, since most seeds are transported only short distances, but the likelihood of such seeds establishing is low. A common feature of both types of kernel is that there is decreasing probability of a seed being dispersed greater and greater distances from the parent plant. Thus a graph of dispersal distances will generally show monotonic decline to zero with increasing distance, whereas one of likelihood of seed establishment may initially increase with distance before declining to zero again (see Figure 3.1). It is much easier to empirically characterize the short-distance part of the curve: because many seeds are involved, stochasticity is less important, mechanisms of dispersal can often be observed, and seeds are easy to track. None of these hold true for attempting to characterize the long-distance tail of the distribution. This is particularly problematic, since it seems that (although relatively uncommon), long-distance dispersal events are particularly important ecologically (Nathan 2006).

Let us now consider what qualities make a good dispersal agent. Although a good dispersal agent transports the seed away from the parent, there may not be selection for ever-further dispersal. Wenny (2000) followed the success of bird-dispersed seeds of the tree *Beilschmiedia pendula* in Costa Rica. Those

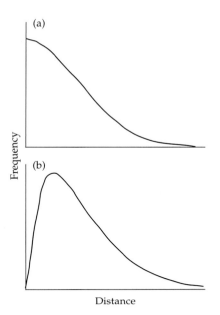

Figure 3.1 Characteristic shapes of many seed dispersal kernels. In (a) we plot the frequencies of seeds being dispersed to different distances; in (b) we plot the frequencies of dispersal and subsequent successful establishment at different distances.

transported less than 10 m from the parent suffered high mortality through predation and fungal diseases. However, those transported more than 30 m had lower survival than those transported 10–20 m. This effect likely arises because the environment close to the parent is more likely than more distant environments to be suitable to this species (since it was demonstrably suitable to the parent). That is, since seeds start out in an environment that has been successful for their parent, sites nearby to the parent may be more likely to offer suitable conditions to offspring than more distant sites. The greater the homogeneous spatial scale of habitat, the less this selection pressure for limited dispersal distances will be.

Another desirable property of a dispersal agent is that it transports the seed in a way that does not damage its future ability to develop into a plant. Dispersers can even enhance germination abilities if passage through the gut thins the hard seed coat. The dispersal agent should also deposit the seed in a suitable micro-habitat to aid germination and subsequent plant viability. Finally, there may be

selection pressure on a dispersal agent to distribute the seeds of a given parent widely: both to reduce between-sibling competition or as a strategy to spread risk. The next six sections of this chapter (3.2–3.7) investigate the implications of these aspects of dispersal agent 'quality' on selection for different forms of animal-based dispersal. In Section 3.8 we explore whether one can predict general trends in the types of plants and types of habitats that seem particularly suitable to different forms of animal-based dispersal. Section 3.9 will turn from ecology to consider the evolution of different animal-based modes of dispersal. Finally, in Section 3.10, we discuss how different forms of dispersal can be combined.

3.2 Dispersal by terrestrial vertebrate frugivores (endozoochory)

In order to increase the uptake of seeds by animals, some are enclosed in fleshy fruits. The signalling functions of fruits will be considered in greater depth in the following two chapters. Here it is enough for us to note that production of fruit may be cheaper physiologically than a seed of similar size, and may act to increase the apparent size and thus ease of detection of the seed. This ease of detection can also be further enhanced by conspicuous colouration of the fruit (see Section 4.4.2). Fleshy fruits provide pulp that attracts a wide array of consumers and thus is likely to impact strongly on the likelihood that a seed is transported by an animal. Use of fruits is widespread: Howe and Smallwood (1982) estimated that 51–98 per cent of canopy trees and 77–98 per cent of sub-canopy trees and shrubs in neotropical forests used fleshy fruits in seed dispersal.

Internal dispersal by frugivores results often in a higher uptake of seeds than transport of seeds on the outside of animals. It is important to note that uptake of the fruit need not result in transport of the seed, since some frugivores (particularly primates, but also birds that manipulate fruits in their beaks) may remove the seeds and only consume the fruit flesh. Such seed removal can result in very low dispersal distances, if the fruit is processed in or under the parent tree. However, useful dispersal can still occur if the frugivore carries the fruit some distance before processing it. The decision to delay processing of the fruit will be influenced by environmental factors, with transport prior to processing being more likely if the fruiting plant simultaneously attracts a number of frugivores that could potentially steal the gathered fruit, or if predation risk is high in or near the fruiting plant and a safer site is available nearby. The likelihood of transporting the fruit away from the parent tree is also increased by fruit characters, such as fruit size and the ease with which flesh can be accessed. Larger, more valuable fruits and those that require longer to process are more likely to be carried prior to processing.

The likelihood that flesh and not seeds will be consumed by a particular frugivore will be influenced by its dexterity. It will also depend on a number of physical aspects of the fruit: such as the number and size of the seeds, their position within the fruit (e.g., scattered through the flesh or concentrated at the centre), and the ease with which seeds separate from the flesh. However, sensory aspects can be important if separation is done in the mouth (or bill) on the basis of different tactile properties of seed and flesh (with seeds being spat out). From the plant's perspective, removal of the seeds prior to consumption may not be strongly detrimental if at least some frugivores transport fruit prior to consumption. Separation of seeds from flesh may be of no interest to those fruit consumers that can gain nutritional benefit from both, but such consumers destroy the seed and will be detrimental to plant fitness.

Another property of a successful dispersal mechanism is that the seed remains viable after the dispersal process. This is complex for dispersal by frugivores. If the seed is discarded rather than consumed by the frugivore then it is more likely to survive the process of dispersal. However, it is well known that seeds can pass through the digestive system of frugivores and remain viable. Indeed the germination rate of seeds with toughened exteriors can be increased by scarification during passage through the digestive systems of frugivores (e.g., Tewksbury et al. 2008b). However, there will be great variability between frugivores, and a given seed type may pass unharmed through some species of frugivore but not others. Properties of the seed itself are likely to affect post-transport viability:

with small size and the presence of a hard coating both increasing the likelihood of surviving passage through an animal. There may be a trade-off related to gut passage time, with short gut passage causing less damage to the seed but also reducing dispersal distance. Some fruits may contain chemicals that have a laxative or constipatory effect on seed passage (Wahaj et al. 1998), suggesting that plants can to some extent manipulate the time seeds spend inside the digestive tract of seed dispersers. However, gut passage time is also affected by gut fullness Thus the number of fruits simultaneously available on one plant, and the local density of such plants can affect the subsequent viability of ingested seeds and dispersal distance (Carlo and Morales 2008). Birds often regurgitate seeds rather than defecating them. Regurgitation often leads to shorter dispersal distances but less damage to seed viability. Large seeds are preferentially regurgitated, but are also more likely to be damaged during mastication by mammals.

Dispersal distances by frugivores are highly variable, but long-range dispersal over several kilometres does occur (see Figure 3.2). Some large frugivores have long gut retention times. Birds and primates can be very fast moving, covering distances of hundreds of metres in a few minutes. Thus dispersal distances of up to tens of kilometres are certainly possible by this dispersal mode, and distances of tens or hundreds of meters are commonplace (Hardesty et al. 2006, Seidler and Plotkin 2006, Jordano et al. 2007). Dispersal distances will be very different for different frugivores and will be affected by seasonal and daily movement patterns, as well as by environmental factors. For example, while mammals were primarily responsible for long-distance dispersal of seeds of the cherry tree *Prunus mahaleb* (Jordano et al. 2007), birds and bats achieved longer dispersal distances than mammals in a neotropical forest (Seidler and Plotkin 2006). Fruit eaten by birds on migratory stop-over sites might achieve very long range dispersal. Similarly seeds eaten at dusk just before birds return to a roosting site might be transported further than seeds eaten earlier in the day. If birds remain feeding on the parent plant for long, then seeds may be defecated under the parent. Frugivore behaviour will be affected by fruit abundance and distribution, not just on the parent

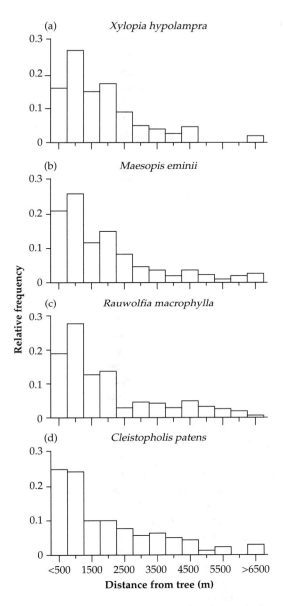

Figure 3.2 Seed dispersal distances resulting from frugivory by the hornbill *Ceratogymna atrata* on four fruit types. Redrawn from Holbrook and Smith 2000.

fruiting trees can increase the attraction of frugivores to a focal plant, but they also compete with that plant for frugivores; generally it seems that the first factor outweighs the second (Blendinger et al. 2008). As one would expect, as fruiting plants become more aggregated, dispersal distances generally decline (Carlo and Morales 2008).

Defecation or regurgitation by frugivores often results in seed clumping (Poulsen et al. 2001, Potthoff et al. 2006), particularly at sleeping sites (Russo and Augspurger 2004). Thus seed dispersal by animals may increase competition between seeds. However, this will sometimes be mitigated by secondary dispersal (see Section 3.10). Further, several different frugivores will generally consume the simultaneously available fruits from a single plant, reducing competition between seeds from the same parent if these frugivores differ in microhabitat selection.

At least some frugivores may deposit seeds non-randomly in ways that benefit subsequent germination. Wenny and Levey (1998) followed individuals of the five avian species dispersing seeds of the *Ocotea endresiana* tree in Costa Rica, monitoring the success of seeds through to the seeding stage a year after fruits were consumed. For four of the disperser species, dispersal distances were low (less than 20 m) and so were seed survival rates. However, over half the seeds transported by three-wattled bellbirds (*Procnias tricarunculata*) travelled over 40 m and often ended up under gaps in the canopy where seedling success was high (see Figure 3.3). This was linked to male bellbirds commonly displaying to females from exposed perches where they are well-lit and can be seen from a distance. Similar non-random dispersal might occur in other dispersers such as manakins (*Pipridae*). Generally, perching sites may be important for concentrating defecation and thus seed deposition (Pratt and Stiles 1983, Gosper et al. 2005).

Mistletoes grow parasitically on other trees, and are often specialists on one or a few different tree species. This habitat specialism might be expected to give rise to a strong need for directed dispersal, since only seeds landing on (or very near) host individuals will flourish. There is indeed evidence of non-random dispersal. It has been suggested that the seeds of one mistletoe (*Viscum album*) stick to the beaks of birds during consumption of the fruit, leading the birds to wipe them off against trees,

plant but also on those in the immediate locality (Sun et al. 1997, Morales and Carlo 2006), as well as by competition from other frugivores. Carlo (2005) demonstrated how the density and identities of surrounding plants can affect the seed dispersal kernels of two fruiting plant species. Surrounding

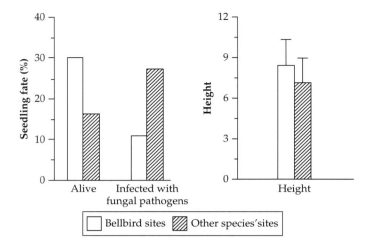

Figure 3.3 For seeds of *Ocotea endresiana*, dispersed by either bellbirds or other bird species, we present the fraction of seedling surviving the first year, the fraction of year-old seedlings infected by fungal pathogens, and the height of year-old seedlings. Redrawn from Wenny and Levey (1998).

increasing the likelihood of deposition in a suitable site. In the same species, it has been suggested that excreted seeds often stick to the rump of the bird, and have to be wiped off on a perch, again increasing the likelihood of deposition on a suitable site (Cousens et al. 2008b). Even controlling for age, taller trees are more likely to be infected with mistletoe, and this can be linked to birds' preference for perching on tall trees (Roxburgh and Nicolson 2008). Within a local population of suitable hosts, mistletoe plants tend to be aggregated, and this can be linked to preferential perching of frugivorous birds on already-infected trees in order to feed on mistletoe fruits (Ward and Paton 2007, Roxburgh and Nicolson 2005). Also, *V. album* ripens fruits in winter, at a time when the evergreen mistletoe provides birds with cover in the deciduous trees that have lost their foliage. Similarly, in a study of the host tree *Cecropia schreberiana*, Carlo and Aukema (2005) found that female trees where more likely to be infected with mistletoes than male trees. The authors suggested that the fruits of the female tree attracted frugivores and thereby increased the likelihood that frugivores that recently fed on mistletoe fruit will deposit seeds as they feed. Experimental removal of the mistletoes from their host was observed to reduce both visits from frugivores and mistletoe seed deposition (Aukema and Martinez

del Rio 2002, Medel et al. 2004). Mistletoes and their hosts often share frugivores, and this may require to some extent a reinterpretation of the mistletoe as a parasite. Van Ommeren and Whitham (2002) investigated seed dispersal in the one-seed juniper (*Juniperus monosperma*) and its associated mistletoe (*Phoradendron juniperinum*). They found that sites of juniper with high densities of mistletoe attracted more frugivores than those with lower densities, and that the number of juniper seedlings was greater in stands with more mistletoe (see Figure 3.4). These interesting results deserve to be followed up in a manipulative study that manipulates infection with mistletoe, to ensure that these results cannot be explained by third variables. More integrative studies could attempt to compare the physiological costs of parasitic infection with the mistletoe (in terms of loss of water and nutrients) with any dispersal benefits to the host, so as to explore under what circumstances (if any) the 'parasite' might provide a net benefit to its host.

3.2.1 Endozoochorous seed dispersal by vertebrate herbivores

It is increasingly clear that herbivores can sometimes eat seeds alongside their target foliage. Such seeds can pass the gut of herbivores undamaged, and thus herbivores can be important in seed

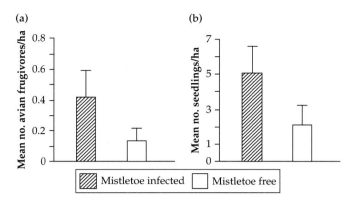

Figure 3.4 In a study of seed dispersal in the one-seed juniper (*Juniperus monosperma*) and its associated mistletoe (*Phoradendron juniperinum*), it was found that (a) stands of juniper with mistletoe attracted more frugivores than those without, and (b) that the number of juniper seedlings was greater in stands with more mistletoe. Redrawn from Van Ommeren and Whitham (2002).

dispersal (Malo and Suárez 1995, Pakeman et al. 2002, Eycott et al. 2007). If seeds are small enough that the herbivores do not avoid them, and tough enough to pass through the digestive system, then it should be no surprise that vertebrate herbivores can be effective seed dispersers. Their large individual size can offer long gut retention times and fast movement through the habitat, that together offer long-range seed dispersal. The plant foliage is attractive to them, so uptake rates may be relatively high; and individual animals can carry many seeds. Such animals spend a large part of their day grazing, and so are likely to excrete the seeds preferentially in areas that should be suitable for plant growth. The excreta may also provide valuable fertilizer for the seeds, especially in nutrient-poor soils. The interesting question from an evolutionary viewpoint is whether this form of dispersal exerts considerable selective pressures on plant traits. Janzen (1984) postulated what he called the foliage-is-the-fruit hypothesis:

We hypothesize that, for a number of species of small-seeded herbaceous plants, a normal and selected model of dispersal was through consumption of the seeds by large herbivores while these were eating the foliage of the parent plants. We also hypothesize that the big herbivores ate the plants, at least partially, because plant traits were selected for consumption since this mode of dispersal was advantageous for plants.

This hypothesis thus suggests that plants may be selected to enhance their seed dispersal by herbivores. Janzen listed ten traits that he would expect in plants dispersed this way:

1. The plant is edible to the herbivore.
2. The foliage is of sufficient nutritional value to be attractive to the herbivore.
3. If the foliage is not always edible and attractive, it should change to become so as the seed crop matures.
4. Seed maturation is timed to coincide with peak herbivory.
5. Seeds are intermingled with foliage or in its immediate vicinity.
6. Mature seeds are retained on the plant
7. Seeds are small, tough, hard, and inconspicuous
8. Seed coatings can resist digestion.
9. If seeds are chemically protected from seed predators, they are so at doses that would not be aversive to large herbivores.
10. The species does well where there are large herbivores.

The problem with testing this hypothesis lies in demonstrating that it is the vertebrate herbivores that have selected a particular combination of traits in the plants. For example, the timing of seed maturation might appear useful for testing the hypothesis in plants that are subject to strong seasonal variation in large herbivore pressure because of migrating herds. Imagine if there was evidence of geographic variation in the timing of seed maturation in a given species that could be linked to the arrival of migrating vertebrate herbivores in each location. This could be interpreted as support for the foliage-is-the-fruit hypothesis, but we can imagine a similar trend occurring in a wind-dispersed plant, simply because there is evolutionary pressure on the plant to produce its seeds before the plant is destroyed by the migratory herd. Similarly reduced investment in defences at the time of seed maturation could be driven by a shifting of limited resources into reproduction. Perhaps the most compelling test would be to look for strong linkage between the traits listed by Janzen. For example, if we compare those species whose seeds are still viable after passage through a herbivore's digestive tract with those species whose seeds are not, and find that the other traits (e.g., reduction of defences at seed maturation, seed retention on the plant, seed maturation coinciding with peak in herbivory) occur more strongly in the first group than the second; then this would be strong evidence for the hypothesis. Another possible approach would be to compare traits between populations of a species where large herbivores have become extinct owing to hunting many decades or even centuries ago (e.g., bison in parts of the US) to populations where these herbivores are still present. If the ten traits listed by Janzen occur in populations that still experience grazing to a greater extent than in the others it would suggest that the lack of selection by large herbivores has led to micro-evolution in that species.

Thus, the foliage-is-the-fruit hypothesis remains plausible, but has not been subject to strong empirical exploration, not because it is unimportant, but because its predictions are difficult to distinguish from those of other selection pressures.

3.3   Dispersal by vertebrate seed predators

At the beginning of Section 3.1, we discussed seed dispersal by animals that are primarily attracted to the fruit in which the seed was embedded. In this section, we are interested in cases where the seed is the primary target of the animal. If such an animal eats the seed, then the plant does not benefit at all.

However, a dispersal benefit may arise if the seed predator does not consume the seed immediately but transports it some distance before caching it for later consumption (a process often called hoarding), providing that later consumption does not occur. Caching occurs in dry fruits that do not rot quickly. It can be seen as a strategy for the seed consumer to cope with seasonal or daily variation in food availability. The plant only benefits from its seed being cached if the seed can germinate before it is discovered and eaten (either by the original cacher or by a thieving granivore). The original cacher may not return to the seed if it dies beforehand, or if it forgets the location of the seed, or if (perhaps due to a milder weather or high availability of better or easier-to-find food) it has less need for cached seeds than anticipated (Jansen et al. 2004). Such environmental effects are also likely to affect the risk of the seed being discovered by other pilfering animals. Thus, it can be expected that milder weather and increased availability of other food generally lower the likelihood of granivores searching for cached seeds.

To evaluate the effectiveness of caching as a dispersal mode, we must consider the probability that the seed is detected by a potential caching individual. This is a function of the size and appearance of the seeds. Seeds that experience caching are generally not as colourful as fleshy fruits. This is most likely because seeds are less perishable than fleshy fruits and so experience less selection pressure to be found and picked quickly. Moreover, the seed dispersal services of animals that cache seeds may be even less reliable than those consuming fleshy fruits because cachers often consume a large proportion of the cached seeds.

When a seed is discovered by a granivore, the key issue will be whether it is eaten immediately or cached. This decision will be influenced by many factors. For example, greater predation risk or risk of having the meal stolen by kleptoparasites increases the attractiveness of caching, for later consumption under less risky conditions. The size and nutritional value of the seed can also influence whether to cache or not (Tamura and Hayashi 2008), as do the time and energy required to break through any protective coating and extract the nutritive value of the seed. Increased local availability of seeds (perhaps triggered by masting by plants) also acts to encourage caching (Zhang et al. 2008). The effect of seed size can be complex: Tamura and Hayashi (2008) found that squirrels hoarded larger seeds more frequently and transported them further than smaller seeds; whereas the opposite effect was found in sympatric mice. For squirrels the dominant factor to explain this behaviour is likely the increased value of larger seeds, justifying longer-distance dispersal (reducing risk of pilferage). For mice the dominant factor is likely to have been the awkwardness of carrying larger seeds, slowing down their food gathering. Chemical defences can also influence caching, with there being a tendency to immediately consume low-tannin nuts and cache high-tannin ones (Xiao et al. 2008). This is consistent with less-defended seeds rotting more quickly. Rodents can use chemical cues from the pericarps of acorns to identify preferred (longer-dormancy) red oak over white oak seeds (Steele et al. 2001).

Mammals and birds can carry seeds considerable distances before caching them. Distances of several tens of metres are commonplace (Jansen et al. 2004), but caching behaviour is unlikely to result in very long-range dispersal of several kilometres. However, seeds can often be recached either by the same or by different individuals, which can increase dispersal distances further. Properties of the seed can influence distance, with larger seeds often being carried further. Aspects of the environment can also affect caching distances, with higher local granivore numbers (which are likely to correlate with risk of pilferage) encouraging more wide-ranging caching (Uemura and Sugiura 2007), and high local seed abundance (and thus lower value of cached seeds) having the opposite effect (Jansen et al. 2004). Widely dispersed caching reduces the risk of seed discovery by pilferers (Daly et al. 1992, Waite 1988).

Caching will generally lead to spatial heterogeneity of seed deposition from a mother plant. Many caching species adopt scatter hoarding: where seeds are stored at individual locations rather than in an aggregate. Whilst this tactic increases the cognitive, time, and energy costs of retrieval, it also makes searching less attractive to potential pilferers (Male and Smulders 2007), as well as working as a spreading-of-risk strategy against pilferers, fungal infec-

tion, or natural disasters causing the loss of cached food in a specific location. Some hoarders use a single larder where they store an aggregate of seeds; however, even in such circumstances the seeds of an individual plant can end up in the larders of several hoarding individuals. This is particularly likely because hoarding species are generally of small body size and so local population densities can be high. Caching may sometimes provide directed transport to suitable sites for seedling establishment. Vander Wall (1993) observed that yellow pine chipmunks (*Tamias speciosus*) preferentially cached in open scrubland rather than nearby closed-canopy forest. Within scrubland, caching occurred preferentially under the canopies of bitterbush shrubs, a micro-site found to be advantageous for seedling development (Vander Wall 1992). Wenny (1999) describes another example of scatter hoarding in open sites that might be advantageous for any subsequent seedlings. In this study, seeds were initially dispersed by birds and secondarily by scatter hoarding rodents. This secondary dispersal was biased towards sites with lower vegetation density.

Germination of the seed whilst cached incurs loss of nutritional value to the hoarding animal, which will thus be selected to minimize the risk of this. They do this in part by preferentially caching seeds that have long dormancy periods. Sometimes the embryo is removed prior to caching to prevent germination (Fox 1982), rendering the dispersal of the seed worthless to the parent plant. Additionally some animals that guard a larder over the winter monitor the development of seeds in the larder and immediately eat those seeds that show signs of germination. This even occurs in some scatter-hoarding species (Jansen et al. 2006). The other disastrous thing that can occur from the viewpoint of the plant is the successful discovery of the cached seed by another granivore. Olfactory cues are used to locate cached seeds (Jorgensen 2001, Vander Wall et al. 2003, Winterrowd and Weigl 2006), and it is likely that seeds are selected to minimize the ease of such detection. This is unlikely to deter the original animal from caching the seed, since a low ability to detect the seed by olfaction should be to the advantage of the original hoarder compared to potential pilferers. The original hoarder can use its spatial memory of stored seeds to relocate them in the absence of olfactory cues. Indeed the requirements of the seed and the hoarder might coincide to the extent that the animal might preferentially cache those seeds that have low production of olfactory cues once cached, thus giving the cacher an advantage over potential pilferers, although we have no empirical evidence for this conjecture. Another, speculative, possibility to reduce olfactory detection would be preferential caching near to plants with strong olfactory emissions that might mask the smells of cached seeds.

Birds appear not to have the olfactory capability to detect buried seeds (Dally et al. 2006 and references therein), but can still pilfer since they find seeds either by random probing or by observation of the initial act of caching. Hoarders can, however, reduce the risk of being observed. Bugnyar and Kotrschal (2002) observed that ravens (*Corvus corax*) preferentially cached in positions where there was a large obstacle between themselves and conspecifics. Several species increase their distance travelled before caching in the presence of conspecifics (reviewed in Dally et al. 2006). Dally et al. (2004) observed that western scrub jays (*Aphelocoma californica*) preferentially cached in shady rather than well-lit sites only when they were observed by conspecifics. These same birds have been observed to more readily retrieve and recache seeds whose original cache location was observed by a conspecific. Finally, rooks (*Corvus frugilegus*) have been observed to cache inedible items like stones only when observed by conspecifics (Dally et al. 2006).

In comparison to other forms of dispersal, caching offers medium-range dispersal distances and relatively low aggregation of seeds from the same plant. However, the chance of a seed germinating may not be high. The survival rate of naturally scatter hoarded seeds has been estimated to be as low as 0.02 per cent in some species (McAdoo et al. 1983), but has also been recorded as being as high as 10 per cent in others (Forget 1992). For seeds that are regularly cached by hoarding species, empirical work that explored what fraction of seeds were cached and compared the ultimate fate of cached and uncached seeds (in terms of success of the germinating seed) would be welcome. Perhaps the biggest attraction of this mode of transport is that seeds are commonly buried in the substrate by some cach-ing species. This should reduce their exposure to granivores and environmental stressors and may provide ideal circumstances for germination. There is also evidence of micro-habitat choice by cachers (Pearson and Theimer 2004). Since micro-habitat choice and depth of burial will have strong influences on the ability of cached but unretrieved seeds to germinate and flourish, further investigation of such choice would be very welcome.

3.4 Inadvertent transport on the outside of mammals and birds (epizoochory)

Some seeds have specialized structures on their exterior that aid the attachment of the seeds to the coats of mammals, and perhaps also to the feathers of birds, that brush past the parent plant. Such traits are seen on the seeds of only around 6 per cent of plants, but can be common in low-growing species in some environments (Sorensen 1986, Vibrans 1999, Hughes et al. 1994). However, it is important to remember that especially if small seeds are wet, they can adhere without obvious adaptations. Even medium-sized seeds may be carried within a coating of mud picked up by an animal, say in a wallow.

In terms of the desirable characteristics of an effective agent of dispersal, the least attractive aspect of this mode of dispersal is the likelihood of a seed actually being dispersed. The animal is not attracted to the parent plant and seed uptake is inadvertent. One would expect higher rates of contact with animals if the parent plant lies on the edge of an established animal track through the environment, however we know of no study that has explored whether plants that disperse in this way are non-randomly distributed with respect to animal tracks. While transport by an animal logically results in a high probability of ending up on a track that animals frequent, such tracks may not be fixed in space over long enough timescales for such an association to develop. It also seems possible to us that the plant could make itself more attractive to grazers around the time that its seeds are ready for transport, trading-off increased tissue damage for increased seed uptake. For example, we might predict that species with chemical or mechanical defences against herbivores might seasonally reduce these defences, in order to encourage greater contact with potential seed dispersers (as in the foliage-is-the-fruit hypothesis discussed in Section 3.2.1). However, we know of no empirical support for our conjecture. In general, we would expect epizoochorous seeds to have long periods of availability on the plant (to counter low encounter rates with dispersers), and for seed availability to be timed to take advantage of migratory herds where they are an important feature of the local ecosystem. Once again, we know of no formal tests of these ideas. There is likely to be little selective pressure on seeds dispersed this way to manipulate the senses of the potential agent to boost uptake rates: since the seed is not of interest to the animal, there would be no value to advertising them. Obtaining close inspection by deception, for example if such seeds mimicked fleshy fruits, might appear a plausible strategy, but we know of no species that employs it. Likewise, we know of no species that uses a mixture of fruit-encased seeds and adhering seeds, although such a strategy seems again to be logically plausible as a means of enhancing uptake of adhering seeds. The presence of adhering seeds is likely to be inconsequential (or of very low cost) to the animal, who are unlikely thus to have developed traits that serve to avoid picking them up. Thus we would not expect strong selection pressure on adhering seeds to be camouflaged (except as a defence against seed predators).

Adhering to the outside of animals is unlikely to impair ultimate seed viability. The only likely exception to this is if seeds are groomed out of the hair or feathers (either by the animal itself or by another individual in social species) and subsequently eaten. It might be that there is selection pressure on the appearance of seeds to avoid standing out from the coat or plumage of the bearer, but we know of no evidence that contrasting seeds are more likely to be groomed out. It may be that the best defence against being groomed out, and against simply being inadvertently knocked or blown out, is for the seed to rest deep within the coat or plumage. There may be a sensory trade-off here, since seeds deeper in the coat may be more likely to contact the skin, and such contact is more likely to cause the animal discomfort and induce grooming.

This mode of dispersal can offer the potential for directed dispersal towards suitable habitats. For example, for aquatic plants, adhesion to wildfowl provides the potential not just for long-distance dispersal, but directed dispersal to another aquatic site. Figuerola and Green (2002) inspected 47 individuals of 6 different waterfowl species in Spain and found the seeds of 15 plant species. Thus, adhesion of seeds to habitat-specialist animals might be an effective way for a plant that relies on a localized habitat type to disperse. Even for dispersing agents that are less habitat specific than waterfowl, final deposition sites may be non-random. For example, it may be that adhesive seeds are commonly knocked loose when the animal brushes through dense vegetation. Such a situation may lead to seeds being preferentially directed towards shaded sites that already have substantial vegetation growth with which the seedling would have to compete. Similarly, grooming out of seeds may preferentially occur in resting-up sites and these may often be shady locations in hot regions. However, very little is known about non-random aggregation by microhabitat in adhering seeds, and much will depend on the mode by which the seed leaves the coat. This may occur simply by the force of gravity as the seed is slowly worked loose from the hair or feathers by the motion of the moving animal, in which case we may expect to see more of a spread of dispersal sites than if removal occurs through grooming. In mammals that have established paths through their territory or established denning or displaying sites, we might expect that most seeds are deposited in such sites, but we know of no study that has explored this. The only data comes from an experiment briefly reported by Sorensen (1986) of seeds artificially attached to captive snowshoe hares (*Lepus americanus*), where it appears that the overwhelming majority were removed as a result of grooming. In terms of spreading of risk and reduction of between-sibling seed competition, variation in final destination of seeds from the same plant is likely to be dependent on whether deposition occurs passively or by grooming. No matter how deposition from a single animal occurs, all the seeds from a single plant will rarely end up on one animal, since the animal will likely only contact part of the parent plant momentarily, and thus variation in deposition sites between the seeds of one plant are likely to occur readily under this mode of dispersal. However, clumping is also possible: it has been claimed that species with adhesive seeds commonly grow around the entrances to rabbit burrows, because rabbits groom themselves there and/or because seeds are knocked off as the rabbit enters the burrow (although specific testing of this hypothesis is lacking).

Carriage on the outside of animals seems particularly suited to long-range dispersal, and Manzano and Malo (2006) determined experimentally that seeds attached to nomadic sheep were transported hundreds of kilometres. Burger (2005) demonstrated that the distribution of the tree *Pisonia gradis* on small tropical islands coincides with the presence of seabird colonies on these islands. The seeds of this tree are enclosed in an extremely sticky resin that adheres strongly to feathers. More generally, species with adhesive seeds seem disproportionately common on islands (see Sorensen 1986 for a review). Increasingly, transport on the shoes, clothes, animals, and vehicles of humans are being acknowledged as sources of long-range seed dispersal (see Mount and Pickering 2009, Wichmann et al. 2009).

3.5 Transport by invertebrates (myrmecochory)

The seeds of over 3,000 plant species (from more than 80 families) have a lipid-rich attached body (the elaiosome) that is attractive to many ant species. These ants can transport both seed and elaiosome to their nest before consuming the elaiosome and discarding the seed. In order to provide seed dispersal, the relative size of elaiosome and seed and their mechanical attachment must be such that it is more attractive for the ants to remove the elaiosome after having carried both back to their nest, rather than separating them immediately on discovery and only transporting the elaiosome. Not surprisingly, ants need larger rewards to transport larger seeds (Edwards et al. 2006a). Further, it must be possible for the ants to harvest the elaiosome without damaging seed viability. It may be that competition between ant species (and the risk of robbery by dominant ant species) is a key driver of

the deferment of splitting the elaiosome from the seed until the safety of the nest has been reached (Mesler and Lu 1983). Where ants achieve transport of the intact seed, the process is called myrmecochory. This relationship shows no specificity, with ant-dispersed plants having their seeds removed by a range of locally available omnivorous ants. Leaf-cutter ants also remove small fleshy fruits (along with leaves) from plants and thereby may disperse the seed of these plants (H.M.S., pers. obs.).

Many ant species are granivorous, but these can still be valuable seed dispersers if seeds are lost by the ants during carriage back to the nest, or if some are stored in the nest but never eaten (because the colony collapses or moves location without removing all the stored food; Levey and Byrne 1993). Storage of seeds in the colonies of granivorous ants is commonplace, as a means to cope with annual variability of seed production. Additionally, some seeds may be discarded by mistake on the refuse pile after being returned to the colony of a granivorous species. In the study of Retana et al. (2004) it was observed that 16 per cent of seeds being transported to the nest of a granivorous ant species were dropped and not recovered during the journey, with a further 1 per cent being (apparently mistakenly) rejected into the refuse pile. Thus myrmecochory may be important even to plant species that do not have an ant-attracting elaiosome. It may be that rejected seeds are not a 'mistake' but result from variation between seeds of a given plant species in profitability (likely driven by variability in chemical defences). If seeds with too high levels of defence to be attractive to ants are sufficiently uncommon and sufficiently time-consuming to detect, then the best strategy for the ants may be to return all such seeds to the nest, and perform evaluation there before rejecting the highly defended individuals (Rissing 1986). This theory makes the straightforward prediction that those seeds that find their way into the refuse pile have higher levels of defensive compounds than would be expected based on the distribution of seeds returned to the nest. This prediction has not received empirical investigation.

Ants do not offer long-distance dispersal, with seeds only being carried at most a few tens of metres to the nest, and often much less (Ness and Bressmer 2005; see also Figure 3.5). However, the sometimes

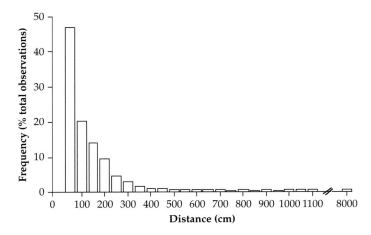

Figure 3.5 Frequency distribution of seed dispersal distances by ants from a study by Gómez and Espadaler (1998). Redrawn from that publication.

very high densities of ants in many environments, particularly tropical forests, promise very high rates of seed removal from under the parent plant. Generally, ant-based dispersal will not be particularly effective at reducing competition between seeds from the same parent, since seeds found by ants in close proximity will often be returned to the same ants' nest and then to the same store or refuse pile. However, if seeds are dropped over an extended period, the position of local ant colonies may change during that time. The site of refuse piles changes as a result of changes in the location of the entrance to the ants' nest, which may happen frequently throughout the year (Hughes 1991). There might also be differential change over time in the numerical strength or activity patterns of local ant colonies, especially if these are of different species, in a way that increases variation in the destinations of seeds from a given plant. However, increased clumping of the seeds of the ant-collected species may be more than compensated for by reduced competition from those plant species not dispersed by ants and thus much less likely to end up on the refuse piles of ant colonies.

An attraction to the plant of this form of dispersal may be movement of the seed to a safe site, since carriage underground will greatly reduce exposure to mammalian and avian seed predators. Some plants may exploit this by dropping seeds early in the day to allow them a full day's exposure to ants before the emergence of nocturnal rodents (Ness and Bressmer 2005). Dispersal by ants is certainly highly directed and this may be to the benefit of some plant species. The refuse site outside an ant colony may be nutrient enriched and seeds may be quickly buried by subsequent ant refuse in a way that enhances germination and decreases discovery by seed predators. Alternatively the refuse sites of ants may preferentially attract vertebrate granivores. Seeds that are stored underground and escape predation may also find a favourable microsite for germination: safe from fire and vertebrate granivores (Heithaus 1981, Gibson 1993, Ruhren and Dudash 1996, Boyd 2001). Whether inside the nest or just outside, the ant colony can have very substantial effects on soil chemistry and physical properties in ways that may influence the germination of seeds.

There have been ample demonstrations that seeds without an elaiosome are rejected by ants that collect intact combinations of elaiosome and seed. Further, a larger elaiosome increases the likelihood of a seed being transported to the nest (Hughes and Westoby 1992, Mark and Olesen 1996). Variation between individual plants in the chemical composition of elaiosomes can affect dispersal success (Boulay et al. 2007). Hughes et al. (1994) suggested that the chemical compositions of the elaiosomes of some plant species have converged to provide a close match to the invertebrate animal prey of many ant species. This could be a potential example of exploitation of pre-existing sensory biases, however it may instead be selection for the nutrients that ants are most able to eat. That is, the involvement of sensory aspects (if any) in this suggested convergence has not been explored. Pfeiffer et al. (2010) present the results of some ingenious experiments that suggest that some seeds without elaiosomes produce a similar chemical signal to elaiosomes and that this increases their chance of dispersal by ants. This was elegantly demonstrated by experimental manipulation of the chemical signatures of different seeds and of inedible dummy seeds. Oleic acid was demonstrated to have a strong effect on ant attraction and seed uptake. It may be relevant that oleic acid is also emitted by dead and decaying ants and triggers carrying and removal of these from the colony. Pfeiffer et al. also demonstrated that while the elaiosome-like signature increased seed uptake, it did not increase it to the levels experienced by similar seeds with real elaiosomes, suggesting that the 'elaiosome mimicry' was imperfect. We look forward to a follow-up study that explores the fate of such seeds after uptake by ants, so that the consequences for plant fitness can be explored more fully.

Davidson and Morton (1981) suggested that the colour of elaiosomes may be selected to be attractive to ants. This hypothesis has not been tested, but seems unlikely in view of the greater use of vision by vertebrate granivores than ants. It is currently unclear how important olfaction is for ants to detect and evaluate seeds and elaiosomes from a distance, but fine-grained selection is achieved on the basis of sampling the surfaces of these objects with the antennae where olfactory and gustatory receptors are located. The large numbers of ants and their highly developed general foraging skills may make specialist adaptations of the seeds to improve detection at a distance unimportant. Gammans et al. (2006) demonstrated that the elaiosomes of two gorse species, *Ulex minor* and *U. europaeus*, featured surface chemicals that increased the seed uptake rate by a mutualistic ant species *Myrmica ruginodis*, but not the granivorous *Tetramorium caesitum*. This observation seems very much worth exploring, both to explore the generality (and sensory basis) of

differential attraction of mutualistic species, and to explain why the granivore has not developed the ability that would allow it to exploit this food source.

Youngsteadt et al. (2008) demonstrated that olfaction was important in detection of the seeds of epiphyte species by ants that cultivate them within the walls of their nests in ant gardens. It may be that because only a small fraction of the seeds that such ants encounter in their local environment are likely to be of the appropriate epiphyte species, long-distance attraction to suitable seeds through detection of vegetative volatiles is selected in order to combat the low spatial density and temporal variability of suitable seeds. There seems to be an obligate specialism on both sides: the small number of plant species involved that are found growing in ant gardens, and the ant species involved seem to require such plants for the physical integrity of their nests. Although this is an obligate relationship, once again there is no tight coupling of two-species plant–animal pairs, but rather each plant species interacts with a range of 'gardening' ant species, and each ant species utilizes a number of different plants in their gardens. However, the interaction between these guilds has apparently led to the evolution of chemical communication over greater distances than ant-seed interactions in other circumstances. The suggestion of Seidel et al. (1990) that the chemical cues emitted by such seeds may be similar to those emitted by ant broods is worthy of further exploration.

Reports of seed dispersal by other invertebrate groups are very uncommon. Wallace and Trueman (1995) report the dispersal of seeds of the tree *Eucalyptus torelliana* by the stingless bee *Trigona carbonaria*. The fruits of the tree produce a resin that the bees collect. Seeds adhere to the resin and are taken up by the bees, transported to the nest, and subsequently discarded. Such discarded seeds can be hundreds of metres from the parent plant, and were demonstrated to retain their ability to germinate after handling by the bees. Dispersal by bees leads to the aggregation of seeds and subsequent competition among seeds. There is also an exceptional case of insects dispersing fleshy fruits in New Zealand. Wellington tree wetas (*Hemidenina crassidens*) are giant, flightless grasshoppers that consume the fleshy fruits of a number of plant species. They destroy the seeds of some of the species but disperse the seeds of others. In particular, some seeds show higher germination abilities when they have passed through the digestive tract of the wetas (Duthie et al. 2006). Lastly, recent evidence suggests that slugs may have been unjustly neglected as seed dispersers. Here an important issue may be that passage through the digestive system of a slug can leave a seed still able to germinate but leave it covered in mucus that makes it unattractive to mammalian seed predators (Türke et al. 2010).

3.6 Transport by fish (ichthochory)

Consumption of fruits and seeds has been documented in almost 200 species of freshwater fish (reviewed in Correa et al. 2007). In terms of sensory effects, nothing really is known about how fish detect fruits or seeds. Vision is sometimes important, since some fish will jump from the water to collect fruits still on the parent plant (Correa et al. 2007). Fleshy fruits consumed by fish seem to sport a diversity of colours, although our knowledge of the fruits that are actually transported this way is not sufficiently complete to allow a comparison with the distribution of colours of fruits dispersed by terrestrial vertebrates. Araujo-Lima and Goulding (1997) suggest that olfaction may be important in seed detection. This is very plausible, especially for turbid waters, and for dull-coloured seeds and fruits that are negatively buoyant. It is likely to be less effective if the seed or fruit is positively buoyant and there is sufficient current that the location of the fruit rapidly changes. However, to date there is no evidence for use of olfaction in ichthochory. The falling of fruit into the water may cause sufficient noise and disturbance of the water to alert fish to this food source (Correa et al. 2007). The presence of bright colouration in fruits consumed by fish cannot be taken as evidence that it evolved because fish utilize these colours in fruit detection, rather the fruit colouration may function primarily to attract terrestrial vertebrates to pluck the fruit from the parent plant or collect them after they have been shed.

The huge advantage of ichthochory compared with water-based dispersal is that it allows dispersal against the direction of the current. There may be advantages too in comparison to dispersal by terrestrial vertebrates. Terrestrial vertebrates generally have lower seed retention times that fish (because of the lower metabolism of fish compared to birds and mammals). Retention for longer than a day seems commonplace in fish (Horn 1997, Correa et al. 2007), during which time the fish can easily cover several kilometres (Pollux et al. 2007), although dispersal distances are probably more often in the range of 100–2,000 m (Horn 1997). Correa et al. (2007) cite a number of studies that demonstrate that seeds can pass through fish intact and are able to germinate. Apparently there can also be subtle intraspecific variation in the effectiveness of seed dispersal. Pacus (*Piaracrus mesopotamicus*) feed to a large extent on fruits during the wet season. Larger pacus transport a higher proportion of seeds intact in their guts than smaller ones (Galetti et al. 2008). As yet, there is little evidence of non-random dispersal by fish, but for plants that have a strong affinity with water, excretion by a fish guarantees that the seed is dispersed to an aquatic environment. If the aquatic environment is spatially constrained (say small streams in a forest) then this aquatic deposition is much less likely with terrestrial vertebrate dispersers. Anderson et al. (2010) reported that fruit-eating fish in seasonally flooded Amazonian forests remained in these temporarily flooded areas continuously throughout the period of flooding. This should mean that defecated seeds are more likely to successfully germinate after the flood waters recede than if they had been excreted onto the permanently submerged main river bed. Nothing is known about variability between seeds from the same parent in terms of final deposition site, but the shoaling tendency of many ichthochorous fish is likely to mean that seeds from the same parent end up in different fish, and between-individual difference in gut passage times (say driven by between-individual differences in gut fullness) are likely to cause wide variation in deposition sites.

Examples of seed dispersal by fish are widespread taxonomically and geographically; this may be related to the lack of morphological specialism required by either the plant or the animal in this interaction. Currently the ecological importance of dispersal by fish is unclear. We would welcome a study that followed the seeds of a plant whose seeds are known to be consumed by fish, in order to explore how much dispersal can be attributed to fish, how much to movement by water currents and how much by terrestrial vertebrates. Care must be taken in differentiating between these modes of dispersal. Kubitzki and Ziburski (1994) suggest that negatively buoyant seeds transported 20 m from the parent must have been moved there by fish, but we would like confirmation (perhaps using artificial objects that are unattractive to fish) that movement along the substrate by water currents does not occur.

3.7 Transport by reptiles

Reptiles have rarely been considered in their role as seed dispersers. Yet, they are important agents of seed dispersal, particularly on islands, where up to 98 per cent of pellets of some species contain fruit remains (Rodriguez et al. 2008), and reptiles may be the only disperser for some plants (Olesen and Valido 2003). Fruit eating is taxonomically widespread in lizards and known from at least 280 species (Valido and Olesen 2007). Strong and Fragoso 2006 demonstrate that the long seed retention times of ectothermic vertebrates (a mean of 1.6 days for red-footed tortoises, *Geocheclone carbonaria*, but

there are also records of a mean of 4 days in lizards; Iverson 1985, and see Figure 3.6) mean that dispersal distances may be further than might otherwise be expected (up to several hundred metres).

3.8 General trends in the use of animals as dispersers

Air and water currents are widespread alternatives to animals as agents of seed dispersal. One would expect air currents to only be an attractive means of transport in open environments: such as grassy plains, deserts, sand dunes, high-canopy species in forests, and areas recently affected by fire; compared with environments where penetration of wind is reduced (e.g., understory species in woodland). This is exactly the pattern seen (van der Pijl 1982, van Rheede van Oudtshoorn and van Rooyen 1999). Of course, water-based dispersal must be restricted to habitats that experience at least periodic surface water.

Seed size seems to strongly covary with dispersal mode; however selection on seed size is a complex issue. Large seeds have a proportionately higher volume to surface area, and so air resistance will have less of a counterbalancing effect to the force of gravity. Thus, post-dispersal selection for large seed size (generally to provide a greater energy store for

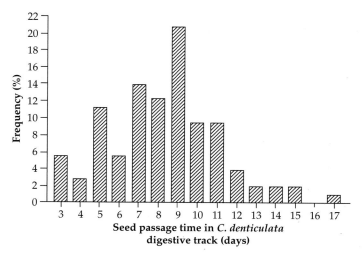

Figure 3.6 Frequency distribution of seed retention times in the digestive tract of the tortoise *Chelonoidis denticulate*. Redrawn from Jerozolimski et al. (2009).

the young seedling; allowing more rapid initial growth) will make wind-borne dispersal less attractive. Although positive buoyancy in water is likely to aid water-borne dispersal, negative buoyancy does not necessarily rule out such dispersal. Denser-than-water objects can be moved along the substrate by flowing water, particularly if the substrate has a downhill slope. Turbulence can lift objects into the water column (increasing dispersal distance) for periods of time. However, as a generality, large and heavy seeds are dispersed by animals (Hughes et al. 1994, Tiffney 2004).

Fleshy fruits are widely distributed taxonomically. They generally tend to be associated with larger seed size. The frequency of fleshy fruits declines with latitude, altitude, and aridity and with decreasing soil fertility; in all these cases this can be seen as a response to the increasing challenge of meeting the physiological cost of fruit production (Willson and Whelan 1990, Westoby et al. 1990). The response to increased aridity is particularly interesting, since another response to high aridity is increased seed size.

Seeds with large nutrient reserves are generally more attractive to hoarding species. From the plant's viewpoint, a large seed should allow more growth before the seedling is reliant on gathered light and soil nutrients, so we would expect to find large seeds (and thus dispersal by caching animals) particularly in low-light (woodland) environments.

Movement by adhesion on the outside of animals requires the animal to pass by the plant. Since many potential dispersers are ground-living, this mode of dispersal is mainly restricted to relatively low-growing plant species. However, generalization by habitat type is less obvious, although deserts may have low numbers of animals that could act as dispersal agents, the relative lack of plants may cause animals to be attracted to those that do occur (e.g., when seeking shade). In the largest comparative study so far, Sorensen (1986) found no strong effect of habitat type on the proportion of plant species using adhesive dispersal.

Myrmecochory is common wherever ants are common. However, it has been argued that the cost to the plant of provision of an ant reward is generally less than that of the fleshy fruit required for vertebrate dispersal (Davidson and Morton 1984), and thus myrmecochory is particularly characteristic of infertile soils where shortage of potassium particularly may limit the production of fleshy fruits (Hughes et al. 1993). However, plant stature may be a confounding factor in this relationship (Westoby et al. 1991). Small plant stature is more common in low-quality soils, and as a further complication ant density and species composition will also likely be related to soil quality. Myrmecochory may also be more attractive to plants of small stature because the relatively small dispersal distances involved may reduce parent – offspring conflict more for smaller plants than larger ones.

The form of dispersal has important demographic consequences for plants. Characteristic dispersal distances show a general pattern with dispersal type: with increasing distances from ballistic to wind to animal-based dispersal in tropical forests (Seidler and Plotkin 2006). While habitat fragmentation generally reduces the dispersal of plant species that are dependent on animal as vectors (Cordeiro and Howe 2003), animal-dispersed species are more resilient to fragmentation than wind-dispersed species (Montoya et al. 2008). This landmark study was based on data from 90,000 forest plots across Spain, and demonstrated that dispersal agent was a better predictor of resistance to habitat fragmentation than any other variable, including seed size, and that the results could not simply be explained by closely related species having similar dispersal strategies. This effect is likely to be due to non-random movements of animal seed dispersers.

3.9 The evolution of different forms of animal-based dispersal

The trend in animal-based seed dispersal is for a scarcity of tightly co-evolved relationships between species (see Section 1.5 and Section 1.6). A given plant will generally have its seeds transported by a diversity of animals, and each animal species will transport seeds from a diversity of plants. This can be understood in part in terms of timescales. Herrera (1985, 1998b) argued that plant species will go extinct at a substantially lower rate than disperser species, and thus one would not expect plants to be closely matched to a more rapidly changing disperser environment. Further, avoidance of competition between sibling seeds might select plants to encourage a diversity of seed transporters that are each associated with distinct seed shadows (Jordano et al. 2007). Unpredictability in the local availability of different vector species would also act against formation of an obligate mutualism with a single dispersing species.

Unlike pollination, there is no obvious selection pressure to act against selection for generalism. For pollen dispersal, there is a definite target for pollen grains (the stigma of a conspecific) which has driven the evolution of delayed payment of the transporting animal until the pollen has been delivered. This is not possible for seed dispersal, and so the incentive to transport seeds must be paid up front, with the plant having no control of the subsequent actions of the transporting animal. Further, the complexity of flower morphology is testament to the ability floral plants have to restrict access only to particular pollinators; in comparison, selection of frugivores (mostly on the basis of fruit size and composition) is much more relaxed (see Chapter 5).

Janzen and Martin (1982) suggested that some fruits may have evolved traits that improve seed dispersal by very large birds and mammals, and may consequently have been adversely affected by the widespread extinction of such 'megafauna' in the last 50,000 years. There is evidence that the largest extant terrestrial mammals are fruit consumers (rhinos: Dinerstein and Wemmer 1988; elephants: Guimarães et al. 2008); and the suggestion that some fruits in areas without extant megafauna share similar traits to those consumed by extant megafrugivores elsewhere (Pantanal: Guimarães et al. 2008). The elephant (*Loxodonta* spp.) seems a particularly important disperser for the African tree *Balanites wilsoniana*. Babweteera et al. (2007) found no evidence for dispersers other than elephants at several study sites. Further, they found juvenile trees established away from adult trees only in forests with elephants; in comparison, forests without elephants featured seedlings only under parent trees. This builds on previous work (Cochrane 2003) that demonstrated that seeds of this species eaten by elephants were much more likely to germinate than those not eaten by elephants. We know of no other examples of obligate reliance on a single frugivore,

although such reliance may occur in species-poor communities, such as at high altitudes. The often repeated story that the dodo (*Raphus cucullatus*) and the tambalocoque tree (*Calvaria major*) had an obligate mutualism now seems unlikely to be true (Witmer and Cheke 1991). Although nutcrackers are very important dispersers of several northern hemisphere pine species (e.g., Hutchins and Lanner 1982, Lanner 1988), there is no evidence that nutcracker birds (genus *Nucifraga*) are essential for the dispersal of any particular species, given that other birds and mammals also disperse the seeds of each species.

From the viewpoint of a frugivore, there is strong selection pressure to exploit a range of different plant species, since fruits of a given species rarely address all the nutritional needs of the frugivore and are generally only available for short periods relative to the lifespan of the frugivore. Thus, frugivores may complement their diets with distinct fruits that together fulfil their nutritional requirements (Whelan et al. 1998; see Section 5.3.2). However, there may be reasons for a frugivore to specialize over short timescales. For example, vertebrates may form search images for locally common fruit types. Individuals may also show dietary conservatism, whereby they focus on a subset of available foods that they have previous experience of and know to be acceptable. However, variability over time and space in fruit availability will select against overly restrictive specialism. Even though individual ants are much more short-lived than most vertebrate frugivores, these too should experience selection for dietary flexibility, since the local availability of fruit will change over generational timescales and also over the spatial scales relevant to the dispersal ranges of the winged queens.

There is strong evidence of seed dispersal by animals from the Permian (see Tiffney 2004 for a review). The earliest angiosperms probably had fleshy fruits (Donoghue and Doyle 1989). Tiffney argues that fleshy coverings to seeds likely originally evolved for reasons other than as a reward to seed dispersers. They may have originally been a repository for chemicals involved in the deterrence of seed predators, bacteria, or fungi (see also Mack 2000; Cipollini and Levey 1997a), or as a mechanical barrier against such antagonists. This covering may then have been co-opted to provide the rewards and sensory traits required for attraction of dispersers. Frugivory occurs in about 36 per cent of avian species spread across 135 families and 20 per cent of mammal species from 107 families. This spread strongly suggests that frugivory has evolved on many separate occasions. This likely reflects the lack of behavioural or physiological adaptation required to make at least some use of fruits as a resource.

The evolution of caching behaviour is relatively under-studied. Scatter hoarding would be very difficult to detect in the fossil record. Larder hoarding is more detectable, and also easier to understand though potential evolutionary stages. Hoarding of seeds has occurred at least since the early Palaeocene (Vander Wall 2001). The precursor of storage of food in a burrow that a mammal uses as a refuge is easy to understand. Seeds may have originally been brought in inadvertently with bedding material. Alternatively seeds may be brought back to the burrow as a safer place to eat them, but be subsequently discarded through a loss of appetite or some distraction. Seeds may also have originally been carried into the burrow to feed offspring, who may have rejected some. By whatever mechanism seeds were originally brought in and not consumed immediately, they could have been readily discovered and consumed at a later time. It is not difficult to imagine how these processes could move through intermediate stages over the course of evolution leading to larder hoarding. The initial evolution of scatter hoarding is much less obvious, and worthy of further research. One promising avenue is the suggestion of Vander Wall (1990) that it may have developed from the tendency of some birds to wedge seeds and nuts firmly in cracks in tree bark or other substrates in order to have them held firmly during processing with the beak to penetrate external layers (such as acorn woodpeckers do).

It is relatively easy to see how specialist structures that allow seeds to adhere effectively to mammalian coats and avian plumage evolved. As discussed earlier, small wet seeds will readily adhere to animals brushing against the parent plant even in the absence of any specialist structures. This form of dispersal would have occurred prior to specialist structures existing. However, mutations that altered seed shape in such a way as to enhance initial attachment and/or retention would provide an immediate selective advantage and naturally lead to elaboration and finally the sometimes-complex structures seen today. Because animals rarely brush past a particular plant, the low likelihood of adhesion probably placed high selection pressure on traits that improve uptake by the animal from such an encounter. Although there may be some costs to the additional material required by specialist structures, these will often be quite low and there is little obvious disadvantage to the seed of having such structures. Indeed these structures may serve to discourage seed predators and/or to anchor seeds to the substrate to allow the radial to penetrate the soil. As such, some of these structures may have evolved under these selection pressures and their use in aiding dispersal may be best seen as an exaptation. It has been repeatedly suggested that adhering dispersal evolved from wind-borne dispersal (Sorensen 1986 and references therein), however this conjecture seems to be based on no more than alleged visual similarity of some structural adaptations to the two dispersal modes.

The initial evolution of elaiosomes and ant–plant mutualism involving seed dispersal occurred along with the rise in relative abundance of ants seen during the Eocene (Dunn et al. 2007), with elaiosomes evolving at least twenty times within the monocots. Again, the precursors of elaiosomes may have had a different initial function and only later been co-opted as an inducement to ants (Forest et al. 2007). The evolution of elaiosomes as an ant attractant may only be possible where ant-worker size is sufficiently small relative to the seed that the seed cannot be consumed in situ (Rissing 1986).

Correa et al. (2007) presented a phylogenetic approach to exploring the evolution of seed dispersal by fish. They suggested that frugivory and granivory evolved as part of an omnivorous diet allowing exploitation of a seasonably abundant food type. Since no plant has been demonstrated to have particular adaptations to seed dispersal by fish, there are no real challenges to explaining the evolution of this mode of seed dispersal.

Hollander et al. (2010) use phylogenetic methods to explore the evolution of seed dispersal strategies among *Ephedra* shrubs. There are fifty species in this genus; of these species some show dispersal by wind, some by frugivorous birds, and some by seed-caching rodents. These authors conclude that dispersal by birds appears to be the ancestral state with the two other strategies evolving from it. Exploitation of modern phylogenetic techniques will certainly be more widely applicable to exploration of the evolution of seed dispersal. More resolved phylogenies, which allow estimation of the timing of particular changes in dispersal strategy, might allow further insight into the ecological factors that underlie evolutionary shifts in dispersal strategy.

3.10 Combination of different forms of dispersal (diplochory)

So far, we have discussed seed transport mechanisms as if they happen in isolation. However, it seems relatively commonplace for a single seed to be transported sequentially by more than one of these mechanisms (a process called Diplochory: see Vander Wall and Longland 2004 for a review). For example, seeds deposited in excreta may be subject to secondary dispersal by animals. The dung itself may be the attraction, or the seeds within it. In the second case, the dung can be more easily detected than the seeds alone, and can offer a concentrated packet of seeds to granivores. Vertebrate excreta can attract dung beetles that inadvertently bury seeds. This can have the double advantage to the plant of moving the seeds to a preferred micro-habitat for germination and making them less accessible to potential seed predators. Similarly, the working of aquatic sediments by invertebrates may bury seeds: Luckenbach and Orth (1999) demonstrated that more seeds were trapped on sediments that were actively bioperturbed by worms. Rodents too may forage for seeds in dung, which could contribute to seed dispersal if the seeds are stored and not consumed or are consumed but pass intact through the digestive system. Seeds in excreta may be washed away by flood waters or taken up by the wind after the excreta has dried.

Fleshy fruits can also be subject to other dispersal forms: fruit can be blown off a plant, and subsequently moved by rolling downhill, by strong winds, or by water currents. The complexity of seed dispersal in general can be seen by examining the fates of seeds consumed by lizards in the study of Nogales et al. (2007). Inspecting bird and reptile pellets, the authors found that the majority of seeds were found in the droppings of birds that had preyed upon the frugivorous lizards. The predatory birds (the Mediterranean shrike *Lanius meridionalis* and the kestrel, *Falco tinnunculus*) did not feed on fruits but were important secondary dispersers that – owing to a predictable difference in habitat choice between each of them and the lizard – significantly altered the seed shadow.

Although hoarding can lead to long-range dispersal, many (if not most) seeds collected by scatter hoarders will be consumed before they can germinate. Thus, it may be best for the parent plant if only some of its seeds are found by hoarders. It may be that one function of masting by trees is to satiate local granivores to reduce the probability that a given seed is consumed directly or cached (Zhang et al. 2008). Those seeds that avoid discovery by granivores may still be moved by winds, or water currents, or simply by rolling downhill. Caching can interact with other dispersal modes, since caching species have been observed to collect seeds from the ground and from faeces as well as directly from plants (Vander Wall et al. 2005; Siepielski and Benkman 2008). After scatter hoarding, seeds are unlikely to be moved by any other dispersal mechanism except under the restricted circumstances where seeds buried in substrate are moved by flood waters or landslide.

Adhesive dispersal too may be combined with other modes. Sticky seeds may be transported inadvertently on the head or bill of a bird that fed on the fleshy fruit in which the seed was originally encased. Similarly, seeds may be initially consumed but stick to the vent or surrounding feathers of the bird after defecation, thus increasing the extent of their dispersal. Sorensen (1986) suggests that dwarf mistletoes (*Arceuthobium vaginatum*) disperse their seeds primarily explosively, but the seeds remain sticky for some minutes afterwards and readily stick to foraging birds. Further, because of the problems associated with the rarity of initial contact with an animal, many species with adaptations to dispersal on the outside of animals may also exploit dispersal by wind.

Diplochory may impose interesting selective pressures, with traits that are advantageous in one phase of dispersal being costly in another. For example, Johnson et al. (2003a) found that pine seeds from species that are characteristically gathered from the ground by scatter hoarding animals have larger seeds than those that are not gathered. This larger seed size reduces the distance travelled in the initial wind-borne phase of dispersal.

Seeds with adaptations to encourage removal by ants may first be separated from the parent ballistically. In another potential example of a trade-off across different dispersal phases, Nakanishi (1994) suggests that ant-dispersed seeds that are initially dispersed ballistically tend to have a smaller elaiosome than those simply shed under gravity or collected by the ant from the plant. After reaching the ground, such seeds may be moved by wind or water, or eaten by rodents or birds, before being discovered by ants. These processes may improve the likelihood that seeds from a single plant are carried to different ant nests. Even after being collected by ants, those seeds that are dumped by the ants on surface refuse piles may be secondarily dispersed by wind, water, or vertebrate seed eaters. Granivorous ants may gather seeds that are adapted to be dispersed in other ways, and since consumption is generally not immediate on seed discovery, there is potential for movement by the ants coupled with survival of a viable seed (see Section 3.5). An interesting case is the removal of fleshy pulp from seeds in fallen fruit by ants (Byrne and Levey 1993). This has been demonstrated to improve germination success of the seeds (Leal and Oliveira 1998), perhaps by reducing the risk of fungal infection.

3.11 Conclusions

The advantages of animal-based dispersal are clear from the studies already discussed. Frugivores and granivores will often be a reliable source of seed transport, since the animals are actually attracted to the seeds or their associated fruit. They can also offer long-range dispersal, variation in deposition site between seeds from the same parent, and preferential dispersal to favourable micro-habitats. However, they may impose significant costs on the parent plant that must offer sufficient rewards to attract the

animals. These costs will often require that fewer seeds are produced per plant, since investment per seed is higher than for wind- or current-based dispersal. The costs associated with frugivory may be particularly unattractive since frugivores and granivores may often damage a large fraction of the seeds that they transport. Thus, we might expect such animal-based dispersal to be commonplace in situations where there are non-transport selection pressures for large seed size and/or air or water currents are weak or non-existent.

In contrast, the main problem with transport by attachment to the outside of birds and mammals is low uptake rates of seeds, since the dispersing agent is not attracted to the seeds or anything associated with them. However, those seeds that are transported this way have the potential for very long-range transport and also for transport between widely scattered specialist habitats. Ants are very important seed dispersers in many parts of the world. In comparison to vertebrate dispersers, they offer much less in terms of dispersal distances, but may be attractive because of their ability to swiftly remove seeds away from granivores and towards physical situations suitable to seedling establishment.

We have identified several aspects of animal-based dispersal that are very much in need of further research. We feel that many aspects of animal-based dispersal are still not well understood. The study of Montoya et al. (2008) suggests that animal-dispersed trees are more robust to habitat loss and fragmentation than wind-dispersed ones. This result suggests than dispersal mode can be a very important ecological predictor of the effect of climate- and land-use-driven changes on plant range and population viability. The rapid current changes to landscapes globally should be a further driver of fundamental research into animal-based dispersal.

Seed dispersal is characterized by looser associations between particular plant and animal species than can be the case for pollinator systems. Thus there are not the extremely exuberant display characteristics of many pollination systems. However, there is a great deal about fruit dispersal to interest evolutionary ecologists, with fleshy fruits having been predominantly studied. We will focus on these in the next two chapters.

CHAPTER 4

Visual communication in fleshy fruits

4.1 Introduction

In this chapter we will explain how the evolutionary and ecological setting of multi-species interactions that we outlined in Chapter 1 can shape the visual communication between fleshy-fruited plants and frugivorous animals. It has long been hypothesized that the colours of fleshy fruits are a visual advertisement that evolved to facilitate detection by seed-dispersing animals (Allen 1879, Kerner 1895). This hypothesis is intuitively plausible, since the majority of fruits that are consumed and dispersed by diurnal animals endowed with colour vision change colour during fruit ripening (such as blackberries do, which are dispersed by birds and mammals). In contrast, the majority of fruits that are consumed and dispersed primarily by nocturnal animals (such as bats), that rely on other senses to locate their food, remain green throughout their life or show only slight colour changes (e.g., many figs). Instead of colour, many bat-dispersed fruits develop strong scent that apparently guides fruit consumers to the food source (Korine and Kalko 2005, Hodgkison et al. 2007, Lomáscolo et al. 2010). Another line of evidence for the evolutionary and ecological importance of fruit colour is that fruit consumers often use colouration as a criterion for choosing among fruits once these are detected. Captive frugivorous birds prefer the darkest fruits available, probably because the darkest fruits once these are detected are usually the ripest ones (Willson 1994). Similarly, free-ranging birds foraging among differently coloured fruits of a neotropical tree prefer the darkest and ripest fruits available, as these are the most nutritious ones (Schaefer and Schaefer 2006). The same tendency in choosing the darkest fruits on offer is observable in the fruit-buy-ing decisions of humans (Crisosto et al. 2003). It is thus plausible to assume that many fruit-eating animals select fruits on the basis of colour. However, there is as yet very little consensus on how frugivorous animals selectively influence fruit colouration and whether selection is consistent enough (among species as well as in space and time) to influence the evolution of fruit colours. This is partly due to the fact that seed dispersers are not the only selective agents on fruit colouration. Abiotic factors (such as temperature and light intensity) and organisms that consume fruits but do not disperse their seeds (such as fungi, microbes, and seed predators) also affect fruit colouration, but their selective pressures on the evolution of fruit colouration are even less well understood.

In this chapter, we will first discuss the changes in fruit colouration that occur during ripening of fleshy fruits, and identify the proximate mechanisms that are responsible for such changes (Section 4.2). Thus if not specified otherwise, the term 'fruit' is used in this chapter to refer to fleshy fruits. We will then describe the patterns of fruit colouration seen within and among plant communities and ask whether specific types of seed dispersers are consistently associated with certain fruit colours (Section 4.3). Since colour is, in fact, consistently associated with specific disperser types, a result predicted by the dispersal syndrome hypothesis, we will then review the ultimate and proximate causes of colour selection by seed dispersers. We structure this part by first discussing whether seed dispersers have innate sensory biases for certain colours (Section 4.4). Then, we will consider whether fruit colours function as signals to seed-dispersing animals (Section 4.5). In general, fruit signals are characterized by two main properties, their covariance with the nutritional fruit quality and their design (the sensory stimulus). The latter influences the likelihood that a signal is detected and perceived. To evaluate whether fruit colours function as signals, we will thus examine how animals select fruit colours based on the fruit's detectability and the specific sensory abilities of different types of seed dispersers. This part is thus mainly concerned with selection by animals. We then switch sides and review whether plants pursue different signalling strategies (Section 4.6). In this part, we will evaluate whether any properties of fruit colours are associated with fruits' nutritional contents. That is, we will ask whether variation in fruit colours can convey information about fruit quality, and if so, whether this association is a by-product of selective forces other than communication or selected for by the animals that select fruits visually. Next, we will follow up the alternative hypotheses that fruit pigments are mainly selected as a protection against fruit pathogens (Section 4.6) or abiotic factors (Section 4.7) rather than communicating with mutualistic animals. This part is linked to the extensive physiological literature on the protective function of plant pigments (see Chapter 7). Finally, we assess whether the colours of vegetative structures that are consistently associated with fruit displays offer a solution to the evolutionary conflict of diverging selective pressures acting on fruit colours (Section 4.8). This chapter thus introduces some of the key factors influencing the evolution of fruit traits more generally. We will explore these factors in more detail in Chapter 5 on non-visual fruit traits.

4.2 Fruit ripening

When unripe, almost all fruits are green because they contain considerable amounts of chlorophyll. There are two plausible (and potentially complementary) explanations for unripe fruits being green. First, early in fruit ripening the chlorophyll within the immature fruit contributes to the energetic requirements for fruit growth and development, and this contribution becomes less important throughout subsequent fruit development (Ashan and Pfanz 2003). The other explanation is that seeds are often (but not always) unripe in green fruits;

that is, they are not ready to germinate. Given that the plant has no interest in losing seeds prematurely, unripe fruits contain very few nutrients, they are hard and difficult to digest, and their green colour makes them relatively inconspicuous against the green foliage of most plants. This has led people to hypothesize that unripe fruits retain their green colour for a communicative function, as means of camouflage within foliage. Unfortunately, no systematic study has aimed to differentiate between these two hypotheses. However, we can conclude that non-green fruit colouration developing during ripening functions as an indicator that a fruit is attaining ripeness (see Sumner and Mollon 2000a), and that a plant offers some kind of nutritional reward to animals that consume these fruits and, often, disperse their seeds.

During the ripening process, nutrients are transported into the fruits, mainly from the foliage, and to a lesser extent from the adjoining green bracts and stems that fruits are attached to. Concomitant to the accumulation of nutrients in the fruit, fruits start changing colour. At this stage, chlorophyll is often replaced by other pigments, but it is also masked by them. A reduction of chlorophyll obviously leads to lower photosynthetic activity, masking of chlorophyll also reduces the light available for photosynthesis, since pigments absorb light at specific wavelengths. The masking of chlorophyll can be illustrated by reflectance spectrometry, where chlorophyll induces a dip in the reflectance of red and orange fruits at around 670 nm (see Sumner and Mollon 2000a). Many ripe fruits retain chlorophyll because it can be used to refix internally generated CO_2 (Ashan and Pfanz 2003). Thus, although ripe fruits often only fix external CO_2 at a rate as low as 1 per cent of that of leaves, their photosynthetic performance may be comparable to that of leaves owing to very efficient internal recycling of CO_2 that is produced during the metabolic activity of fruit ripening. Fruit photosynthesis may contribute significantly to the carbon requirements needed for reproduction. Photosynthesis in young peaches contributes up to 15 per cent of the weekly requirements for growth, and sun-lit fruits may produce up to 9 per cent of the total carbon requirements throughout the fruiting season (Pavel and DeJong 1993). Similarly, net CO_2 uptake of young fruits accounts for up to 10 per cent of the dry mass gain in that period (Nobel and delaBarrera 2000). Fruits that remain green when ripe are likewise characterized by higher photosynthetic activity than fruits of other colours, since they can obtain a positive carbon balance at high light levels. At low light levels, however, they have apparently greater respiration rates than fruits of other colours (Cipollini and Levey 1991). As such, green colouration in ripe fruits might entail an ecological cost in habitats of low light availability such as forest understory or cloud forests. We know of no study that has addressed whether green fruits are more prevalent in the forest canopy or in open habitats than they are in the forest understory.

Given non-green fruit colouration entails costs of reduced photosynthetic abilities (at least in habitats of high light availability), we have to ask what are the benefits associated with bright non-green fruit colouration? Before we attempt to answer this question, we will provide a short summary of the biochemistry of fruit colouration.

4.2.1 Biochemistry of fruit colouration

Fruit colours result from the reflectance of the fruit surface, the ambient light illuminating the fruit, and the sensory physiology of the viewer. Fruit reflectance in turn is influenced by a mix of different fruit pigments and the specific properties of the epidermis and of epicuticular waxes (see Box 4.1 for comparison to floral colouration). The main fruit pigments are carotenoids producing yellow, orange, and red colours, anthocyanins imparting red, purple, black, and blue colours, and chlorophyll resulting in green colouration. Colour variation of anthocyanins is caused by factors such as the chemical structure of the anthocyanins, co-pigmentation, variation in the pH content of vacuoles, and the quantity of anthocyanins present in plant tissue (see Section 2.6.3). Red fruits may thus be either coloured by anthocyanins (as in cherries or strawberries) or they may be coloured by carotenoids (as in peppers). A commonly observed colour change during ripening is that unripe green fruits change their colour to a red mid-ripe stage, which finally becomes black when the fruits are ripe (as in blackberries). This appearance of red colour is associated with the initial accumulation of anthocyanins. Consistent with the photoprotective role of anthocyanins in vegetative tissue (see Section 7.3.2), they also often accumulate in the peel of apples and pears in response to low temperatures and high light levels (Steyn et al. 2009). The colour change

Box 4.1 Colour patterns in fruit vs. flowers

In contrast to flowers, fruits are almost always uniform in colour. They rarely sport stripes or spots (but see, for example, the fruits of some *Elaegnus* spp.). This difference between flowers and fruits is probably related to important constraints and possibly also to the lower selective pressures on fruits to be recognizable for mutualists. First, there are distinct developmental constraints related to fruit and flower architecture. Flowers comprise several ontogenetically differentiated structures that serve multiple purposes (e.g., signalling to pollinators, nectar production, and reproduction), whereas most fruits develop from a single flower structure, the ovary wall. This proximate difference alone may explain why flowers are more diversely coloured than fruits. Yet, flowers also need to be recognizable for pollinators because efficient pollen transfer is achieved if pollinators tend to visit conspecific flowers often, a behaviour termed flower constancy. Flower constancy obviously requires certain species-specific cues such as colouration, colour patterns, and olfactory cues. In seed dispersers, fruit constancy has not been described, although it is well known that seed dispersers do not consume fruits randomly (Jordano 1988). Indeed, plants would not increase their seed dispersal rate if seed dispersers repeatedly visit conspecific plants and thus disperse seeds under such plants; because seed predation risk underneath conspecifics will often be higher and germination probabilities lower.

from red to black is mainly caused by further accumulation of anthocyanins; it is thus based upon a quantitative change in anthocyanin concentrations, not a qualitative one.

In contrast to the pigment-based colours discussed so far, UV reflectance in fruits is produced by epicuticular waxes that represent a physical barrier between the fruit and the environment. To human eyes that are insensitive to UV reflectance, these waxes appear bluish-grey (as for example in grapes and plums). Peak reflectance in the UV can vary from up to 80 per cent to approximately 20 per cent, and is influenced by the thickness of the wax layers and by structural differences in the waxes that are involved. Because the waxes are a physical barrier, they may influence interactions between the fruit and the environment and thereby be influenced by abiotic factors (see Section 4.7). White fruits are primarily coloured by polyphenols (Harborne 1976).

4.3 Patterns of fruit colours

4.3.1 Bird-dispersed fruits

Visiting a local market almost anywhere on the planet may give the impression that fruit colours vary endlessly, with many gaudy colours prevailing (see Plate 3). This impression is slightly misleading for two reasons. Obviously, fruits available at local markets are the product of man-made selection and, since fruit colours affect buyers' decisions, they have been the target of selection. Thus, the choice on offer does not necessarily mirror local fruit colour diversity as experienced by natural frugivores. Second, fruit colour diversity is fairly constant across different sites. In a large-scale quantitative comparison Burns et al. (2009) did not find pronounced geographical differences in fruit colour diversity among bird-dispersed fruits of different areas in South America, North America, and Europe. Wheelwright and Janson (1985) demonstrated that colour diversity of bird-dispersed fruits was remarkably similar across four different regions: Florida, Europe, Costa Rica, and Peru. In all these regions as well as in Illinois (Willson and Thompson 1982), fruits were predominantly red and black; these colours accounted for approximately two-thirds of all

fruits. Later studies by Lee et al. (1994) on *Coprosma* species in New Zealand and by Corlett (1996) and Schmidt et al. (2004) on bird-dispersed fruits in Hong Kong and Venezuela, respectively, confirmed this pattern (Figure 4.1). Importantly, the study by Lee et al. (1994) and Schmidt et al. (2004) also provided a proximate mechanism for why these fruit colours were common. Red and black fruits were the most contrasting fruit colours against a background that consists primarily of leaves. In an experimental study on the pecking rates of artificial fruits, Alves-Costa and Lopes (2001) found that red and black fruits were more rapidly detected than white fruits. The authors used fruits made out of modelling clay to assess experimentally how fruit colours affect fruit consumption. This method works well because different frugivores leave distinct bite or peck marks in the soft modelling clay. Employing the same method, Galetti et al. (2003) also found that red and black fruits were pecked most. They also found that pecking rates depended on the type of habitat, with a pronounced edge effect. The probability of attempted fruit consumption was lower inside the forest compared to the forest edge. The relative abundance of black vs. red fruits may vary, though. Red fruits are more common in Europe, Alaska, and New Zealand (Turček 1961, Lord and Marshall 2001, Traveset et al. 2004, Valido and Olesen 2007), while black fruits dominate in Illinois (Traveset et al. 2004) and the other areas mentioned above. Importantly, both red and black colours are produced by the same biochemical pathway – the phenylpropanoid pathway – and even the same pigments, anthocyanins.

Apart from red and black colours, UV reflectance in fruits is also associated with dispersal by birds or by rodents, which are both sensitive to UV light (Burkhardt 1982, Willson and Whelan 1989, Altshuler 2001). Geographic variation in UV-reflecting fruits is poorly documented, partly because objective quantitative colour assessments of fruit colouration were rare until recently. Willson and Whelan (1989) reported 30 per cent of fruits (a total of 53 species) in central Illinois having UV reflectance, while Altshuler (2001) reported 61 per cent out of 39 species in Barro Colorado having UV reflectance. In contrast, in the tropical forests of Brazil, Venezuela, Kenya, and New Guinea,

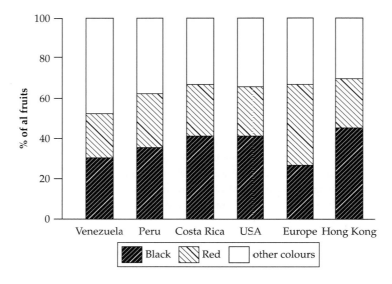

Figure 4.1 Approximately 60 per cent of bird-dispersed fruits are either black or red in different vegetation communities. Black and red colours are imparted by anthocyanins or betalains. Data from Wheelwright and Janson (1985), Corlett (1996), and Schmidt et al. (2004).

UV-reflecting fruits were very rare (Schmidt et al. 2004, Flörchinger et al. 2010, Cazetta et al. unpubl. data, Lomáscolo unpubl. data). The predominance of red and black fruits is, however, limited to fruits that are mainly dispersed by birds. Fruits consumed and dispersed by other animal groups consistently sport different colours (Janson 1983, Gautier-Hion et al. 1985, Korine et al. 2000, Lomáscolo et al. 2008). For example, in savannah habitats that are occupied by large seed-dispersing mammals in Africa and that have been occupied by the large extinct mammals in South America (e.g., the Pantanal), there is a higher proportion of orange, brown, green, and yellow colours than in bird-dispersed fruits (Donatti et al. 2007). The central question is thus, why do fruit colours vary according to disperser type?

4.3.2 Association between visual fruit traits and disperser type

Studies on fruit–frugivore interactions adopted an adaptive framework based upon earlier work during the 1970s and early 1980s. The adaptive framework posited that plant–animal interactions were mainly driven by co-evolutionary processes between fruiting plants and frugivorous animals.

The cornerstone of the framework is that mutualistic animals consistently select plant traits, but that animal groups differ in their selection of these traits and thus in their selective pressures upon plants (McKey 1975, Janson 1983). The distinctness and the consistency of selective pressures should result in trait convergence among plants sharing a similar suite of dispersers (or pollinators in flowers). This hypothesis has been termed the syndrome hypothesis (Janson 1983, Knight and Siegfried 1983, Gautier-Hion et al. 1985). It is important to remember that this hypothesis does not necessarily predict exclusive use of a fruit resource by only one type of seed disperser. Even plant species whose fruits are consumed by more than one seed disperser group may evolve traits that match one specific type of seed disperser, if it consistently disperses the seeds more effectively than other fruit consumers.

The syndrome hypothesis is not focused only on colours traits, but assumes that a combination of plant traits evolves under the selective regime of a given seed disperser type (for other traits, see Chapter 5). Among morphological and chemical fruit traits, fruit colour was one of the characters that best fitted the assumption of the syndrome hypothesis (Voigt et al. 2004). Analysing a large data set from the Peruvian tropical forest, Janson (1983) found that all fruits can be categorized into two classes according to fruit colour, size, and protection. One class consisted of large yellow, brown, or green fruits with a protective husk (a distinct, stiff, non-nutritious layer, such as orange peel). This husk may protect against unwanted fruit consumers but is probably also involved in maintaining the physical integrity of large fruits. The second class consisted of small red, black, blue, or white fruits without a husk (like cherries). Janson argued that the two classes of fruits correspond to consistent differences in size, jaw morphology, and visual abilities between mammals and birds, respectively. The dull, large fruits represent a mammal fruit syndrome, while the colourful small fruits represent a bird fruit syndrome. The evolution of size differences is explicable because the average frugivorous mammal has a larger individual size than the average frugivorous bird, and mammals can therefore on average handle larger fruits. The evolution of differences in smell is explicable because mammals apparently use their sense of smell more than birds, since many mammals are nocturnal. Finally, birds lack the manual dexterity, teeth, and long gut passage times of mammals that could be used to remove protective husks. These conclusions were corroborated by the comparison of fruit characters in savannah habitats inhabited by large mammals. Donatti et al. (2007) found that fruits in these habitats are often very large and that larger fruits have a stronger smell than small fruits. Similar results were obtained for figs that are dispersed by either birds or bats and that differ consistently in their size and colour, with bird-dispersed figs being small and conspicuously coloured (predominantly red), while bat-dispersed fruits are larger, scented, and mainly green (Kalko et al. 1996, Korine et al. 2000). Similarly, a recent study on geographical fruit colour variation within a species showed that fruits of the South American mistletoe *Tristerix corymbosus* were green in areas where they were dispersed by the nocturnal marsupial *Dromiciops gliroides*, whereas they were yellow in areas where they were exclusively dispersed by birds (Amico et al. 2011). Taken together these results suggest that plants may use different sensory modes, odour and colour, to advertise the presence of fruits to seed dispersers that, like mammals

and birds, differ in the relative importance of smell and vision. The sensory abilities of animals would thus present a functional mechanism for explaining dispersal syndromes.

In a large study on fruits in tropical Africa, Gautier-Hion and co-workers (1985) found slightly distinct dispersal syndromes. These authors found that fruit traits including colour diverged mainly between red, orange, and purple monkey- and bird-dispersed fruits on one hand and yellow and brown fruits consumed by rodents and ruminants on the other hand. They argued that the colourful fruits correspond to a monkey – bird syndrome because both these disperser groups rely on vision to identify fruit resources, while other mammals, characterized by poorer dichromatic colour vision abilities, rely primarily on scent to find fruits. Again, animals' distinct sensory abilities are among the key factors explaining fruit syndromes. Taken together, studies consistently show that fruits dispersed by birds tend to be small and colourful, while fruits dispersed by mammals tend to be larger and less colourful, thus supporting the existence of fruit syndromes. However, we should envision dispersal syndromes not as a dichotomy, but as a ends of a continuum between plants that are mainly dispersed by one type of animal (e.g., birds) and plants that are mainly dispersed by another type of seed dispersers (e.g., mammals).

Whether there are fruit syndromes that correspond to seed dispersers other than mammals and birds is not well known. Valido and Olesen (2007) present data on the colours of fruits dispersed by lizards on three different archipelagos. Not surprisingly, many of the fruits eaten by lizards are colourful since these reptiles possess excellent colour vision. However, just like mammals or birds, lizards are also catholic in their colour selection and eat a wide range of differently coloured fruits.

During the 1980s, interspecific comparisons were mainly done without controlling for the effects of phylogeny. When syndromes were scrutinized in phylogenetically controlled studies, the syndrome hypothesis was often not supported (Herrera 1987, Fischer and Chapman 1993). Consequently, the dispersal syndrome hypothesis remains contentious for fruit traits, particularly for visual traits because earlier studies categorized fruit colour according to human colour vision, which is not representative of the colour vision of most frugivores. Fruit colour was rarely included as a variable in phylogenetically controlled studies. As such, the question of whether communication to frugivores in addition to shared phylogeny explains why plants dispersed by distinct frugivores sport different colours (at least, as seen by humans) remains largely unresolved.

Only recently have researchers employed rigorous phylogenetic analyses on fruit colours, either as measured according to human colour categorization (Voigt et al. 2004) or measured quantitatively by spectrometry (Lomáscolo et al. 2008, 2010). The work by Lomáscolo et al. clearly supports the dispersal syndrome hypothesis because fruit colour and fruit size evolve in correlation in figs (*Ficus* spp.), and small red or purple figs have evolved independently in different clades, while large fruit size is correlated with inconspicuous, mainly green, fruit colouration. Interestingly, fruit colouration does not only differ between fruits consumed by birds and bats, which differ extremely in the relative importance of colour vision for foraging; fruit colouration also differs between distinct types of diurnal seed dispersers. For example, regional variation in fruit colours was associated with the relative importance of primates and birds as seed dispersers (Voigt et al. 2004), even though both seed disperser types are known for excellent colour vision abilities. This study concluded that, independent of phylogeny, fruit colour reflects differential selection by these distinct types of frugivores more than other morphological or chemical fruit traits. These results are corroborated by a large-scale comparison of the colours of fruits eaten either by primates or birds. Analysing a data set of more than 400 species from different regions worldwide, Lomáscolo and Schaefer (2010) found that both birds and primates can discriminate among the fruits typically dispersed by them and by the other animal group. Fruits dispersed by primates have a higher reflectance in the green part of the visible spectrum, while fruits dispersed by birds have a higher reflectance in the blue part of the spectrum.

The consistent association between fruit colour and disperser type may be explicable by distinct evolutionary mechanisms. It may result from directional selection on fruit colour variation imposed by seed dispersers or it may result from correlated selection. If, for example, two fruit traits are strongly integrated amongst each other (owing to developmental, physiological or genetic linkage), directional selection that induces changes in one trait will be associated with directional changes in the other trait even though there is no selection acting directly upon the second trait. At present, we have almost no information on whether and to which degree colour traits are integrated with other fruit traits. The multiple functions of fruit pigments make such correlations very likely. The only study to date (Valido et al. 2011) shows that integration of hue, chroma, and brightness (the colour traits that humans perceive) is weak and much less than that found within morphological or biochemical traits. Importantly, however, colour and nutritional fruit traits are consistently associated throughout the phylogeny of fruits in southern Spain. The brightness of fruits is consistently associated with the contents of non-structural carbohydrates, while the hue and chroma of fruits is correlated with the lipid contents of fruit pulp. If these results hold more widely and do not pertain only to the Mediterranean area, we thus expect colour to evolve fairly independent of morphological fruit traits but in concert with nutritional fruit traits. This result entails that owing to the consistent link between colour and nutritional traits fruit colour may serve as reliable indicator of fruit nutritional rewards (Section 4.5).

In another study, fruit size covaries with fruit colouration (because smaller fruits were more contrasting to primates; Sumner and Mollon 2000a). Biochemical constraints of pigment production may be a mechanism that could lead to a consistent association between fruit colouration and disperser type without direct selection by dispersers. For example, black colours produced by high concentrations of anthocyanins are very rare among large fleshy fruits. This may be explicable by metabolic costs associated with producing large quantities of anthocyanins. The metabolic costs are related to enzyme biosynthesis and activity, to the conjugation of cyanidine molecules to a monosaccharide molecule, and to the transport of the cyanidin-3-glucoside into the vacuole (Gould 2004). Information on the costs of pigment production is extremely rare, partly because pigments fulfil multiple functions in most plant organs, thereby complicating the assessment of their costs and benefits. We would therefore welcome studies that test the hypothesis that costs of pigment production explain the negative relationship between conspicuous colour and fruit size against the ecological hypothesis that colour cues become less important for fruit detection with increasing fruit size. That is, the larger a fruit, the easier it may be detected by alternative cues such as differences in the size and shape of a fruit compared to the size and shape of objects in the background.

Under the fruit syndrome hypothesis, variation in the size of fleshy fruits is traditionally seen as an adaptation to different frugivores. However, phylogenetic analyses yielded that the evolution of fruit size is correlated to that of leaf size, suggesting that a large proportion of variance in fruit size is an indirect consequence of variation in leaf size (Herrera 2002). Likewise, the association between protective husk and large fruit size could also be explicable by economic reason. Plants producing large fruits have typically smaller fruit crops than those producing small fruits (e.g., Flörchinger et al. 2010). Consequently, the total surface area of husks increases disproportionately in small fruits relative to the production of fruit tissue.

4.4 Selection by frugivores

Here we consider two proximate mechanisms for how seed dispersers may select fruit colours. First, animals may have innate biases or learned preference for certain colours. Innate biases are well known for a number of foragers including seed dispersers (Schmidt and Schaefer 2004) and insect pollinators (Naug and Arathi 2007). Consequently, fruit colours that match animals' innate or learned biases would be at a selective advantage. Second, certain colours may be more contrasting against the background and thereby be more effective in facilitating fruit detection (Kerner 1895, Willson and Whelan 1990, Schmidt et al. 2004).

4.4.1 Sensory biases

The idea that the evolution of fruit colouration has been driven by frugivores' innate biases or learned preference for certain colours dominated research in the 1980s and 1990s. This hypothesis predicts that seed-dispersing animals of a given group have consistent colour choices, with little variation between and within individuals (Willson and Whelan 1990). The extensive work by Mary Willson on frugivorous birds showed that, overall, these predictions were not supported. While frugivorous birds may choose fruits based upon their colour, their colour choices were often transient and inconsistent between individuals (Willson et al. 1990a, Willson and Comet 1993, Willson 1994). Individual birds differed in initial colour preference as well as in the transitivity and temporal stability of the preference (Willson et al. 1990a). Few individuals preferred the most commonly encountered fruit colours: red and black. Likewise, birds did not discriminate among conspecific fruits that either reflected UV (having a bloom) or that did not (waxy bloom removed) (Willson and Whelan 1989, Allen and Lee 1992). In North-western crows (*Corvus caurinus*), adults as well as hand-raised juvenile birds showed pronounced individual variation in fruit colour choices, suggesting that innate biases were not strong (Willson and Comet 1993). Similarly, a more recent study reported spatially fluctuating selection by frugivores on different-colour morphs in a polymorphic shrub (Whitney 2005). Overall, Willson concluded that the behavioural evidence for birds' preference to influence fruit colour evolution was weak (Willson and Whelan 1990).

Willson's conclusions are based upon the behavioural flexibility that most fruit-consuming animals show. In particular, many frugivorous animals typically consume a large diversity of fruits of varying colours. For example, in one study, more than 100 species of fruits were consumed by single individuals of Salvin's Curassow (*Mitu salvini*) over the course of 14 months (Santamaría and Franco 2000). Variation in seasonal abundance of fruit resources is key to inducing behavioural flexibility in consuming fruits of different colours; an effect that is likely to reduce the extent of innate preferences for certain colours. Indeed, garden warblers (*Sylvia borin*) that were caught in the wild were quicker to discount colour preferences in foraging when confronted with varying associations between colour and reward than same-aged garden warblers that were hand-raised (Schaefer et al. 2008c). The behavioural flexibility to consume many different fruit species is thus a prerequisite for frugivorous animals, given that the availability of specific fruit resources is often short and unpredictable. As such, there are few incentives for animals to rely upon strong colour preferences when consuming fruits. On the contrary, it would probably be maladaptive if animals were very conservative in eating only fruits of a given colour.

However, since the seminal work by Willson, some studies have reported consistent colour preferences in birds for red fruits, although it is unknown how widespread such preferences are. The first study was that by Puckey et al. (1996) documenting consistent preferences for red fruits in captive Australian white-eyes (*Zoosterops lateralis*). Importantly, maintaining birds on a particular diet of a different colour did not alter the colour preferences for red. The preference for red can be reversed in that species, however, if red fruits are only half as abundant as fruits of other colours (Giles and Lill 1999). A preference for red fruits was also found in a field study of the dimorphic red or orange fruits of salmonberry (*Rubus spectabilis*; Gervais et al. 1999). The authors cut canes bearing either red or orange ripe fruits and placed them 10–20 cm apart from each other. They found that red fruits from that species were removed more quickly from the experimental sites. The paired study design suggests a strong preference for red. The naturally occurring geographic variation in fruit colour frequencies of the two morphs of salmonberry (*R. spectabilis*) is thus probably maintained by other selective agents in addition to selection by frugivores. Finally, two studies showed consistent age-dependent differences in fruit colour selection. Adult redwings (*Turdus iliacus*) consistently preferred UV-reflecting over non-UV-reflecting blueberries, while inexperienced hand-raised birds did not show such a colour preference (Siitari et al. 1999). Schmidt and Schaefer (2004) found the opposite result in blackcaps (*Sylvia atricapilla*) where hand-raised juvenile birds preferred red artificial fruits over those of other col-ours, while this colour preference was lost in adult birds. Similarly, the related garden warbler likewise shows an innate preference for red over orange fruits (Schaefer et al. 2008c). The adaptive significance of these innate sensory biases is thought to help naïve individuals to find and identify food resources. In summer, when inexperienced juvenile birds start to forage on their own, the majority of fruits are red in Europe. Later in autumn, when migratory blackcaps encounter differently coloured fruits, the colour preferences is lost, implying a low potential for directional selection upon fruit colouration (Schmidt and Schaefer 2004). The loss of colour preferences is probably explicable by the higher experience of adult birds that allow them to quicker discount colour preferences when confronted with differently coloured fruits of varying nutritional contents (Schaefer et al. 2008c).

While most experimental studies were focused on birds, behavioural flexibility in foraging and the transitivity of colour preferences are also known for many other taxa, such as fish (Rodd et al. 2002, Smith et al. 2004). Innate sensory biases for certain colours are also known from pollinators where they may vary regionally (e.g., Raine and Chittka 2007). Such innate biases may enable bumblebees to forage more efficiently, but there is little evidence as yet that innate biases influence the evolution of flower colours. Likewise, the adaptive significance of innate colour biases remains poorly known.

Yet, some neotropical and European birds possess innate sensory biases (Figure 4.2) to prefer foraging from more contrasting fruit displays even in captive situations where all alternative fruit displays are clearly visible to the birds (Schmidt et al. 2004, Schaefer and Braun 2009). Given that colour plays a dominant role in avian foraging, the sensory system of birds may be tuned to respond strongly to the most contrasting displays. Similar sensory biases are also known for some pollinator groups (Naug and Arathi 2007). However, it is unclear at present whether sensory biases for contrasting colours are widespread among seed dispersers and pollinators or whether they occur only in a limited range of taxa. Honkavaara et al. (2004), for example, did not find evidence for sensory biases in redwings to respond more strongly to contrasting fruit displays

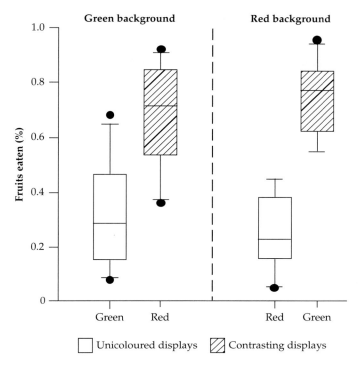

Figure 4.2 Four neotropical frugivorous bird species consumed more artificial fruits from contrasting displays than from uniformly coloured displays. The colours on the x-axis denote the colour of the artificial fruits that were presented on a specific background. The preference for contrasting displays was independent of the colour of the fruits. Shown are medians, 2nd and 3rd interquartiles and 10th and 90th percentiles as well as outliers by dots. Redrawn from Schmidt et al. (2004).

We can conclude that learned colour preferences are apparently not strong and consistent enough among species to exert strong directional selective pressure upon fruit colouration, and that there is little evidence as yet that innate biases for certain colours influence plant colour evolution more strongly. However, it is possible that innate biases for contrasting displays are strong enough to affect the evolution of contrasting displays in plants.

4.4.2 Effect of fruit colouration on ease of detection

Apart from actively preferring certain colours, animals might select fruit colours by consistently finding fruits of a given colour more easily. In general, searching and identifying targets depends on the background against which these targets are usually seen. Unlike animals, which move, plant colours

are usually displayed against an unchanging and predominantly green background. Apart from being green (see Chapter 7 for consideration of non-green leaf colouration), the background of leaves also varies in another visual dimension, in brightness. Brightness measures the intensity of light that is reflected from a given surface. The intensity reflected by leaf surfaces varies dramatically (3,000-fold, see Section 2.3) depending on whether leaves are illuminated by direct sunlight or whether they are shaded by other leaves resulting in a mosaic of bright and dim surfaces. The brightness of a fruit may thus be a poor signal because there is considerable variation in the brightness of the background. The colour of a non-green fruit, however, may be a good signal because it can stand out from the predominantly green background.

It is a long-standing hypothesis that fruit colours are optimized to allow animals to detect fruits from a distance (Allen 1879, Kerner 1895). This fruit contrast hypothesis remained untested as long as fruit colours were not analysed according to the visual systems of frugivorous animals. The earliest studies that modelled fruit colour according to consumer vision focused on primate vision. These studies suggested that the chroma of fruit colours, that is the saturation of the colour, allows primates to distinguish ripe fruits reliably from unripe fruits and from the background of leaves (Sumner and Mollon 2000a). Indeed, the need to reliably identify fruit resources may have been a driver in the evolution of trichromacy in primates (see Box 4.2) because the ability to distinguish fruit targets from background distractors is a prerequisite for foraging on fruits.

If frugivores exert selective pressures on fruit colouration as a signal to distinguish fruits from background, we can expect that the colour difference between fruit targets and background items should be large. This is because the larger the colour differ-

---

Box 4.2 Fruit colours as drivers of primate colour vision

As far as it is known, primates are unique among eutherian mammals in possessing trichromatic colour vision. The origin of primate visual pigments is ancient (Goldsmith 1990), but during the long history of mammals in the Mesozoic – where they were mostly small and nocturnal – the capacity for photopic colour vision in mammals declined. The adoption of nocturnal habits in the early mammalian history probably reduced the number of cones to two, one absorbing in the blue part of the spectrum and one in the green part. Thus, because primates, but not other diurnal mammals, secondarily gained a third cone, the factors selecting for trichromacy have attracted much debate. Because we lack molecular evidence for similar shifts in sensory abilities in most other taxa (except possibly for the two types of avian vision that differ in the peak absorbance of the short-wave receptor, UVS type and VS type), the debate on sensory adaptations has mostly focused on primate vision. It is a long-standing and popular hypothesis that primates evolved trichromatic vision to distinguish fruits from the background of leaves (Allen 1879, Polyak 1957). This hypothesis is consistent with the difficulties that red – green colour-blind humans have when searching for red fruits among the green foliage (Mollon 1989). If a target (the fruit) differs in colour from the distractors in the background (the foliage), the 'pop-out' effect allows quick identification, and visual search is performed in a pre-attentive search mode, in which the entire background is scanned for targets in parallel (Treisman and Gelade 1980). In contrast, if colour differences between target and distractors are small, a slow, serial search modus is employed, where items are scrutinized one after the other to discriminate targets from detractors. Vision-modelling studies demonstrated that the position and sensitivities of cones in primates are indeed optimized for detecting fruits, and also edible young leaves, against the foliage (Sumner and Mollon 2000b). This result is true for the two visual systems that are found in primates. Old world primates and apes, the catarrhine monkeys, possess a uniform trichromatic colour vision system with peak sensitivities at 424–34 nm, 531–9 nm, and 562–8 nm. In contrast, there is remarkable polymorphism in the visual abilities of New World primates, the platyrrhines. In howler monkeys (*Aloutta* spp.) colour vision is based upon an opsin gene duplication on the X chromosome that placed two allelic forms of the opsin gene on the same chromosome (Surridge et al. 2003). The same duplication also occurred in Old World primates. It has thus been suggested that colour vision based upon the duplication of opsin genes arose independently in Old World and New World monkeys (Hunt et al. 1998). The fixation of the duplication suggests that colour vision was advantageous and stabilized by positive selection. While this conjecture is intuitively plausible, it makes it difficult to explain the intraspecific colour vision polymorphism found in other neotropical primates.

In many neotropical primates there is only a single, polymorphic photo pigment gene on the X chromosome encoding for different sensitivities in either the long-wavelength range (approx. 563 nm) or the mid-wavelength range (approx. 535 nm; Jacobs and Blakeslee 1984, Surridge et al. 2003). As such, heterozygous females possess trichromatic colour vision, while homozygous females and males are dichromats. In most primates the polymorphism consists of three alleles resulting in 67 per cent of females being heterozygous. It is usually assumed that a heterozygote advantage favours females with trichromatic vision because they are better able to detect fruits (e.g., Regan et al. 2001, Riba-Hernandez et al. 2004). Indeed, there are experiments documenting that trichromatic individuals had an advantage when competing for orange food on the floor of their enclosure (Caine and Mundy 2000). Likewise, trichromatics were more efficient in selecting ripe fruits under semi-natural conditions (Smith et al. 2003). However, when the fruit and energy intakes of trichromats were compared to that of dichromats, there was no difference between the colour vision phenotypes (Vogel et al. 2007). How can this puzzling result be explained? Dichromats and trichromats forage together in groups. Thus, if trichromats are more efficient in detecting ripe fruits from a distance, feeding rates on the smaller spatial scale within a tree may not be different as dichromats may detect fruits based upon shape. Alternatively, the advantage of higher colour contrasts may be less obvious in the bright light environment of the deciduous forest studied by Vogel et al. (2007) but more apparent under the dimmer light conditions that typically occur underneath the canopy of tropical rain forests (Osorio et al. 2004). Consequently, the maintenance of the polymorphism in different lineages is inconsistent with a clear heterozygote advantage and may rather mirror situations where people are unaware of their red – green colour blindness if not specifically tested (Osorio et al. 2004). The polymorphism is consistent with a scenario of balancing selection, but the exact mechanisms that would favour dichromats are unknown. It may occur if colour vision phenotypes forage in different niches, leading to frequency-dependent selection. Thus the underlying mechanisms contributing to the colour vision polymorphism are still under debate.

While the hypothesis that foraging for fruits drove the evolution of colour vision in primates is popular, it is difficult to prove. In particular, many trichromatic Old World primates forage primarily on leaves and unripe fruits rather than on ripe fruits. This led to the suggestion that trichromacy evolved to distinguish nutritious and not well defended, but often differently coloured young leaves from better defended mature leaves, based upon a red – green channel in trichromats that does not occur in dichromats (Dominy and Lucas 2002). These authors showed that fruits differ from the background by both a red – green channel, and a blue – yellow channel that is also available to dichromats. The perceived red–green ratio in leaves was not only correlated with variation in leaf protein contents but also distinguished consumed leaves from leaves that were not consumed (Dominy and Lucas 2004). The authors found, however, no colour axis that separated fruits that were consumed from those that were not. They concluded therefore that folivory (rather than frugivory) could explain the evolution of trichromacy in primates. The situation, however, is more complex than the dichotomy between frugivory and folivory suggests. Analysing the food of Old World primates in Uganda, Sumner and Mollon (2000a, 2000b) found that the primate visual system is optimized for detecting food targets that are not mature leaves against the background of foliage. Thus, any object that differs from the background, be it a young nutritious leaf or a fruit or even an insect will be well detected. This is in accordance with the general predictions on the evolution of sensory systems, which usually are optimized for detecting various targets (e.g., food, potential mates, and predators) and not solely for detecting specific targets such as fruits or leaves. This hypothesis, termed sensory drive, is further in accordance with another hypothesis by Allen (1879) that once colour vision evolved in primates, it can and was used for different purposes. A recent analysis supported this view, showing that the evolution of trichromacy predated that of red colouration in sexually selected signals in primates (e.g., skin colours and pelage; Fernandez and Morris 2007). Consequently, once colour vision evolved, and was likely associated with an attraction towards colourful stimuli in a foraging context, it represented a pre-existing bias that promoted the evolution of colourful traits in a context of mating as well.

ence between targets and distractors in the background, and the smaller the colour difference between different distractors, the faster a given target can be identified. This mechanism is particularly important in long-distance signalling (e.g., that of a tree advertising fruits to entice fruit-consuming animals to visit the tree). If the size of a fruit is constant, a larger contrast is more easily seen against the background. There are few experimental data to assess whether frugivorous animals indeed select for more conspicuous fruit colours against the background. Those that are available, however, support this view. Testing the rate of fruit detection on artificial fruits of constant size, Cazetta et al. (2009) found that fruits with larger contrasts are more readily detected. Interestingly, they found that the rate of fruit detection is a sigmoid function of fruit contrasts, which is consistent with the prevailing paradigm that increasing contrasts do not increase the conspicuousness of a target above a certain threshold (see Figure 2.1).

Behavioural observations of avian foraging under controlled conditions are congruent with this result of selection for increased conspicuousness. Birds find more contrasting fruits from a larger distance, and they attend primarily to differences in colour and not to differences in brightness between fruits and their foliage. For example, if UV-reflecting blueberries (*Vaccinium myrtillus*) were tested on two backgrounds against control blueberries lacking the UV reflectance, the UV-reflecting fruits were only removed more quickly on the background against which they had significantly higher contrast than the control fruits (Schaefer et al. 2006). Red fruits were detected from a larger distance against foliage than identically sized black fruits. This occurs because red represents are strong colour contrasts to the green background, whereas black represents primarily a contrast in the brightness of reflected light. We could thus hypothesize that red fruits are primarily adapted to increase the conspicuousness to seed dispersers, while black fruits may either be selected for high contents of anti-oxidants (see Section 4.5.3) or as defence against fruit pathogens (see Section 4.6).

Thus, the hypothesis that seed dispersers select fruit colours based upon the ease of discriminating them from the background seems plausible. We would thus expect fruit colour evolution to be driven towards increased conspicuousness for the primary type of seed dispersers. Since seed dispersers differ in their abilities to perceive colour, with birds and reptiles being tetrachromatic, primates being trichromatic, and other mammals being dichromatic (Section 2.5), we might expect that the persistent differences between the colours of fruits that are dispersed by distinct disperser types represent adaptations to their respective colour vision abilities. However, the available evidence suggests that this is not the case. While fruits dispersed by birds are more contrasting to birds than fruits dispersed by primates, primates also perceive higher contrasts of bird-dispersed fruits compared to those of primate-dispersed fruits (Lomáscolo and Schaefer 2010). Similarly, modelling the visual abilities of birds, Schaefer et al. (2007) investigated the contrasts between fruits and their species-specific background, the foliage. If the contrasts between fruits and the background they are presented against are optimized for detection, fruits should have higher contrasts against their own leaves than against the leaves of other species. This prediction was not supported, suggesting that conspicuousness against the prevailing background is not optimized in fruits (also not in flowers; see Chittka 1997).

Optimization of fruit contrasts may not occur because the colours of fruits and leaves (as seen by birds) were correlated in a large sample of bird-dispersed plants (Burns et al. 2009). According to this study, any directional change in fruit colouration would concomitantly entail a change in the same direction in the leaves and thereby maintain fruit contrasts relatively constant. As such, the assumption of contrast optimization is obviously simplifying as it ignores the role of a number of important selective agents. However, it provides a good starting point to ask which factors may explain the lack of optimization by influencing fruit colouration. First, fruit detection is not independent of fruit size. The larger a given target, the easier it is to discriminate this target from the background independent of colour properties. For example, a large fruit may stand out from the background primarily by having a different shape than background items (leaves). The lower importance of visual contrasts in large fruits was suggested by a negative relationship between fruit contrasts and fruit size in fruits consumed by primates (Sumner and Mollon 2000a). If detection of large fruits is less dependent on visual contrasts than that of small fruits, the way fruits are presented will be an important factor influencing their detectability. For example, fruits that grow singly may need to have larger contrast against the background to secure detection than fruits which grow in clustered infructescences because the infructescence will appear as a large uniformly coloured patch. As yet, data are missing to test this hypothesis that fruit contrasts are contingent on fruit presentation. Other plausible hypotheses are that there are distinct signalling strategies in plants fruit (Schaefer and Schmidt 2004), that fruit pigments are also selected by seed dispersers for their biochemical properties (Catoni et al. 2008b, Schaefer et al. 2008b), or that fruit colours are selected by fruit predators (Whitney and Stanton 2004, Schaefer et al. 2008a) or are pleiotropically linked to abiotic factors (Traveset and Willson 1998). In the followings, we will discuss the evidence for each of the hypotheses, but first we will discuss evidence that fruit colours are indeed signals adapted to consumers.

4.5 Fruit colour as signal

If we ask whether fruit colours are an adaptation towards signalling to fruit-eating animals, it is important to clearly differentiate between cues, indices, and signals (Chapter 1). As a quick reminder, only signals evolve in order to communicate to other organisms (Otte 1974), whereas cues are informative but are not primarily selected for their communicative function and indices are specific types of cues that necessarily indicate a specific quality of the sender. Are fruit colours thus cues, indices, or signals? This depends at least partly on the type of fruit quality that we examine. For example, the colouration imparted by anthocyanins is necessarily correlated with – and hence an index of – anthocyanin contents in the skin of the fruit (see Section 4.5.3). Since the pulp of some fruits can also be coloured by anthocyanins, the relationship between the total anthocyanin contents in fruits and fruit colouration may be more variable than the example of fruit skin suggests. Fruit colours are probably no index of fruit nutritional contents,

mainly because the relationship between colour and nutrients is sufficiently variable between different species. At present, we probably lack sufficient knowledge to distinguish between cues and signals in fruit colouration. We should therefore retain cautionary statements on fruit signalling. Nevertheless, signal theory represents a very useful framework to trace the evolution of fruit colour because it provides predictions on the patterns we would expect if fruit colours are signals adapted to increase the communication efficiency to mutualistic animals.

4.5.1 Signalling strategies

If we examine communication systems, there are two different ways to optimize the signalling efficiency between organisms. On one hand, signallers can optimize signalling behaviour by increasing the conspicuousness of their signals and thereby reducing the costs that receivers pay to search for and discriminate among signals. On the other hand, signallers may increase the reliability of their signals. Models on signal evolution proposed a trade-off between the conspicuousness and the content of signals (Schluter and Price 1993). Let us consider how such a trade-off may arise. The concept of optimization entails that certain colours are better than others at attracting animals because they are more conspicuous. If individuals are selected to increase conspicuousness, variability in colouration will be reduced (unless conspicuousness is costly, for example through enhanced attraction of seed predators; and may therefore represent a handicap signal) so that conspicuous colours are less suitable (than other colours) to indicate variation in the quality of fruit rewards among individuals. This logic applies to comparisons within populations but also to interspecific comparisons on the possible strategies that plants might pursue within a vegetation community. For example, if we assume that plant species converge on the same, most contrasting fruit colours, it is easy to understand that the more different species converge on these colours, the less likely is colour variation among fruits to indicate specific information on fruit nutritional contents. This is simply caused by the high inter-specific variability in fruit nutritional contents of species converging on that colour. Thus, signal convergence may actually conflict with the fundamental principle that organisms receiving signals select for an increased reliability of information transfer. As yet, the relative importance for signal reliability and signal conspicuousness for the evolution of signals is not well resolved (Schaefer and Schmidt 2004, Darst et al. 2006), although it certainly represents one of the most fundamental and most exciting areas of current research on communication.

Apparent trade-offs between the reliability and the conspicuousness of signals have only been found in fruit colours and in aposematic colours of poison frogs. The most conspicuous fruit colours in a Venezuelan rainforest, red and black, were unrelated to the biochemical contents of fruits, while less contrasting fruit colours indicated macro-nutritional and allelochemical fruit contents (Schaefer and Schmidt 2004). Yellow to orange fruits were characterized by high protein contents and low levels of tannins, while blue fruits had high levels of sugar and tannins. Fruit consumers are thus able to evaluate the nutritional contents of less conspicuous fruits by just looking at a fruit. The relationship between fruit contents and fruit colouration is blurred in red and black fruits, possibly because so many different fruit species converge on these highly conspicuous fruit colours. A similar negative relationship between the content and the conspicuousness of a signal has been found in the warning colouration of poison frogs. The conspicuousness of aposematic colouration was inversely correlated to the toxicity of these frogs (Darst et al. 2006). In this signalling system, increasing either conspicuousness or toxicity induces avoidance reactions in chickens. These examples indicate that the diversity of colours in communication systems may be at least partly explicable by different signalling strategies.

The hypothesis about different plant advertising strategies is attractive but remains untested owing to the paucity of data. If advertising the contents of a fruit is a stable signalling strategy for plants, we expect a consistent relationship between fruit contents and fruit colours. Only two studies analysed this relationship in fruits (Valido et al. 2011, and Cazetta et al., unpubl.). As mentioned previously, Valido et al. found persistent covariation between

lipid contents and colour (hue) and chroma of fruits independent of phylogeny. Likewise, fruits' levels of carbohydrates covaried with fruit brightness. While these consistent patterns of covariance between nutritional and colour traits demonstrate the evolutionary potential for a signalling strategy to reliably indicate fruit contents to fruit consuming animals, they are open to alternative interpretation. In particular, shared biosynthesis rather than adaptations to communication may explain the consistent covariance between visual and nutritional fruit traits (see below).

4.5.2 Signal honesty

If fruit colours are adapted to reliably indicate fruit contents, we need to consider the mechanisms that could maintain the evolutionary stability of this putative signalling system. In theory, fruit colours could be handicaps if the physiological costs of signal production are high enough that only high quality individuals could produce both sufficient colour and nutrients. As yet, there is no evidence for such a scenario. Indeed, there are generally few demonstrations of cost-enforced honest signals in the communication between mutualists (Edwards and Yu 2007). Because the fitness costs associated with consuming a low-quality fruit are not very high for frugivorous animals, we do not expect them to strongly select for honest signalling in fruits.

An alternative scenario is that signals can be cost-free and reliable. This occurs under the very restricted conditions that receivers can verify each signal (Lachmann et al. 2001) and punish cheaters so that these incur higher costs than honest signallers of low-quality resources. On one hand, fruit consumers can verify the relationship between fruit colour and fruit reward when consuming the fruit. For example, they can discriminate 1–2 per cent difference in sugar contents (Schaefer et al. 2003a). Even if they avoided consumption of fruits from plants that cheat and promise more rewards than they offer, it is doubtful that the costs incurred by the plant surpass those of plants indicating low rewards honestly. In essence, although instant verification is possible in frugivory, we do not expect it to result in honest, cost-free signalling.

On the proximate basis, covariance between colours and nutrients is at least partly explicable by fruit biochemistry. Biosynthesis of the most important fruit pigments, anthocyanins, is strongly upregulated by sugars inducing the gene expression of key enzymes such as chalcone synthase (Takeuchi et al. 1994, Solfanelli et al. 2006). This functional link between pigment production and sugar contents may result in a consistent link between colouration and the sugar contents of a fruit. Likewise, covariance between lipids and the hue and chroma of fruits throughout the phylogeny of fruiting plants in Spain may be related to the presence of the other major fruit pigments, carotenoids, which are lipid-soluble (Valido et al. 2011). A major limitation for understanding the relationship between colour and contents in fruits is insufficient knowledge on the proximate biochemical mechanisms – and possible functional links – of pigment production and the production of nutritional rewards. This may come as a surprise given the major advances in molecular and genetic studies focusing to increase nutritional contents of commercially grown fruit crops. These artificially selected species have a high potential to become ideal model systems for revealing the underlying proximate mechanisms that likely contribute to the evolution of fruit–frugivore interactions and that would allow us to distinguish between cues and signals.

4.5.3   Secondary compounds

We have asked whether fruit colour traits reveal an underlying invisible quality of the fruit to seed dispersers. By doing so, we have focused on nutrients because fruit consumers ingest fruits to obtain nutrients that fuel their metabolism. However, macronutrients are not the only nutritional resource that fruit consumers may obtain from fruit consumption. Fruits contain many secondary compounds, some of them being deterrent for a certain range of fruit consumers (see Section 5.1), but some secondary compounds have long been acknowledged to possess beneficial effects on fruit consumers. The best known are probably plant anti-oxidants that reduce reactive radicals and thereby alleviate oxidative stress. Since many of these plant anti-oxidants retain their quality if consumed by animals or humans, these dietary anti-oxidants have become a major focus of medical and behavioural studies in the last two decades. Fruits are a principal source of dietary anti-oxidants because the concentration of these compounds is often much higher than in other plant tissue (Ames et al. 1993, Catoni et al. 2008a). This is also one of the main reasons why the consumption of fruits is considered to be healthy for humans.

Fruits contain a cocktail of different dietary anti-oxidants, the most important groups are Vitamin E (tocopherol), Vitamin C (ascorbate and derivates), polyphenols (including anthocyanins), and carotenoids. Anthocyanins and vitamin C are the most common anti-oxidants in fruits (Catoni et al. 2008a). Given that anthocyanins and carotenoids play a dual role in imparting colour and functioning as anti-oxidants, it has been postulated that fruit colouration may indicate the contents of anti-oxidants (Schaefer et al. 2004). Yet, by just looking at a fruit humans are unable to visually evaluate the concentrations of the major pigment groups, carotenoids, chlorophyll, and anthocyanins in a limited number of commercially available fruits (Lancaster et al. 1997). Similarly, birds are unable to evaluate the carotenoid contents of a fruit by just looking at it, as revealed by a study on 60 wild fruit species (Schaefer et al. 2008b). This is attributable to the fact that carotenoid contents may be masked by anthocyanins, which are 16 times more concentrated in fruits than carotenoids. Also, anthocyanins absorb more light per molecule than carotenoids do. However, birds are able to evaluate anthocyanin contents in the fruits of 60 species visually (Schaefer et al. 2008b) – humans seem to have a similar capacity (HMS, unpubl. data).

Thus, fruit colour can be a cue (possibly even an index) of the contents of anthocyanins, which belong to the flavonoids, a group of strong anti-oxidants. However, it is important to note that anthocyanin contents can vary independently of the skin colouration. Blueberry varieties are probably the best example because they may either possess dark coloured fruit flesh or whitish fruit flesh. In both cases, the skin colouration appears similar, but comparing the anthocyanin contents of these fruits likely yields strong differences.

While increasing the contents of anti-oxidants in food consumed by humans has become a major enterprise, the effects of fruit anti-oxidants on natural consumers' health have received surprisingly little attention in evolutionary ecology. Only one recent study showed that fruit anti-oxidants indeed influence health parameters in fruit consumers (Catoni et al. 2008b). This study first established that captive blackcaps (*Sylvia atricapilla*), a common frugivorous European passerine and seed disperser, preferred food with anthocyanins over food without anthocyanins (Figure 4.3). They found further that anthocyanins circulated in the blood of these birds after consumption. Thirdly and most importantly, blackcaps that were fed anthocyanin extracts from fruits were more likely to mount a humoral immune response than birds that had no access to anthocyanins but were maintained on an otherwise identical food (Catoni et al. 2008b). This study argues that fruit consumers might select fruit pigments for health reasons and not solely because of the colour they impart on the fruit. Consequently, we can conclude that, even if fruit consumers exert consistent selective pressures on fruit colouration, the selective pressures may vary according to the relative importance of long-distance signalling (conspicuousness) and the relationship between

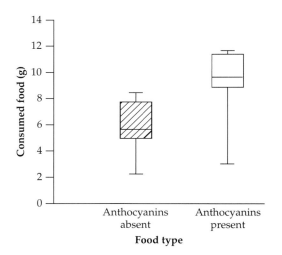

Figure 4.3 Blackcaps, a frugivorous bird, consume more artificial food if it contained anthocyanins compared to identical food without anthocyanins. Shown are medians, 2nd and 3rd interquartiles and 10th and 90th percentiles. Redrawn from Catoni et al. (2008b).

colour traits and macro-nutritional or micro-nutritional contents such as those of anti-oxidants. An important corollary is thus that – even without considering the relative influence that fruit predators may have on fruit colouration – seed dispersers may exert distinct and sometimes divergent selective pressures upon fruit colouration. In other words, if birds selected fruits solely based upon their conspicuousness, they would probably select mostly red-coloured fruits over black-coloured fruits (Schaefer et al. 2006). If, however, they selected fruits based upon their contents of anti-oxidants, they should select black fruits over red fruits.

If plants employ a signalling strategy of indicating fruit contents through visual or olfactory signals, then this strategy is likely to be invaded by cheaters. This is arguably the main argument of why reliable signalling is not widespread. So far evidence for cheating is not very strong, but admittedly only a few studies ever focused on this topic. One type of cheating occurs if coloured seeds mimic fleshy fruits or unattractive fleshy fruits mimic more attractive ones. We discuss these types of fruit mimicry in Section 8.5.

4.6   The roles of fruit predators

Many fruit consumers are detrimental to plant fitness because they consume fruit pulp without dispersing the seeds, or because they consume and destroy the seeds and discard the fruit pulp. Here we refer to both groups as fruit predators. The relative role that fruit predators may have on the evolution of fruit colouration is poorly understood. Given that fruit predators include birds and primates that eat or destroy seeds, we can safely assume that at least these types of predators use the same visual signals to identify fruits as seed-dispersing birds and primates do. The situation is likely to be different if we turn to different types of fruit predators such as insects, which possess visual systems quite different from those of vertebrate seed dispersers. Here, there is the evolutionary potential to reduce conspicuousness to one type of fruit consumer, the insect predator, while maintaining high conspicuousness to the intended receivers of seed dispersers (Willson and Whelan 1990). Red is an ideal candidate for this type of colour because many insects lack a red receptor (Briscoe and Chittka 2001) and are therefore less sensitive to colour variation in the red part of the spectrum than seed dispersers such as birds and primates. Whether differential conspicuousness to predators and seed dispersers indeed represents an evolutionary trajectory in fruit colouration is as yet unexplored.

We can probably assume that the evolution of most fruit colours should represent a trade-off between the relative risk of fruit predation on one hand and the benefits of attracting seed dispersers by highly conspicuous fruit displays on the other (Greig-Smith 1986). This is akin to the situation in sexually selected colours in animals that represent a trade-off between increased conspicuousness in mates and reduced conspicuousness to predators (Endler 1980, Håstad et al. 2005). Unfortunately, the relative role of fruit predators as eavesdroppers on the communication between fruiting plants and seed dispersers is mostly unknown. Importantly, however, even fruit predators that are colour blind or that depredate unripe fruits before the colouration of ripe fruit appears can exert strong selective pressures upon fruit colouration. The best studied system is the fruit colour polymorphism in *Acacia ligulata*, an Australian shrub that produces red, yellow, and orange fruit morphs. Morphs differed in the production of viable seeds per ovule owing to differential seed predation by heteropteran insects (Whitney and Stanton 2004). Because insect predators fed on the unripe fruits prior to the appearance of colour differences between the morphs, the differential predation is evidently explained by pleiotropic effects of fruit colour alleles. This study demonstrates that the selective pressure of seed dispersers, the agents that we assume to predominantly influence fruit colouration, may be overridden by selective pressures of less obvious, but often numerically more important, selective agents. Unfortunately, there are almost no other studies that simultaneously assessed the selective role of fruit predators and seed dispersers on fruit colouration in the field. There is, however, evidence that fruit pigments may also be selected by fruit predators. A recent study showed that fruit extracts consisting mainly of anthocyanins considerably reduce fungal growth in vitro and in grapes detached from the plant. The moderate contents of

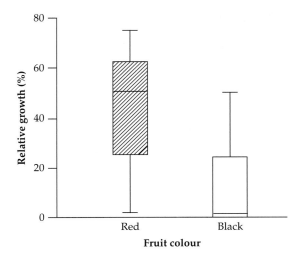

Figure 4.4 Anthocyanin contents in fruits reduce fungal growth. The lower anthocyanin contents of red mid-ripe blackberries reduced the growth of nine fungal species by 50 per cent compared to the growth on control agar plates, whereas the higher anthocyanin contents in black ripe blackberries reduced the growth of nine fungal species by 95 per cent. Shown are medians, 2nd and 3rd interquartiles and 10th and 90th percentiles, redrawn from Schaefer et al. (2008a).

anthocyanins in mid-ripe red blackberries reduced by 50 per cent the growth rate of nine fungi species that were isolated from fruits compared to that of the control. The higher anthocyanin contents in black fruits reduced fungal growth rate by 95 per cent compared to the control (Figure 4.4; Schaefer et al. 2008a). As such, fungi as fruit predators may select for higher anthocyanin contents. The characteristic colour changes during ripening from red to black that many fruits (like blackberries) show may thus be related to the increased risk of fungal infection with increasing nutritional contents in ripe fruits and decreasing contents of deterrent secondary compounds. This hypothesis is further supported by the fact that black fruits are less conspicuous to seed-dispersing birds than red fruits.

4.7 Influence of abiotic factors on fruit colouration

As yet, the influence of abiotic factors on fruit colouration is not well understood. Given that the colouration of ripe fruits only appears at the end of the ripening period, it has been argued that pigments have only a minor role in the thermoregulation of fruits (Willson and Whelan 1990). While this is probably true, there are good reasons to expect other correlated effects associated with fruit colouration. In general, pleiotropic effects are predicted to be common in colour polymorphisms in plants and animals because correlated selection promotes the physical integration of colour and other traits in multi-trait co-evolution processes (Forsman et al. 2008). This is because in polymorphic species, selection may result in the modification of the genetic architecture and the developmental pathways leading to increasing divergence in colour morphs. Colour morphs may thus occupy distinct peaks in the adaptive landscape that represent alternative trait–value combinations (Wright 1988). These adaptive peaks may not always be due to selection by visually-oriented animals but can be explicable by pleiotropic effects (see Section 6.7).

While balancing selection of herbivores and frugivores contributes to the fruit polymorphism in *Acacia ligulata*, abiotic factors may also affect the colour polymorphism in that species. Sampling along an almost 600 km long transect revealed clinal variation in morph frequencies (Whitney and Lister 2004). Frequencies of the yellow morph decreased in favour of the red morph with decreasing temperature and increasing rainfall. Interestingly, the morphs did not differ in seed mass, aril mass or the profiles of fatty acids and flavonoids, but they showed consistent differences in carotenoid contents. These results are comparable to that of the fruit colour polymorphism in *Myrtus communis*, which sports either the common morph of bluish-grey fruits (to the human eye) that are UV-reflecting to birds, or the rare morph of white fruits. Colour morphs of that species did not differ in morphological or nutritional traits, and they were equally consumed by frugivores (Traveset et al. 2001). Seeds' propensity to germinate from both morphs did not differ under controlled conditions, but seeds from the more common blue morph germinated faster in outdoor conditions. Faster germination of the more common morph was not found in salmonberry *Rubus spectabilis* (Willson and O'Dowd 1989) but, here, the type of soil affected germination probability; suggesting pleiotropic effects similar to those

reported for the *Linanthus* flowers by Schemske and Bierzychudek (2007; see Section 6.7).

The mechanism that produces UV reflectance in most UV-reflecting fruits is also probably influenced by abiotic parameters. Waxy epicuticular layers reflect in the UV and are found on many fruits such as plums, kakis, grapes, and some figs. Studies on grapes revealed that the thickness of epicuticular waxes increased with increasing illumination, so that sun-exposed grapes had stronger UV reflectance than grapes of the same plant grown in shade (Rosenquist and Morrison 1989). Since the waxy layers reflect light, they may effectively protect the fruits from heating up in direct sunlight, which would elevate water losses by the fruit and could lead to rapid desiccation. Indeed, sloe (*Prunus spinosus*) fruits whose epicuticular layers had been removed showed a higher desiccation rate relative to control fruits (Schaefer and Schaefer 2007). Increasing thickness of epicuticular layers further provides a physical barrier between the fruit and the environment. Consistent with this conjecture, the thickness and structure of epicuticular waxy layers may further modulate the susceptibility of fruits to fungi (such as *Botrytis*) (Rosenquist and Morrison 1989). In the shade of the canopy or in densely clustered infructescences, the epicuticular layers are more likely to be rubbed off by adjacent plant tissue than those of fruits that are not in direct contact with either foliage or other fruits (Rosenquist and Morrison 1989 and pers. obs.). Given that the thickness of the waxy bloom is thus influenced by abiotic factors and by factors such as the abrasion by adjacent plant tissue, it is not surprising that studies investigating avian fruit choice among fruits that either sport the waxy bloom or lack the bloom have generally found no consistent selection by birds on the presence of the bloom (Willson and O' Dowd 1989, Allen and Lee 1992, Honkavaara et al. 2004, but see Siitari et al. 1999). Also it is feasible that some strongly UV-reflecting fruits that stand out from the foliage (and therefore have lower abrasion of the waxy layers) attract birds to the plant, and that these birds then show no selection against other fruits whose waxy layers are removed. As such, a measure of the 'attraction power' of UV could be to calculate the average reflectance of all fruits in a plant, since selection occurs on the level of whole plant.

4.8 Signalling role of secondary structures

Given the different selective pressure on fruit colouration, one solution to mitigate the various trade-offs is to transfer the signalling role of fruit colours to adjacent tissue that is consistently associated with fruit displays (see Plate 4,5). Secondary structures that are associated with fruit displays may include the bracts and pedicels of fruit displays and colourful capsules enclosing the fruit. Secondary structures are common in fruit displays, occurring in 30 per cent of trees and shrubs in Illinois, US (Willson and Thompson 1982). The colouration of these structures shows a general trend: usually the secondary structures are coloured red and the ripe fruits are black or blue (Willson and Thompson 1982). This combination is also the most common bicoloured display in two neotropical areas, where more than 70 per cent of black or blue fruits are associated with red-coloured secondary structures or unripe fruits (Wheelwright and Janson 1985). A prolonged display of red colouration in unripe or mid-ripe fruit stages is often also seen as being adaptive to increase the conspicuousness of the entire fruit display; a situation that is termed temporarily bicoloured fruit display (Willson and Thompson 1982). Given that red colour is a better long-distance signal than black colour (Schaefer et al. 2006) and that, conversely, black colour provides a better protection against fruit-rot fungi (Schaefer et al. 2008a), we can hypothesize that the trade-off between long-distance signalling and defence against fruit predators is mitigated by a spatial separation of signals. The colour of secondary structures (or partially ripe fruit) would thus function primarily as a long-distance signal, whereas the colour of the fully ripe fruit may primarily be a consequence of selection for defence against pathogens and hence does not represent a signal.

If secondary structures function as long-distance signals, they should display strong visual contrasts. Modelling the contrasts of fruit displays according to the visual abilities of birds indicated that the contrasts created by secondary structures indeed increased the contrasts of the entire fruit displays (Schaefer et al. 2007). There is also evidence that the contrasts created by secondary structures influence fruit detection. More fruits are removed from displays with red secondary structures compared to displays lacking red colouration – independent of whether the secondary structure consisted of mid-ripe red fruits or of red stems (Willson and Melampy 1983; Burns and Dalen 2002). Moreover, there is evidence that the signalling function of accessory structures indeed promotes long-distance attraction. When Morden-Moore and Willson (1982) compared the fruit removal rate of ripe black cherries in the presence and absence of mid-ripe red cherries, the effect depended on the spatial arrangement. In isolated infructescences, more black fruits were eaten in the presence of mid-ripe red fruits than when displayed alone. However, this effect disappeared when infructescences containing mid-ripe red fruits and those that did not contain mid-ripe red fruits were presented in close proximity. This difference is explicable by the fact that once red fruits attracted frugivores to a given site, they consumed black fruits independent of the presence of secondary structures within this plant. This is because fruit displays with only black fruits are visible at shorter distances, but they are less effective in attracting frugivores from far away.

Experiments with captive frugivores showed that even if alternative displays within a cage are equally visible to birds, they still prefer to consume ripe fruits from the more contrasting displays (Schmidt et al. 2004, Schaefer and Braun 2009). As discussed earlier, the preference for contrasting displays may be explicable by innate sensory biases to respond more strongly to strong contrasts. Likewise, a preference to consume ripe fruits from infructescences that still contain some unripe fruits (Figure 4.5) may help consumers to avoid ripe fruits that are starting to rot because they have been exposed to fruit predators for long periods. This hypothesis is based upon the findings that a lesser proportion of damaged fruits have been found in blackberry (*Rubus fruticosus*) racemes that contained unripe fruits compared to those that contained only ripe fruits (Greg-Smith 1986). Although the hypothesis is plausible, there is currently no evidence that seed dispersers avoid infested fruits by feeding on infructescences that

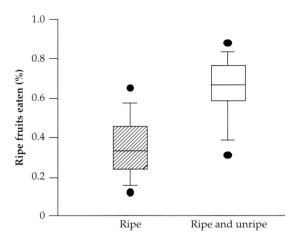

Figure 4.5 In captivity birds of various species preferred consuming ripe fruits from displays that contained contrasting unripe fruits, which were not eaten, relative to displays that only contained ripe fruits. Shown are medians, 2nd and 3rd interquartiles, 10th and 90th percentiles, and outliers. Redrawn from Schmidt et al. (2004).

also contain unripe fruits. On one hand, the retention of different ripening stages on the plant may serve an adaptive function owing to the increased contrasts of fruit displays. On the other hand, it may simply result from differential flowering times or from resource limitation of the plant. As yet, the relative importance of each of these mechanisms has not been fully addressed.

A recent study by Amsberry and Steffen (2008) yielded no effect of bicoloured fruit displays compared to control displays. In their study on *Ardisia nigropunctata*, ripe fruits were red and unripe fruits were either white or pink. Given that red is a more contrasting colour against green background than the colours of unripe fruits in this species (Lee et al. 1994, Schmidt et al. 2004), this result, which seems counterintuitive at first, demonstrates the importance of measuring the contrasts of the different parts of fruit displays in order to analyse their possible adaptive function. It is likely that an alternative mechanism, such as the effects of differential flowering time, explain the sequential fruit ripening in this species (Amsberry and Steffen 2008).

Secondary structures, such as conspicuously coloured bracts and stems of infructescences, are ideal for studying signalling in fruit-seed disperser com-

munication. First, the colour change of these morphological structures is often associated with the time of fruit ripening. Second, because non-green colouration develops in photosynthetically active tissue and likely reduces photosynthesis, some of the alternative explanations that may partly explain fruit colouration are difficult to reconcile with this colour change in vegetative tissue.

Secondary structures can also indicate fruit nutritional contents. Schaefer and Braun (2009) investigated the polymorphic black elder (*Sambucus nigra*) that sports either red or green pedicels associated with infructescences of ripe black fruits. They found that the occurrence of both phenotypes is dependent on the habitat; the phenotype with red pedicels occurs mainly in gaps, whereas the phenotype with green pedicels occurs mainly within the forest. Modelling avian vision suggested that only the red phenotype increases the conspicuousness of the fruit displays of black elder, as the green pedicels do not contrasts strongly against the background. Consistent with these results, fruits were more quickly removed from isolated, artificially coloured red pedicels than from isolated, artificially coloured green pedicels. Importantly, pedicel colouration is not only a trait that increases the conspicuousness of fruit displays, but it also conveys information on fruit quality. Fruits from red pedicels are characterized by higher sugar contents compared to the fruits from green pedicels, which is likely explicable by the higher illumination in gaps that should result in higher rates of photosynthesis (Figure 4.6). Thus pedicel colouration not only increases signal efficacy, it also functions as an indicator of fruit quality. Consistent with this conclusion, captive blackcaps preferentially consumed the sweeter fruits from red pedicels than fruits from green pedicels. They apparently attended to the signal of the secondary structure because fruits from red pedicels were preferred from the start before they had tasted fruits. Most importantly, however, colour variation within the red and the green phenotype also indicated the sugar contents of the fruits. The higher the contrasts created by the pedicels, the richer the fruits were in sugars (Schaefer and Braun 2009). The positive covariance between the contrasts of pedicels and sugar contents are explained by two distinct mechanisms. In

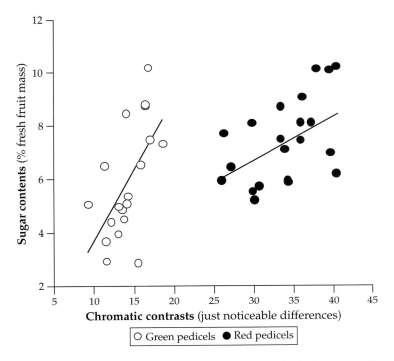

Figure 4.6 Fruit quality is a linear function of fruit advertisement in both alternative phenotypes of pedicel colouration in black elder. Symbols indicate mean values for plant individuals. Red pedicels indicated higher sugar contents and had higher chromatic contrasts against leaves than green pedicels. Redrawn from Schaefer and Braun (2009).

the red phenotype higher contrasts are produced by higher contents of anthocyanins in the pedicels. The concomitant increase of anthocyanins in pedicels and sugar contents in the fruits may both be explicable as a physiological response to high illumination intensity, since anthocyanins often accumulate as a light-stress reaction in vegetative tissue in response to increased illumination (see Section 7.3.2). Given that the link between variation in red pedicel colouration and sugar contents in fruits is probably environmentally enforced, it is unknown whether red pedicels indeed evolved as signals, that is to increase the conspicuousness of the fruit displays or to convey information about fruit quality, or whether they evolved as a physiological stress response that is associated with concomitant changes in nutrient levels in fruits. In black elder, red pedicels can thus be viewed cautiously as cues that indicate fruit quality, whereas the timing of the development of red colouration (being associated with fruit ripening) is a signal.

In contrast to the relationship between red pedicels and the nutritional contents of fruits, the positive covariance between sugar levels in fruits and the contrasts of green phenotypes against the distinctly shaded green background of the leaves is not easily explained as a cue. It is produced by a negative correlation between chlorophyll contents in pedicels and sugar contents in fruits. Thus, lower chlorophyll contents in the pedicels indicate higher sugar contents in fruits, which is not explicable by physiological mechanisms. Instead, it suggests that black elder individuals incur costs of foregoing primary production to create more contrasting pedicels (i.e., with a different shade of green from the leaves) that signal high fruit quality to seed dispersers. Given that the background of leaves is optimized for photosynthesis, it is evident that signals that are contrasting to this background – to indicate fruit quality – can only be achieved in green tissue at the expense of deviating chlorophyll contents away from the optimum. Variation in the green pedicels is thus consistent with a signalling function. Hence, this study documents that the array of plant signals is possibly larger than usually acknowledged, and includes shades of green.

There are two evolutionary scenarios for the evolution of signalling in green pedicels. It is unknown at present whether the reduction in chlorophyll contents is sufficiently costly to prevent cheating by lower-quality individuals in this signalling system. If this is the case, green pedicels would function as handicap signals that are kept honest by the costs associated with signalling. Such a decrease may not be costly in the high irradiance environment of red phenotypes, but it may be costly in the low light environment of green phenotypes, which is considered to be a habitat of lower quality owing to the low shade tolerance of black elder (Atkinson and Atkinson 2002). However, compared to the overall photosynthetic capacity of an individual, the reduction in photosynthesis in the pedicels is likely to be small. Alternatively, costs might be small if the verification of fruit signals by the consumer restricts the possibility for cheating. It would be interesting to explore these alternatives both in models and experiments.

4.9 Summary

The evolutionary ecology of fruit colouration is less studied than that of floral colouration (Section 6.5). Therefore, the ultimate mechanisms that select for fruit colour diversity are not well understood. Nevertheless, there is a consistent association between certain fruit colours and seed disperser types. Small red or black fruits are consistently associated with dispersal by birds; larger fruits that are often green, yellow, or brown are associated with dispersal by mammals. As yet, it is unclear whether this pattern results from direct and consistent selection by seed dispersers or whether it is explicable by correlated selection. The production costs of pigments in large fruits may set an upper limit to the production of colourful large fruits. Colour preferences by seed dispersers are apparently weak and inconsistent, and thus play at most a small part in driving fruit evolution. However, seed dispersers may consistently select fruit colours owing to their conspicuousness. This hypothesis underscores the importance of analysing fruit colours as they are seen by the animals interacting with the plant. Fruit colours may also function as signals of certain fruit contents, but the relationship between fruit reflectance and contents is relatively unexplored. In contrast, the signalling role of secondary structures that are associated with fruit displays seems to be well supported by the few available studies. Finally, studies have only recently started to assess the influence of factors other than seed dispersers on fruit colouration. Fruit antagonists including frugivorous insects and fungi select fruit colouration owing to pleiotropic effects of the pigments themselves or because pigment production is linked to the activation of certain biochemical pathways that may also produce other secondary compounds with known activity against these plant antagonists (Schaefer and Rolshausen 2006), or because colour affects detection by insect antagonists.

Fruit colouration is likely to be selected by multiple agents including mutualistic and antagonistic biotic agents, as well as abiotic factors. Fruit colour polymorphisms represent an ideal study system to assess the relative importance of each, and to study whether balancing selection occurs. It thus represents an opportunity to assess the relative importance of adaptation and constraints in communication because the visual stimuli themselves are fairly simple, constant, and – at least when compared to signals in animal communication – relatively unchanging. We can thus ask whether fruit colour evolution is driven towards increased efficiency in communication to mutualists or whether it is largely constrained by selective pressures of fruit pathogens and abiotic effects. The multiple agents likely involved in shaping fruit colouration suggest that aiming for a unifying hypothesis to explain the evolution of colour in all fleshy fruits throughout the plant kingdom is probably unrealistic. However, the hypothesis that fruit colours function as signals to attract seed dispersers remains attractive but more detailed information on the relative conspicuousness of fruits to distinct types of frugivores is needed. Comparing the relative conspicuousness of fruits for different frugivores with the relative visitation rates of these frugivores would be a correlative study suggesting a pattern, while manipulating fruit reflectance experimentally and assessing fruit visitation afterwards would clarify the mechanism. To understand the evolution of fruit colours, fundamental questions concerning the intraspecific variability of fruit colours have yet to be addressed. While some variation in fruit colours does exist along environmental gradients, studies on the variability within populations are very rare. Finally, fruit colours present an ideal study system to assess the reaction of wild animals towards coloured food stimuli. As such, the study of fruit colouration holds large potential for experimentally assessing how visual signals are selected in complex natural environments.

CHAPTER 5

Evolutionary ecology of non-visual fruit traits

5.1 Introduction

Fleshy fruits are made to be eaten. They look appetizing from a distance, but appearances are often deceiving. Although the plant offers nutrients in exchange for seed dispersal, astringency is the rule, not the exception, and some fruits are outright toxic to humans. The astringency and toxicity of many fruits to humans might occur because other seed dispersers are less sensitive in their taste and less discriminating in fruit choice, although we consider this unlikely given the fine-tuned selectivity demonstrated by some frugivores (Schaefer et al. 2003a). Alternatively, the directed deterrence hypothesis posits that deterrent secondary compounds in fruits screen out unwanted fruit consumers (which may destroy the seeds or consume fruit pulp without dispersing the seeds), while not affecting legitimate seed dispersers. Yet, targeting specific seed dispersers by using biochemistry to screen out other animals does not entirely solve the puzzle of the relative unpalatability of many fruits. Even in the tropics where the availability of fruits is less seasonal than in temperate areas, few mammals and birds depend entirely upon the free lunch provided by plants. At the same time, many plant species would benefit from greater attention from seed-dispersing animals because many fruits remain uneaten, fall beneath the parent plant, and rot. The characteristic smell of rotting fruits attracts a large suite of animals including seed dispersers. Even though fruits that have fallen to the ground can be dispersed, a central question remains: why do so many animals forgo the free lunch provided by fruit?

The answers to this question are related to the key aspects of communication among multiple species that we explained in the previous chapter with the examples of visual fruit communication. There is strong evidence that selective pressures upon communication are often diluted in the various selective pressures of multiple species interacting with the same communicative trait, particularly because some of those species do not attend to communication but interact with chemical substances owing to their biochemical functions. Here, we will extend the issues of multiple species interacting with a plant trait, and of the consequential multi-functionality of those traits to gustatory, tactile, and olfactory communication of fruits

As in the previous chapter, we start this chapter by examining the ripening process in fleshy fruits (Section 5.2). We then describe how fruit pulp biochemistry changes during fruit ripening (Section 5.3.1) and how variable fruit pulp biochemistry is in ripe fruits (Section 5.3.2). Fruit biochemistry features prominently in this chapter because it determines the quality of fruit rewards. We explain how fruit consumers can select for fruit pulp biochemistry (Section 5.4) and how that selection translates into differential fruit removal. The removal rates of frugivores influences fruit longevity and we examine the traits and selective pressures associated with it (Section 5.5). We then examine fruit communication in other senses, namely fruit texture (Section 5.6), fruit scent (Section 5.7), and echolocation (Section 5.8). Finally, we consider the sensory aspects of dry fruits (Section 5.9), before summing up how the geographic mosaic (Section 5.9.2) provides a conceptual tool to account for the spatial and temporal variability of factors shaping non-visual fruit communication.

The most convincing explanation for why so many animals forgo the free lunch provided by the plant is that fleshy fruits are the result of an evolutionary triad between plants, seed dispersers, and fruit predators like insects and microbes (which are the most abundant and ubiquitous frugivores but often the least obvious ones to humans) (Janzen 1977, Herrera 1982, Cipollini and Levey 1997a, Tewksbury 2002). Seed dispersers and fruit predators impose conflicting selective pressures on the design of fleshy fruits. Ideally, fruits should be attractive to the former and at the same time repellent to the latter. Given that fungal spores and bacteria are ubiquitous and can travel by air, there are few means to grant exclusive access to fruits by seed dispersers. One option is a thick fruit husk that covers the fruit but that can be crushed by legitimate seed dispersers. However, this morphological barrier does not come without costs. The husk may make the fruit unavailable to small seed dispersers, or may increase the handling costs of the fruit such that it becomes less attractive. Also, an animal that can crush thick husks can incidentally crush the seeds of the fruit at the same time. The seeds may thus need further morphological protection to ensure successful seed dispersal. Finally we suggest that the husk (and any associated protection of the seed) will have a production cost, as well as increasing the weight of the fruit such that greater investment may have to be made in the structure of the plant to support it.

Likely for the reasons outlined above, biochemistry rather than physical defence is the primary pathway that plants use to mediate their interactions with fruit consumers. Plants produce a bewildering diversity of chemical compounds (estimated at approx. 100,000 known compounds, Hartmann 1996); many of them influence the feeding behaviour of pollinators, seed dispersers, and herbivores. Unfortunately, evaluating the selective pressures shaping the expression of plant chemical compounds is a complex endeavour given that many of these compounds have multiple functions. Multi-

functionality likely arises because some of the many selective pressures acting upon plants interact with each other and become coupled over evolutionary time. The concept of multi-functionality entails the need to analyse *relative* selective pressures associated with each function in order to fully appreciate the evolutionary dynamics of plant biochemistry. Multi-functionality calls for a holistic approach, integrating physiology and ecology (both biotic and abiotic factors) to understand the evolution of plant biochemistry.

Fruit (and nectar) biochemistry differs from the biochemistry of other plant organs (leaf, flower, root) in that the plant must balance the conflicting demands of defence against unwanted consumers and attraction of beneficial consumers (i.e., seed dispersers and pollinators). Importantly, the balance between both demands shifts throughout fruit development. When the seeds are not yet able to germinate in a developing fruit, the fruit is termed unripe. At this stage, any consumption of the fruit is detrimental to plant fitness; there is no need to attract seed dispersers. Only when the seeds attain their ability to germinate, does the fruit become ripe from the perspective of the plant. At this point, however, the fruit is not necessarily ripe in the colloquial sense or from the perspective of a fruit-eating animal. Analysing the changes in fruit quality that occur subsequent to the maturation of seeds thus provides an opportunity to scrutinize which traits might have evolved specifically to attract seed dispersers. Although the balance between defence and attraction shifts towards increasing attraction during fruit ontogeny, fruit traits associated with ripening might also have evolved to screen out frugivores that do not disperse seeds.

The evolutionary triad between plants, seed dispersers, and antagonist such as microbes requires that we approach the issue of communication by non-visual fruit traits with the null hypothesis in mind that some of these traits may not be selected for the purpose of communication. The central aim of this chapter is thus to clarify whether plants evolved traits in order to communicate with frugivores via taste, touch, and smell or whether frugivores respond to gustatory, tactile, and olfactory cues that provide information about fruit quality.

In this chapter we will ask how reliable gustatory, tactile, and olfactory information is and whether some traits specifically evolved in order to communicate to animals. This is asking whether there is any evidence that traits evolved specifically for a communicative function, representing what Marler (1977) termed 'true communication'. Alternatively, communicative traits could be merely by-product of non-communicative functions such as physiological changes in the fruit or adaptations to abiotic factors.

5.2   Fruit ripening

The control of ripening is of paramount importance for the producers of commercially grown fruits. Particular efforts have been devoted to delay fruit maturation to increase shelf life and reduce economical losses during the transport of fruits. This economic interest has resulted in substantial research efforts to understand the genetics of fruit ripening in model species such as the tomato and strawberry. Tapping into the research tools of food sciences holds considerable potential for evolutionary biologists to uncover how plants control fruit ripening (see Box 5.1) and how fruit ripening in

turn affects the interactions between plants and their seed dispersers.

In fleshy fruits, ripening is associated with some or all of the following: (i) the loss of cell wall structure resulting in softening of fruit tissue, (ii) an increase of carbohydrate and lipids in fruit pulp, (iii) the production of volatile flavour compounds, and (iv) increasing non-green pigmentation. Thus, fruit texture, smell, and taste are all important cues that can indicate the ripening stage of a fruit. These sensory modalities act in a hierarchical manner during the decision-making about whether or not to swallow a fruit. Vision and scent are long-distance cues that can attract animals to a given plant. They can also act as short-distance indicators that reveal the status of a fruit (e.g., damaged or intact). Once an animal decides to pick a given fruit it receives information by touch, and finally it can receive gustatory information during the swallowing of a fruit. Additionally, the strength of a fruit's attachment to the peduncle is another cue of ripeness. Firm attachment can be a very effective means of deterring feeding by small animals because immature songbirds that had not yet learned to pick only ripe fruits were unable to detach unripe *Goupia glabra* fruits

Box 5.1 Hormones involved in fruit ripening

At first glance, the ripening process of fleshy fruits and of dry fruits (like nuts) differs dramatically. The obvious changes in colour, smell, texture, and fruit pulp nutritional composition are all restricted to fleshy fruits caused by the up- and downregulation of many suites of genes (e.g., Manning 1998). However, the molecular processes underlying fruit ripening in dry and fleshy fruits are conserved and share important characteristics, such as the key regulatory control by the ethylene signalling pathway. Ethylene (C_2H_4) is the simplest unsaturated hydrocarbon that regulates many developmental and metabolic processes related to growth and senescence in plants (e.g., abscission of leaves, fading of flowers). In general, dry fruits mature in a process akin to senescence and disperse their seeds in an abscission-like fashion including dehiscence and shattering. This contrasts with fleshy fruits that can be categorized as either climacteric or non-climacteric fruits. Only climacteric fruits such as apples increase respiration rates at the onset of ripening and only they require ethylene for maturation (see Section 5.3 for discussion of ethylene as interspecific cue of fruit maturation). Interestingly, however, although ethylene synthesis does not increase through the ripening process of non-climacteric fruits (e.g., grapes and citrus fruits), alterations in ethylene responsiveness might be able to mediate the physiological processes associated with fruit ripening (Adam-Phillips et al. 2004). In addition to ethylene, other plant hormones contribute to the regulation of ripening. In non-climacteric fruits, ripening is delayed by auxins that coordinate growth and development in plants, whereas jasmonate contributes to ripening in the tomato, a climacteric fruit.

from the peduncle (HMS, pers. obs). In the next section, we will explore how the sensory traits change during ripening and ask how reliably they are associated with fruit maturation. We start with changes in fruit pulp chemistry because they determine the reward of a fruit, whereas changes in external aspects determine the stimuli that an animal can associate with the reward.

5.3   Fruit pulp biochemistry

5.3.1   Changes during ripening

As a rule of thumb, nutritional contents of fruit pulp increase during fruit ripening, the contents of deterrent secondary metabolites decrease, while the contents of pigments, aromatic compounds, and volatile flavour compounds increase. Yet, there are many species-specific and component-specific variations to this pattern. Furthermore, a substance that is deterrent to one consumer does not necessarily deter another. Despite these reservations, the trend of an increase in the contents of lipids and carbohydrates is fairly robust. Multi-variate models comparing the effects of different fruit compounds on

the rate of seed dispersal among species suggested that lipids and carbohydrates are the most important fruit contents that stimulate fruit removal by avian seed dispersers (Figure 5.1; Herrera 1998a, Schaefer et al. 2003b).

In contrast, fruit phenolic contents reduce fruit removal rates, and their contents decrease during ripening (Schaefer et al. 2003b, Cazetta et al. 2008). The reduction in phenolic contents follows the trend that defensive compounds present in unripe fruits are catabolized, translocated, or detoxified by chemical reactions (e.g., complexation) during ripening. But, there are some exceptions. In one study, the contents of a specific group of phenols, the condensed tannins, did not change during ripening, but these compounds had no effect upon the fruit removal rates of seed dispersers (Schaefer et al. 2003b). In summary, attractive pulp constituents increase during ripening, whereas deterrent components decrease concomitantly; a pattern that reflects plants' shift in interest from protecting fruits when seeds are immature to promoting their consumption by seed dispersers, once they are mature.

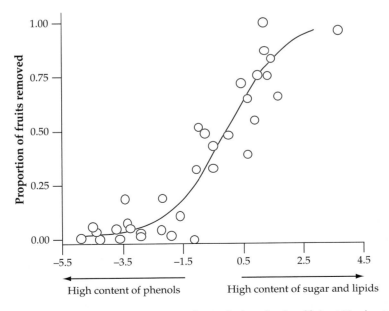

Figure 5.1 The proportion of fruits removed form the parent plant can be visualized as a function of fruit nutritional contents in a Venezuelan rainforest community. Higher contents of lipids and/or carbohydrates in fruit pulp lead to a higher proportion of fruits removed, whereas higher contents of constitutive defences such as phenols lead to lower removal rates. Redrawn and modified from Schaefer et al. 2003b.

5.3.2   Patterns in ripe fruits

The proportions of the three main nutrients (carbohydrates, proteins, and lipids) differ strongly among ripe fruits of different species. Overall, non-structural carbohydrates are the most important nutrients in the fruit pulp of Mediterranean fruits, constituting 67 per cent (range: 26–94) of dry fruit pulp, whereas protein and lipids make up on average 5 per cent (range: 1–27) and 7 per cent (range: 0–59), respectively (Herrera 1987). Similar patterns are described from Asian and neotropical plant communities (Corlett 1996, Schaefer et al. 2003b). The distribution of protein and lipid fractions are both skewed to the right; that is, there are few species with high protein or lipid contents. In contrast, the proportion of non-structural carbohydrates shows the opposite pattern and is skewed to the left (Herrera 1987, Corlett 1996). Compared to alternative food items (e.g., insect prey), the availability of protein is limited in fleshy fruits. Many estimates of protein contents in fruits are troubled by the fact that most researchers analysed nitrogen contents and used a conversion factor of 6.25 to calculate protein contents. This conversion factor is, however, derived from animal tissue and most likely over-estimates protein contents because nitrogen is present in many secondary compounds (Izhaki 1993, Levey et al. 2000). In general, there are distinct types of fruits, which can have high contents of either lipids or of carbohydrates. There is a strong negative correlation between these nutrients, which is explicable in light of their biochemical properties: carbohydrates are water soluble, whereas lipids are hydrophobic.

There are very few studies on fruit biochemistry at large spatial scales. Hampe (2003) documented that the contents of lipids and soluble carbohydrates decrease northwards in Europe, whereas pulp water content remained constant along a latitudinal gradient. Fruit sizes increase with increasing annual rainfall, suggesting that the largest fruits of a given species are found in the wettest climate that that species occurs in. Geographic patterns in fruit secondary chemistry are not well known. There are, however, consistent differences in the contents of phenols and condensed tannins, with higher relative contents at lower latitude (Schaefer et al., unpubl. data). It thus seems as though abiotic factors influence the allocation of fruit nutritional rewards to fruit pulp. Given that seed dispersers prefer nutrient-rich fruits, fruit biochemistry is likely shaped by an amalgamation of biotic and abiotic selective factors.

A corollary of the skewed distribution of nutrients is that the nutrient profile of single fruit species is usually not balanced for the dietary requirements of most frugivorous animals. The adaptive hypothesis is that a skewed nutrient profile induces fruit consumers to search for other prey and thereby facilitates the dispersal of seeds rather than the dropping of seeds underneath the mother plant during extended stays in the plant. As such, animals can either complement their diet by feeding on different fruit species or by complementing a frugivorous diet with other prey items. Indeed, resource complementarity has been suggested in North American fruits that either had low or high sugar and lipids contents, respectively (Whelan et al. 1998). These authors presented fruits in different combinations and relative abundances to free-ranging fruit consumers and found preferences for rare resources. Such preferences indicate that consumers can complement their diets with different fruit resources. These preferences, in turn, can translate into differential plant fitness owing to so-called neighbourhood effects (see Section 3.2).

The complementarity of fruit resources is not only contingent upon the distribution of fruit nutrients but also on the abundance and identity of secondary compounds found in the fruit pulp. For example, fruits of guelder rose (*Viburnum opulus*) contain secondary compounds that make these fruits very acidic. Cedar waxwings (*Bombycilla cedrorum*) consuming these fruits had elevated nitrogen losses consistent with producing bicarbonate as an acid buffer (Witmer 2001). Free-ranging waxwings consume guelder rose fruits only when an additional food source rich in proteins becomes available: the catkins of *Populus deltoides*. Feeding experiments showed that waxwings maintained or increased body mass on a mixed diet of guelder rose fruits and nitrogen-rich catkins even though catkins offered negligible amounts of energy (Figure 5.2). Given that waxwings seem to be the main seed disperser of guelder rose at least in parts of its range, this is a convincing example of how the effective seed dispersal of one plant depends on the

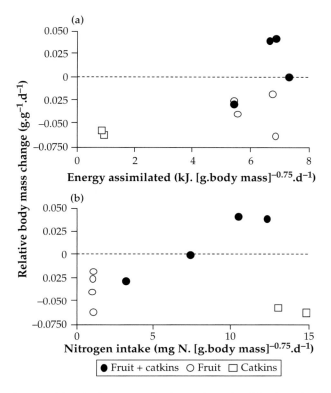

Figure 5.2 (a) Cedar waxwings fed a mixed diet of *Viburnum opulus* fruits and *Populus deltoides* catkins (filled dots) maintained body mass even though the catkins provided minimal energy returns. Waxwings feeding on fruits only (open dots) or catkins only (open squares) did not maintain body mass. (b) Feeding on catkins resulted in high nitrogen intake suggesting that fruits provide sufficient energy returns but lack nitrogen, whereas catkins provide nitrogen but are deficient in energy. Redrawn from Witmer 2001.

presence of another (a neighbourhood effect). More importantly, it illustrates that thriving on a fruit diet can be contingent on the timing and spatial proximity of complementary food items.

Large quantities of deterrent secondary compounds in the pulp of fruits could either inhibit or reduce feeding activity or induce consumers to complement their diet with substances that detoxify the deterrent compounds. Although the presence of deterrent compounds in the pulp of fruits may seem surprising at first glance – since the fruit is meant to be eaten – there are many factors that could select for the presence of these compounds in fruits (see Section 5.3.3). One speculative idea that haunts the literature is that the presence of deterrent secondary compounds or imbalanced nutrients is adaptive *because* it induces fruit consumers to complement their diet, travel more between complementary

plants, and thereby results in a more effective dispersal of seeds (Izhaki and Safriel 1989, but see Mack 1990). The rationale behind this idea is that if a given consumer ingests many fruits in one foraging bout on one plant, the likelihood increases that many seeds will be defecated in a single patch. There is evidence that such aggregated seed dispersal limits recruitment because negative-density-dependent effects (predation on seeds and competition among seedlings) will thin the number of seeds that can successfully grow in a given location (Russo and Augspurger 2004). Further, given the short digestive time of some frugivores, consumption of many fruits from one plant at a single visit may result in some of the seeds being defecated under the parent plant. Although the presence of deterrent compounds can decrease the amount of fruits eaten during a foraging bout, we know of no study that has actually demonstrated that the avoidance of aggregated seed dispersal is an important factor selecting for the presence of secondary compounds in fruit pulp. It may be that the cost of secondary compounds in terms of reduced uptake of fruits overwhelms any benefits in terms of manipulation of the distribution of seeds from those fruits that are taken up. However, this is essentially an empirical question, and we would very much welcome empirical studies of the fitness consequences of secondary fruit compounds that consider the range of ultimate fates of the seeds from a given plant. Genetic modifications that knock out certain fruit compounds would be a powerful tool to address this question.

In the absence of external cues that indicate the ripening stage of a given fruit an animal has to probe each fruit prior to deciding whether or not it is ripe and thus suitable for consumption. This probing behaviour has been filmed by the group of Elisabeth Kalko in fruit-eating bats on Barro Colorado Island that feed on *Ficus* spp. that produce red ripe fruits which are apparently scentless (because they belong to the bird dispersal syndrome; see Section 4.3.2). Such a mechanical inspection of fruit quality damages the fruit by puncturing the mechanical barrier that bars colonization by fungi, bacteria, and other fruit predators. The evolution of communication to reliably indicate fruit quality before the fruit is damaged should thus be beneficial to both plants – because it lowers their loss of seeds – and to animals – because it directs them to the most rewarding food resources. In the case of the fruit-eating bats, the external cue of ripeness, red colour, is not easily perceived during the night. The mutual benefits are conditions that generally favour the evolution of signalling between plants and animals. Yet, before assuming signalling to be the default mechanism that influences the feeding behaviour of frugivores, we need to ask whether there is evidence that external indicators indeed evolved for a communicative role.

5.4   Selection by fruit consumers

The amount of nutrients present in fruit pulp is highly variable among fruit species (see above), but it also varies intraspecifically according to variation in abiotic factors such as light availability and, as evidenced by artificial selection experiments to increase the nutritional contents of economically grown fruits, according to genetic variation. Owing to the genetic variation for nutrient allocation patterns to fruit pulp, we can thus ask whether seed dispersers select for higher nutritional contents in the fruits they consume.

In captivity, fruit-eating birds can exhibit fine-tuned discrimination abilities: preferring the more nutritious food even when differences are limited to 1 per cent in the contents of carbohydrates or 2 per cent in lipid contents (Schaefer et al. 2003a). Similarly, fruit-eating birds discriminate among distinct amino acids when given the choice in captivity (Schaefer et al. 2003a). The variation in fruit pulp biochemistry among plants of a population often exceeds the difference that frugivorous birds can discriminate. For example, a population of black elder exhibited a fourfold variation in the contents of carbohydrates ranging from 2.5 to 11 per cent (Schaefer and Braun 2009). Theoretically, selection by seed dispersers might thus shape the amount of nutrients offered by plants. However, the large intraspecific variation in nutritional contents testifies that stabilizing selection by seed dispersers on fruit nutritional contents may not be very strong. There are several reasons why this might be so. First, even in captivity there is strong intraspecific variation in fruit choice (Willson and Comet 1993, Lepczyk et al. 2000). Further, the relative profitability of exploiting a given food resource is set by its marginal returns, the distances between alternative food patches, the predation risk associated with each patch and with travelling between the patches, and the likelihood of encountering another profitable patch. Experiments in captivity showed that even a moderate distance between alternative fruit resources resulted in the consumption of less preferred fruit species that occurred in the same patch as preferred fruits (Levey et al. 1984). These results suggest that non-preferred fruits may more often be consumed if they grow in the neighbourhood of preferred fruit species, resulting in so-called neighbourhood effects.

Despite these reservations on the transfer of fine-tuned discrimination abilities in captivity to selection in the wild, the general pattern of a preference for nutritious fruits persists in the wild. Not surprisingly, seed dispersers are choosy consumers. They consistently prefer fruits with either high lipid contents or high contents of carbohydrates, but their relative preference can switch between years and change from a lipid-rich fruit to a carbohydrate-rich fruit (Herrera 1998a). Preferences for nutrient-rich fruits have been found in distinct habitats and distinct foragers (e.g., Fuentes 1994, Schaefer et al. 2003b, Cazetta et al. 2008). Despite the robustness of nutritional preferences we cannot infer the selective pressures acting upon plants from those studies. This is because similar nutritional preferences are known for fruit predators that are detrimental to plant fitness. For example, pulp-feeding olive fly (*Dacus oleae*) larvae heavily infested wild olive trees (*Olea europea*) whose fruits were characterized by a relatively high fruit yield (the ratio between fruit pulp and seeds) (Jordano 1987). In the same study, vertebrate seed predators preferred fruits with a relatively low fruit yield because these were less often infested by *Dacus* larvae. This example demonstrates that overall selective pressure upon fruit traits depends on the complexity of multi-species interactions.

Although it seems intuitive to expect that fruit consumers select for nutritious fruits, there can be some important differences. A cleverly designed study revealed differences in the use of saguaro cactus fruits (*Carnegia gigantea*) by two congeneric dove species, the white-winged dove (*Zenaida asiatica*) and the mourning dove (*Z. macroura*). Both doves are seed predators that grind up the seeds of the cacti in their muscular gizzards. Saguaro cactus is a keystone species of the Sonoran desert because it is abundant and produces large amounts of fruits during the hottest period in summer. The cactus shows a distinct isotope signature that allowed Wolf and colleagues (2002) to analyse how feeding on the cacti fruit contributed to the carbon and water budgets of the two doves. Their analysis shows that white-winged doves forage on saguaro fruits to obtain both nutrients from the seeds and water from the fruit pulp, whereas mourning doves obtain only nutrients from the fruit resource. This analysis demonstrates the power of using stable isotopes to study the factors controlling foraging. In the case of saguaro fruits, it is easy to imagine that white-

winged doves select as seed predators against high water contents in fruit pulp, whereas this selection pressure might be countered if seed dispersers select for higher water contents.

To infer selective pressures, we thus need to quantify the selective pressures of each the interacting animals. Given that the distribution of different frugivores will vary between plants in a local population and between populations, there will be temporal and spatial variability in the net selection pressure experienced by an individual (see Section 5.9.2).

5.5 Fruit biochemistry mediates interactions

There is substantial variation in the time that ripe fruits persist on a plant. They can remain as long as several months (over 5 months in guelder roses) without visual signs of any decay, or rot within days (e.g., 6 days in *Alocasia odora*; Tang et al. 2005). In general, two processes determine the persistence time of ripe fruits: fruit consumption, and active abscission by the plant. Active abscission is based on cell wall degradation in the so-called abscission zones and is responsible for such diverse processes as pollen release from anthers, fruit softening, fruit abscission, and the sprouting of radicles from germinating seeds (Roberts et al. 2002). We would expect the timing of abscission to be under selective pressures. If abscission is too late, then birds (for example) have greater difficulty in pulling ripe fruit from the plant, and dispersal is compromised. If abscission is too early, then the fruit is more likely to be blown to the ground by the wind, becoming more accessible to seed-destroying rodents (for example), again at a cost to successful dispersal. Fruit abscission is the proximate mechanism that is ultimately shaped by the relative probability of successful seed dispersal in a given time interval and as such is dependent on the identity of the consumer who eats the fruit. Fruit biochemistry, in turn, controls the relative probability of fruit consumption by legitimate seed dispersers relative to fruit consumption by predators that consume fruit pulp but do not disperse seeds. Here, we will examine how fruit biochemistry influences the ratio of dispersed seeds relative to those that are predated.

5.5.1 Trade-off hypotheses

The likelihood that a seed is removed and transported from the parent plant depends on the relative probability that the fruit is consumed by the intended consumer (a seed disperser) and not the unintended consumer that either consumes fruit pulp without dispersing seeds or that eats and destroys the seeds. Note that the removal of a seed from the parent plant does not equal successful seed dispersal because it does not guarantee that the seed will end up at a suitable site for germination (see Chapter 3). In general, the risk of loosing seeds to frugivores that consume fruits without dispersing the seeds is thought to increase linearly with time (Thompson and Willson 1979). This linear increase is predominantly explicable by the increasing likelihood of fruit loss to ubiquitous fruit consumers such as bacteria and fungi. It may be that fungal spores (for example) are so abundant in the environment that as soon as a fruit becomes vulnerable, it is infested, after which the development of the fungus within the fruit is a relatively deterministic process. At first glance, the probability of fruit consumption by seed dispersers should also increase linearly with time. However, compared to bacteria and fungi, the abundance of seed dispersers, and consequently fruit removal, seems to be more variable, even unpredictable in space and time (see Gorchov 1988, Herrera 1998a). In contrast to microbial attack, the act of fruit consumption by a disperser is an almost instantaneous act, that will happen at a highly unpredictable time after the fruit becomes available, and will often not happen at all before the fruit spoils. Owing to the unpredictability of seed dispersal, the fundamental question is how can plants mediate the contrasting requirements of defending fruits against predators and making them attractive to seed dispersers? The answer to this question is at the same time a route to understanding the pronounced interspecific differences in fruit longevity.

In the introduction to this chapter, we proposed that plants primarily use biochemistry to mediate their interactions with other species. Plants can attract seed dispersers by accumulating nutrients in fruit pulp, thereby increasing their 'payment' for dispersal service; plants can defend fruits against fruit predators by accumulating deterrent secondary compounds that act as feeding inhibitors. The important issue is the optimal balance between nutrients and secondary compounds to produce fruits that are attractive to seed dispersers and, concomitantly, deterrent to fruit predators.

The most obvious way of resolving the trade-off between attraction and defence is to use chemical defences that are only deterrent for fruit predators and are neutral or beneficial for seed dispersers. There are some notable examples of this kind of directed defence. The presence of capsaicin, the substance that brings the pungent taste to chilli (*Capsicum* spp.), screens out mammalian seed predators while having no effect on avian seed dispersers (Tewksbury and Nabhan 2001). Feeding experiments revealed that capsaicin in *C. annuum* deters the feeding activity of two rodent species in Arizona that destroy the seeds of chilli peppers. In contrast, an avian frugivore, the curve-billed thrasher (*Toxostoma curvirostre*) consumed pungent and non-pungent chilli peppers alike and dispersed their seeds. Another example are anthocyanins, the fruit pigments that produce red to black colours, that strongly inhibit fungi growth while increasing the feeding rate of avian seed dispersers (Schaefer et al. 2008a, 2008b; see also Section 4.6). Yet, many, and perhaps most, secondary compounds are not as specific in their action. In particular, so-called constitutive defences such as tannins and phenols are repellent to most consumers, mammalian and insect herbivores, fruit predators, and seed dispersers (Herrera 1982, Schaefer et al. 2003b). Constitutive defences are (in contrast to inducible defences) already present before the attack of herbivores. Broad deterrence is also found in other secondary compounds that can be more bioactive than phenols or tannins, such as glycoalkaloids (Cipollini and Levey 1997b, 1997c). Although a broad, undirected bioactivity of secondary compounds seems to be common, it does not necessarily follow that seed dispersers and fruit predators have similar fruit preferences. Yet, this seems to be the case. Comparing the persistence time of bagged fruits that were inaccessible to seed dispersers with fruits that were accessible, Tang et al. (2005) concluded that species with a low risk of microbial infection were at the same time less attractive to seed dispersers (Figure 5.3). Similarly, a study in a neotropical

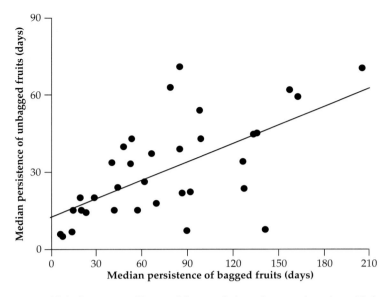

Figure 5.3 The persistence time of fruits that are accessible to seed dispersers (unbagged treatment) correlates with that of inaccessible fruits (bagged treatment; dots represent different species). Since bagged fruits are only consumed by fruit antagonists (fungi, bacteria), this correlation suggests that mutualists and antagonists have similar fruit preferences and are likely to exert conflicting selective pressures upon fruit traits. Each dot represents one species. Redrawn from Tang et al. 2005.

rainforest revealed that pathogens and vertebrate seed dispersers were both attracted by nutrient-rich fruits (Cazetta et al. 2008).

In 1997(a), Martin Cipollini and Doug Levey proposed two alternative models on the conflicting demands on fruit pulp chemistry. The nutrient–toxin titration model predicts a positive relationship between the contents of nutrients and those of secondary compounds. The model is based upon the rationale that more nutritious fruits can better afford higher quantities of chemical defences than less nutritious fruits, simply because their higher rewards would entice seed dispersers to tolerate relatively higher defences. The alternative 'removal-rate' model predicts a negative relationship between the contents of nutrients and those of secondary compounds. The model assumes that attractive fruits are characterized by high levels of nutrients which results in high removal rates by seed dispersers. These attractive fruits could thus require low levels of deterrent secondary compounds, simply because the likelihood of fruit predation is inherently low, because of their short exposure time (Cipollini and Levey 1997a).

Subsequent work on *Solanum* spp. and on the patterns of fruit pulp biochemistry in different vegetation communities found a negative relationship between nutritional and allelochemical contents in fruit pulp, supporting the removal-rate model (Cipollini and Levey 1997a, Schaefer et al. 2003b, Cazetta et al. 2008). These studies highlight the evolutionary triad between fruiting plants, seed dispersers, and fruit predators.

The negative relationship between the contents of nutrients and secondary compounds illustrates that fruiting plants can employ alternative strategies to achieve seed dispersal. Given that nutrient-rich fruits are attractive to both seed dispersers and fruit predators, they are unlikely to last long on the plant because the fruits are either consumed by seed dispersers or by fruit predators. In contrast, less nutritious fruits are more strongly defended and can therefore persist for months on the plant, increasing the likelihood of seed dispersal. A good example is, again, the guelder rose (also often called the European cranberry bush). The fruits of this species ripen in late summer, persist during winter, and last until late spring. Even in late spring, the fruits do not wither, they desiccate on the plant – suggesting that this species has low attractiveness to seed dispersers and fruit predators alike (HMS, pers. obs.). Interestingly, the attractiveness of these fruits decreases over time: cedar waxwings in captivity prefer early winter fruits over fruits collected in spring, when both were presented simultaneously, even though they only consume the latter in the wild in spring when *Populus deltoides* catkins are available (Witmer 2001; see Section 5.3.1). When alternative fruit resources are available in autumn and winter, seed dispersers consume these rather than guelder rose fruits. Interestingly, guelder rose is an invasive species in parts of the US, even though its fruits are not particularly attractive. This example suggests that an extended fruiting period can compensate partly for low uptake rate by seed dispersers.

5.5.2 Community patterns

The example of guelder rose demonstrates that seed dispersal can depend on the availability of alternative fruit resources. Dispersal and pollination are processes that can be thought of as biological markets, where the relative supply of goods (fruits) dictates the prices (attractiveness of fruit) associated with a transaction (seed dispersal). Although biological markets are not a new metaphor, few researchers have explored the dynamics of seed dispersal markets. For example, given the relatively invariant nature of plant–animal networks (Bascompte and Jordano 2008), we could ask whether the proportion of plants pursuing distinct strategies is likewise relatively invariant in space. If plant strategies are dependent upon the community context, we can hypothesize that fruiting plants with highly different biochemical profile are more likely to co-exist than those with similar profiles. For example, the relative abundance of nutrient-rich fruits with low defences and nutrient-poor fruits with high defences may be relatively constant among communities. However, any market dynamics are strongly influenced by the abundance and relative accessibility of fruits (Moermond and Denslow 1983). If competition among plants leads to adaptive differentiation – as has been proposed with regard to flowers (McEwen and Vamosi 2010) – market theory would provide a conceptual framework to study the associated dynamics.

Similarly, and given the apparent impact of fruit removal rates on fruit biochemistry, we might ask whether the likelihood of fruit removal is similar among distinct habitats and geographic areas. There are reasons to believe that they are not. For example, many fruiting plants in temperate areas produce fruits during autumn, at a time where the abundance of seed dispersers is relatively high, because migrating birds are highly frugivorous and often the main group of seed dispersing animals (Thompson and Willson 1979). Yet, the abundance of these migrating animals fluctuates strongly from year to year, presumably driven by abiotic factors in other areas. Consequently, in habitats where plants rely on seasonally available seed dispersers there can be a strong decoupling in the fruit supply and the number of available seed dispersers (Herrera 1998a). Herrera observed that the relative fruit preferences of two common seed dispersing bird species changed between years depending on the availability of alternative fruit supplies (Figure 5.4). This decoupling occurs even though some mobile seed dispersers such as birds can track the abundance of fruit resources over local and regional scales (Moegenburg and Levey 2003, Tellería et al. 2008). Such a decoupling would increase the unpredictability of fruit removal by seed dispersers and might thus increase bet-hedging strategies of plants (such as extending the temporal period of fruit availability, and spreading investment in fruit production over the plant's lifetime) or increase the proportion of plants with strongly defended fruits, compared to less seasonal habitats with a more predictable abundance of seed dispersers. We know of no study analysing the relationship between fruit traits and predictability of fruit removal by seed dispersers across communities and habitats.

A study on geographic variability in pungency in *Capsicum chacoense* in south-eastern Bolivia shows that indeed fruit defences vary geographically. Unlike *C. annuum* in Arizona, *C. chacoense* is naturally polymorphic for the proportion of capsaicinoids and displays pronounced geographic variation in the proportion of plants that produce these pungent secondary compounds (Tewksbury et al. 2008a). Interestingly, this variation is directly linked to variation in the damage caused by the fungal pathogen *Fusarium semitectum* that attacks the seeds. In

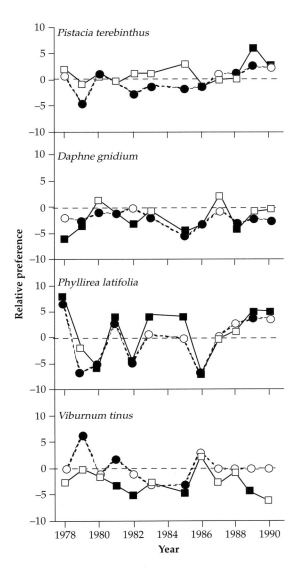

Figure 5.4 The relative preference of two bird species (squares and solid line: blackcaps *Sylvia atricapilla*; dots and dashed line: European robins *Erithacus rubecula*) for different fruit species fluctuates among years depending on the availability of alternative fruit supply. Values around zero indicate no preferences, significant preferences (positive values) or avoidances (negative values) are indicated by filled symbols. Redrawn from Herrera 1998a.

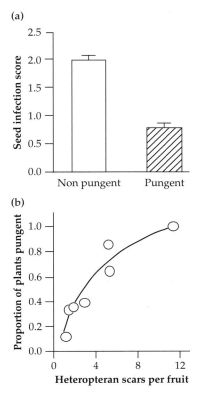

Figure 5.5 *Capsicum chaconese* is naturally polymorphic for the production of pungent capsaicin. (a) Non pungent seeds are twice as likely to be colonized by fungi (*Fusarium semitectum*) than pungent seeds. (b) The proportion of plants producing capsaicin within populations is a function of the predation pressures by Hemiptera which puncture fruit skin and thereby facilitate infection by the fungi. Redrawn from Tewksbury et al. 2008b.

C. chacoense the fungi is the primary cause of seed mortality, but the likelihood of infection is contingent on fruit biochemistry. Fungal infection was twice as high in seeds from non-pungent fruits compared to pungent fruits (Figure 5.5a). The risk of fungal infection may be a selection pressure leading to the evolution of signalling ripeness to dispersers. It may be expensive for plants to have potential dispersers sampling fruit prematurely, especially if any mechanical damage done to the unripe fruit during handling provides an important pathway into the fruit for microbial pathogens. Hemipterans foraging on the chilli facilitate the entry of the fungi into the fruit and its spread between fruits. Most notably, the foraging pressure of hemipterans was a strong predictor of the proportion of capsaicinoid-producing plants in a given site (Figure 5.5b). Taken together, these results suggest that pungency in chillies is an adaptive response to the likelihood of fungal attack. Antagonistic co-evolution between *C. chacoense* and

F. semitectum is thus a likely scenario that is mediated by the production of pungent chemicals (Tewksbury et al. 2008a).

In Chapter 1 we laid out that most interactions between plants and animals are not pair-wise but involve multiple species, as in the insect-mediated fungal attack on *C. chacoense*. In this chapter we have illustrated various examples that demonstrate how the biochemistry of fleshy fruited plants can mediate interactions with both seed dispersers and fruit predators. Given that interactions between multiple species are generally characterized by pronounced temporal and spatial heterogeneity, we expect geographic variation in fruit biochemistry to be particularly pronounced. As yet, the study on *C. chacoense* is one of very few documented examples of both the adaptive value of secondary compounds in fruit pulp and of geographic variation in the profile of secondary compounds. We suggest that research on variation in fruit biochemistry is a particularly promising field to uncover the geographic mosaic of co-evolution (see Thompson 2005a).

5.5.3 Competitive interactions

We have so far examined how fruiting plants mediate their interactions with other organisms through the production of biochemical compounds in fruit pulp. However, plants are not the only players in the evolutionary triad of fruits, seed dispersers, and fruit predators that actively influence the outcome of the interactions by the use of biochemistry. In 1977, Daniel Janzen wrote a seminal article on why fruits rot, seeds mould, and meat spoils. In essence, he hypothesized that microbes compete with vertebrates for fruit resources. He suggested that rather than competing directly for nutrients, microbes would benefit from excluding other consumers – thus gaining a selective advantage – by producing molecules that render a fruit objectionable to other fruit consumers.

We all know that fruits rot, and that rotting quickly decreases the attractiveness of fruit to ourselves and other vertebrate frugivores. The main point, however, is to differentiate whether the exclusion of other frugivorous consumers is a mere by-product of the feeding activity of microbes or whether microbes actively exclude competitors

using biochemical engineering. There is mounting evidence that the latter is the case (see Section 5.6.2 for the role of ethanol and smell). Experiments soon after Janzen's paper were not designed to differentiate between by-products and biochemical engineering. These experiments nevertheless documented that not only do microbes produce bioactive substances but also that many of these substances deter fruit consumption by vertebrates. Vertebrates use vision, smell, and taste to discriminate among fruits. Not surprisingly, if decay is visible, seed dispersing birds discriminate between decaying fruits (which are often brown) and undamaged fruits – preferring the latter (Borowicz 1988). Hence, visual inspection screens out those fruits that are heavily infested. Birds also avoid consuming fruits that look healthy on the surface but that are already infested internally, by dropping them significantly more often than undamaged fruits (Buchholz and Levey 1990). Birds thus use at least two levels of discrimination that are based upon vision and taste, respectively.

Seed dispersers also interfere with immobile insects such as larvae feeding on fruits. Interestingly, the abundance of frugivorous insects on fruits was negatively correlated to the abundance of frugivorous avian seed dispersers in southern Spain (Herrera 1984). Frugivorous insects tend to occur mainly on unripe fruits in areas where avian seed dispersal is common, whereas they are commonly found feeding on ripe fruits in areas where avian seed dispersers are scarce. Fruit consumption by birds thus apparently interferes with fruit consumption by insects. Supporting this scenario, Herrera (1984) found that aposematic insects such as hemipteran bugs are more likely than undefended insects to feed on ripe fruits where predation by birds is presumably high. The research by Herrera suggests that insects avoid feeding on the same fruits that seed dispersers do, likely because the seed dispersers often also consume insects on or in the fruit.

The interactions between fruits, insect fruit predators, and seed dispersers can be intricate. In the tropical tree *Acnistus arborescens*, birds preferentially consumed fruits that were infested by pulp-mining larvae (Valburg 1992). Infested fruits are characterized by small holes and a vinegary smell, so that birds might use either vision or olfaction to discriminate among fruits. Seed dispersers can also preferentially consume infested fruits because of nutritional changes associated with infestation. *Solanum mauritianum* fruits infested by the fruit fly (*Dacus cacuminatus*) contained twice as much protein and essential amino acids as uninfested fruits (Drew 1988). During oviposition, fruit flies insert bacteria together with the eggs into the host fruit. The authors suggested that the increase in proteins and amino acids is likely due to the additional allocation of amino acids by the plant in order to repair tissue damage by the plant. This nutritional change facilitates bacterial growth; the bacteria in turn provide essential nutrients and amino acids to the developing larvae (Drew 1988). Another possible nutritional change following infestation is to increase the contents of secondary compounds that deter microbes or other unwanted fruit consumers that could enter the fruit through the wound.

If the plant allocates additional amino acids to the fruit, this is an extremely interesting situation because it is not clear who ultimately benefits from the increased nutritional content of infested fruits. It seems plausible that bacteria and the developing larvae benefit because they grow more quickly. If they are the only ones profiting from increased nutrient allocation, it remains unclear why the infected fruit is repaired and not abscised. However, assuming that seed dispersers could detect infested fruits in *S. mauritianum* (owing to changes in fruit appearance), the high nutritional contents of infected fruits could lead to a higher probability of successful dispersal. In this case the advantage may be to plants and seed dispersers; a situation that could then select for higher chemical defences by the larvae. This example highlights the complexity of multi-species interactions, calling for further study on the effect of nutritional changes that occur after infestation.

Seed dispersers that prefer infested fruits are obviously detrimental to the fitness of insects laying their eggs into developing fruits or into flowers. These insects usually time their development so that the offspring leaves the fruit before it becomes ripe. Avoidance of interference with seed dispersal might thus be the main selective force shaping developmental time. In other words, developmental time of insects is adapted to minimize the risk of being consumed by vertebrates. In a related vein,

infected fruits often appear to ripen more quickly than uninfected fruits, probably to facilitate seed dispersal before the fruits or seeds become too damaged. The relative abundance of each of the fruit consumers will then shape the selective pressures acting on each of the participants in these multi-species interactions. We would welcome experimental studies addressing the conjecture of more rapid development of infested fruits.

The evolutionary ecology of competitive interactions among fruit consumers in general and that of communication in particular is an exciting research field. This is because the interests between the different players do not coincide. Frugivores inside the fruits like microbes and developing insects have adaptive benefits if they can (i) change fruit biochemistry so that the fruit becomes unpalatable to other fruit consumer, and if they can (ii) communicate those changes through vision, smell, or taste. Communication through vision or smell (even birds are likely to use smell at short distances) is particularly useful for those consumers (like developing insect larvae) that incur high risk of mortality during fruit handling by a seed disperser. It is thus easy to imagine the evolution of communication between the distinct fruit consumers. However, plants should be under selective pressures to conceal such communication because any fruit that is picked up by a seed disperser has a higher likelihood of establishing than fruits that are disregarded on the plant. Our hypothesis that communication could evolve between different fruit consumers has not yet been investigated. We would not expect such communication to always occur. If the infestation makes the fruit more attractive to vertebrate frugivores than pristine fruits, then the invertebrate should be selected to conceal its presence. However, in this situation it may be in the plant's interests to signal the infestation to vertebrate frugivores. This idea is conceptually similar to the idea that plants signal that they are attacked by herbivores, discussed in Chapter 9.

5.6 Fruit texture

Fleshy fruits soften during ripening owing to two different processes. First, there is an active process of cell wall hydrolysis caused by the upregulation of the activity of specific enzymes (e.g., cellulase) (Abeles and Tanaka 1990, Manning 1998). Second, fruit softening occurs as a by-product of water accumulation during fruit ripening. Fruit growth is at least partly achieved by cell expansion owing to increasing storage of water. For example, common myrtle (*Myrtus communis*) fruits increased in mass during ripening from 2.5 to 8.8 g with a concomitant increase in water content from 28 per cent to 72 per cent (Wannes et al. 2009). At present, the relative importance of cell wall hydrolysis and water accumulation for determining fruit softness is not well resolved.

Although the softening of fruit tissue occurs during ripening in most species, the degree of fruit hardness per se is a poor indicator of fruit maturation. Ripe fruits in a Venezuelan rainforest community had a mean water content of 72 per cent, but there was pronounced interspecific variation in fruits, ranging from 32 per cent water content in oily palm fruits to 95 per cent water contents in sugary fruits (Schaefer et al. 2003b). To illustrate this difference think about cheese: camembert cheese has a moisture content of 52 per cent, cottage cheese of 79 per cent, and cheddar cheese of maximally 39 per cent. Obtaining juice from some of the fruits growing in the Venezuelan rainforest is thus more difficult than obtaining juice from cheddar cheese! Similarly, the degree of fruit hardness showed a 100-fold variation in Australian and Papua New Guinean fruits consumed by bats (Dumont and O'Neil 2004). Not surprisingly, bigger fruits tend to be harder (because of the structural challenge of maintaining the integrity of the larger body) and are typically exploited by larger bat species that have a higher bite force than smaller bat species (Aguirre et al. 2003). Consequently, fruit texture can be a reliable cue of fruit ripeness on an intraspecific level, but it is not a reliable cue indicating fruit ripeness on an interspecific level.

Although fruit hardness is not reliably associated with fruit maturation it can nevertheless be informative of fruit contents. In a sample of 99 south-east Asian fruits, fruit hardness was inversely correlated with the sugar contents of fruit pulp (Dominy 2004). This relationship is explicable on the proximate level by the consistent positive correlation between sugar and water

contents in fruit pulp. The author suggests that the reliability of texture as a cue of sugary fruit rewards explains why primates (including humans) commonly inspect fruits by touching them. Owing to the reliability of fruit texture as informative cue of fruit quality in this relatively large fruit sample, we expect animals other than primates to also attend to this cue. However, there are few quantitative demonstrations of such a behaviour. American robins (*Turdus migratorius*) made no within-bill selection when consuming hawthorn (*Crataegus monogyna*) fruits, suggesting that no discrimination occurred in this species once a fruit was picked (Sallabanks 1993). Note, however, that *C. monogyna* fruits are not particularly sweet; rather they are characterized by low water contents and relatively high lipid contents. As such, fruit hardness is probably not a very good indicator of nutritional rewards in this species. Hence, we hypothesize that fruit texture is not a reliable cue of the distinct nutritional rewards provided by fruits. Specifically, we predict that it is less reliable for lipids than it is for sugar rewards.

If animals select fruits based upon fruit hardness because it predicts sugar rewards, this association could theoretically be exploited by free-riding plants that produce soft fruits without investing in sugar. At first glance, the potential for exploitation is not particularly high because animals that consume a fruit necessarily verify the sugar contents of these fruits. These animals could thus directly select against cheaters and thereby greatly limit the spread of cheating. However, a closer inspection of the feeding behaviour of birds reveals important differences in sensitivity to sugar contents.

In general, fruit-eating birds can be classified as gulpers (e.g., thrushes, flycatchers, trogons) that do not mandibulate fruits before consumption but swallow them intact, whereas mashers (e.g., finches, tanagers, and parrots) crush fruits during mandibulation to discard seeds (Levey 1987a). Mandibulation of fruits ensures that mashers taste fruit juice before consuming a fruit. It is thus unsurprising that they discriminated between alternative foods that differed in sugar contents, whereas gulpers did not (Levey 1987b). Consequently, a population of mashers would be more likely to select against cheaters, whereas cheating is more likely to invade a population with fruits consumed by gulpers. Obviously, visualizing the evolution of cheating along a dichotomy of mashers and gulpers is simplifying the complexity of natural frugivore communities. Yet, fruit consumers vary in their nutrient discrimination abilities (Schaefer et al. 2003a), and, conceivably, such variation may affect the degree of cheating. We know of no study that investigated this topic.

Cheating could likewise occur if plants add sweeteners without caloric value to fruit pulp to attract seed dispersing animals. Such sweeteners are known from some plants (Bassoli 2004), although the functions of these compounds are poorly known. Although plausible (Lev-Yadun and Mirsky 2007), we know of no demonstration that sweeteners are employed to influence interactions with animals.

Gulping and mashing behaviour determines the relative importance of sensory input in the interaction between birds and plants. Gulpers regurgitate or defecate seeds; in both cases the likelihood that the seed is removed from the parent plant is relatively high. In contrast, mashers discard the seeds when feeding (to avoid ballast) and thereby more commonly drop seeds underneath the parent plant. Thus, the more reliable dispersing agents use tactile and gustatory information to a lesser extent in fruit choice than mashers. We hypothesize that the different feeding technique between gulpers and mashers is mainly explicable by the comparatively stronger and thicker bills of mashers that probably facilitate separating fruit pulp from the seeds efficiently.

Fruit texture also influences the feeding behaviour of bats that chew fruits. Fruit hardness influences the location and magnitude of bites by different bat species (Dumont 1999). However, fruit hardness apparently does not structure the dietary scope of frugivorous bats in the neotropical savannah community that Aguirre and colleagues studied (2003). In omnivorous and insectivorous bats, maximal prey size, food hardness, and bite force were all correlated, whereas there was no such correlation between bite force and maximal fruit size in frugivorous bats. We hypothesize that this lack of a correlation is explicable because the relatively low energetic content of fruits in general may mean that exploitation of the largest fruits that could physically be chewed by bats is not energetically rewarding.

Fruit hardness is a trait that is characterized by substantial genetic variation, as evidenced by the artificial selection experiments on commercially grown fruits. There is further evidence that fruit hardness can indicate the sugar rewards of the fruit. Although feasible, there is currently no evidence to conclude that fruit hardness decreases during ripening *in order to* communicate the ripening stage to seed dispersing animals. That is, texture might most appropriately be considered a cue used by frugivores, rather than a signal. Supporting this scenario requires careful disentangling the relative importance of physiological processes for fruit softening from ecological ones that are associated with communication to facilitate seed dispersal.

5.7 Communication through fruit odour

Communication by scent differs from visual communication (see Section 2.7) in a number of ways. Communication by scent is not restricted to the presence of light and it does not need a direct line of sight. As such, olfactory communication is particularly suitable for attracting nocturnal foragers or foragers in dense habitats where visibility is limited. In contrast to visual communication, olfactory communication is slow, there is low control over directionality, and the tracking qualities of olfactory cues are relatively poor compared to vision (Endler 1993b). Olfactory communication can be much more specific than visual communication, requiring special receptors (Section 2.7). Yet, many plant odours are not very specific: resulting in interactions among multiple plant species and multiple animal species.

Olfactory communication by fruiting plants is particularly interesting because it allows studying how communication can evolve from precursors. Moreover, many of the molecular mechanisms underlying fruit odour production are particularly well known because odours can influence buyers' decisions. Here, we briefly explain some of the proximate mechanisms of fruit odour production, in order to ask whether odour production is selected to communicate with animals or whether it can be an informative cue that is a by-product of other processes.

5.7.1 Ethylene as cue

Ethylene, one of the main plant hormones involved in fruit ripening (see Box 5.1), is a gaseous signal molecule that diffuses freely from cell to cell but that also diffuses into the air (Bleecker and Kende 2000). Because ethylene regulates a large number of processes in plants that are generally related to senescence and plant growth (e.g., asymmetric growth of stems and petioles), it is primarily selected for its function as a plant hormone. If a plant individual releases a hormone, like many animals do in their urine and faeces, other organisms can obtain information from those hormones about the status and quality of that individual. Because hormones are particularly informative, it is easy to imagine that the incidental release of hormones could be selected towards intentional release of hormones to transmit information about status and quality. As such, the release of hormones like ethylene could evolve into a true communication system where the release of hormones is modified in order to increase the effectiveness of communication (in terms of altering the behaviour of others in ways that increase the fitness of the emitter). While this scenario seems plausible, ethylene is usually produced during the ripening process and thus at times when the seeds may not yet be mature. In this case, we would expect that plants conceal ethylene if animals cue on it.

The scenario that frugivorous animals use ethylene emission to locate ripening fruits has not yet been tested experimentally. It is therefore unknown whether insects perceive ethylene and whether they use it to locate suitable food resources. The dearth of knowledge on the role of ethylene in mediating plant-frugivore interactions mirrors the limited information on the role of ethylene emission in the interactions between plants and herbivores (Chapter 9). Here, it is at least known that ethylene emission increases in response to herbivore attack. The molecular substances inducing ethylene production are apparently sensitive to oral secretions of herbivores because ethylene production in response to insect feeding is significantly higher than that in response to mechanical damage. Ethylene production in vegetative tissue elicits gene expression of different defensive proteins, for example those that form quinines, which are thought to block digestion in the gut of herbivores (reviewed in von Dahl and Baldwin 2007).

Ethylene is probably associated indirectly with the attraction of animal consumers because it stimulates the production of fruit volatiles as well as floral scents. Plant volatiles consist of several groups, most notably hydrocarbons, esters, aldehydes, ketones, acids, and alcohols. Again, the role of fruit volatiles in attracting animals in general and seed dispersers in particular is not very well known. Some fruits that are dispersed by primates, like *Protium* spp., have a strong characteristic smell perceivable by humans from a distance (HMS, pers. obs.). It is, however, not well known whether such volatiles are involved in the long-distance attraction of seed dispersers. Compared to vertebrate seed dispersers, the attraction of insects by fruit odours is slightly better known. Some fruit predators, such as apple moths, that lay eggs in fruits use the volatile profile of fruits for host identification (Bengtsson et al. 2006). A combination of field and laboratory experiments showed that apple moths use a combination of different volatiles to identify their hosts (Knudsen et al. 2008). During years when fruit production of the host tree, rowan (*Sorbus aucuparia*), is low, apple moths switch to alternative hosts which also belong to the Rosidae and which presumably produce similar volatile blends. Yet, even in insects the role of fruit volatiles as foraging cue is contentious. In *Drosophila* for example, it is unclear whether volatiles are used as a cue involved in chemotaxis or whether the presence of volatiles simply increased the probability of non-directed search activity.

5.7.2 Ethanol as cue

Fruit pulp usually contains micro-organisms, which can be either detrimental, neutral, or possibly even beneficial to plant fitness. The high acidity of fruits tends to inhibit the growth of bacteria. Consequently, filamentous fungi, including yeast, are the most important micro-organisms that colonize fruits (Cipollini and Stiles 1992). Yeasts metabolize the carbohydrates found in fruit pulp and produce ethanol and carbon dioxide in the process. As such, ethanol occurs mainly in fruits as a result of the fermentation process by yeasts; however, it can also be produced by the plant, particularly under limited oxygen supply.

Ethanol production is likely to be adaptive to the yeasts that produce it because high ethanol contents have anti-microbial properties and also defend the fruit resource against competitive consumption by vertebrates (Janzen 1977). The effect of ethanol production can be even more complex. During early fermentation, up to ten different yeast species can be found in a single fruit, whereas in later stages one initially rare species, *Saccharomyces cerevisiae*, dominates. This species outcompetes the other species of the yeast community owing to the combined production of heat and high ethanol concentrations via fermentation and because it is more tolerant to high concentrations of ethanol. *S. cerevisiae* thus increases its fitness by chemically modifying the habitat it lives in – the fruit resource – which is a particularly convincing example of ecological engineering as a result of competition (Goddard 2008).

Given that ethanol production is associated with the presence of carbohydrates, frugivorous animals might use ethanol evaporating from fruits as a foraging cue that indicates the presence of fruit resources. Indeed, the fruit-feeding tropical butterfly *Bicyclus anynana* locates carbohydrate resources mainly using ethanol (Dierks and Fischer 2008). This butterfly is not a seed disperser but it competes with the yeast for fruit-derived carbohydrates, demonstrating that ethanol production can also entail ecological costs owing to attracting competitors to fruit resources. Some seed dispersers also use ethanol to guide their food choices, but in a different way. Egyptian fruit bats (*Rousettus aegyptiacus*) that are likely to consider olfactory information during their primarily-nocturnal foraging react to ethanol cues: they avoid fruits with more than 1 per cent ethanol (Sánchez et al. 2004). At ethanol concentrations below 1 per cent in bats and 3 per cent in yellow-vented bulbuls (*Pycnonotus xanthopygos*), both seed dispersers were indifferent to the presence of ethanol (Mazeh et al. 2008). Above these ethanol concentrations bats and birds consumed significantly less fruits compared to alternative foods with lower ethanol concentrations. Both consumers avoided consuming many fruits with relatively high ethanol contents, presumably to reduce the risk of intoxication. Controlled feeding experiments showed that the effects of ethanol on consumers are more complex than hitherto realized. Ethanol decreased in the breath of Egyptian fruit bats after they ate fructose compared to eating either sucrose or glucose (Sánchez et al. 2008). Compared to birds, however, the fruit bats were less efficient in metabolizing ethanol. Taken together these results are consistent with the well-known avoidance of damaged or rotten fruits by seed dispersers and therefore extend the example of ecological engineering to the exclusion of vertebrate foragers.

If yeasts use ethanol production to outcompete other fruit consumers, they could use it as a signal to prevent fruit consumption by vertebrates. This conjecture predicts that ethanol emission by the fruit is a signal from yeasts (not the plant!) to seed dispersers that the exploitation of this fruit resource is no longer profitable (Janzen 1977). The evolution of such a communication system would clearly be advantageous for yeasts that avoid consumption and for vertebrate fruit consumers that avoid ingesting fruits that can potentially be harmful. In this scenario, plants might be selected to conceal the emission of ethanol because they would benefit from any fruit that is consumed regardless of its nutritional state. This conjecture is admittedly simplified. Plants should pursue strategies that maximize the number of seeds transported away from the mother plant. Seed dispersers that are able to discriminate among fruits of different quality may have a higher fruit uptake at a given plant than those that are unable to do so. We know of no studies that investigate whether plants conceal ethanol production. The available evidence suggests that communication between yeast and seed dispersers would be more likely to evolve among yeasts and bats rather than among yeasts and birds.

The avoidance of ethanol in fruits contrasts with the regular ingestion of similar or higher ethanol concentrations by pentailed tree shrews (*Ptilocercus lowii*). The bertam palm (*Eugeissona tristis*) exudes nectar from flowers where the petals are still closed. This nectar contains a high ethanol concentration and is produced by the plant for periods of up to 46 days. Feasibly, this unusual phenomenon facilitates the development of a rich community of yeasts living on the nectar. Not surprisingly, the inflorescences release a strong alcoholic smell that may facilitate foraging by tree shrews that act as pollinators (Wiens et al. 2008). Thus, unlike the situation in fruits, the production of ethanol may actually increase the fitness of bertam palms. In this case, however, the production of ethanol is unlikely to be a signal from yeasts to pentailed tree shrews, but rather a cue that is selected by intense competition among yeasts and that gives away the location of the nectar resource.

At present, the way that multi-species interactions in the evolutionary triads among plants, mutualists, and antagonists shape plant traits in general and communicative traits in particular is not well resolved. Ethanol is a good example of how a single chemical can be used in different contexts, for signalling (yeast to seed dispersers) and as a cue to locate food resources (tree shrews). We predict that similar cases will be found more commonly in the interactions among plants, micro-organisms and seed dispersers. In order to infer the adaptive basis of communication, it is therefore crucial to analyse in each system the fitness benefits and costs associated with communicative traits.

5.8 Echolocation

Many bats have mastered the night skies by using bisonar echolocation to perceive their surroundings. Echolocation helps bats to navigate through complex habitats such as forests without crashing into objects, it also helps to locate and identify suitable prey. Short, downward modulated calls can yield information on the position, size, and texture of objects (Thies et al. 1998). However, relying solely on echolocation to find and identify fruits is difficult because canopies present very complex surfaces where the echoes of multiple objects are scattered and interfere with each other. In other words, the echoes from fruits can be masked by those of leaves and branches, making it extremely difficult to identify fruits that are mingled among the foliage by echolocation only. These difficulties explain why bats apparently rely on a mixture of different sensory information to select the fruits they consume.

Experimental work has shown that bats mainly rely on scent and echolocation during foraging on fruits. A central question is then, how they integrate information from these different sensory modalities in fruit selection. It has been shown that odour-guided foraging is mainly important during an initial stage of foraging as a long-distance cue to locate fruiting trees. Echolocation is then used during the final localization of individual fruits (Hessel and Schmidt 1994, Thies et al. 1998). Yet, recent experiments that systematically explored the use of distinct sensory cues established that olfactory cues are also important in fruit selection, the final stage of foraging. Bats only gleaned scented fruits in flight from the vegetation and did not consume freeze-dried fruits whose original shape was preserved but whose odour was not (Korine and Kalko 2005). This result is likely explicable because the direct identification of the fruit target based upon echolocation is hindered by echoes bouncing off at various angles from the surrounding vegetation.

Interestingly, the echolocation behaviour changed depending on whether or not the fruit emitted odour. In the presence of fruit odour, bats emitted lower mean values of the upper boundary and terminal frequency of echolocation calls. Bats used the same modulation to distinguish ripe undamaged fruits from damaged or infested fruits, which they avoided (Korine and Kalko 2005). The authors therefore propose that bats use mainly odour to detect fruits and to classify them as edible or not. Yet, given that scent is not a precise marker, owing to rapid diffusion in the air, they probably use echolocation in addition to scent to determine the exact position of a free-standing fruit that is not intermingled in vegetation.

Specific adaptations in plants to target foraging bats as seed dispersers are not well known. Morphological adaptations to foraging by bats rely on the production of specific echoes that stand out from the cluttering of the surroundings. Such morphological adaptations are constrained in fruits whose shape is primarily that of a sphere where maximal volume is covered by a minimal surface. Spherical objects such as fruits yield relatively simple echoes that are distinct from the flat surfaces of leaves but that are more difficult to detect than the echoes of some mirror-like flowers (for morphological adaptations to attract bats as pollinators, see Section 6.4). Despite these reservations, there are morphological adaptations in petiolate fruits to bats as seed dispersers. For example, bats are important seed dispersers for *Piper* fruits that are characterized by erect, spike-like fruit stands that stand out physically from the surrounding foliage. However, bats still primarily use odour to locate *Piper* fruits, possibly because these fruits do not stand out sufficiently to prevent interference by foliage. The situation in *Piper* is typical for other important fruit resources used by bats such as *Ficus.* However, given that many flowers that are pollinated by bats stand out from the vegetation (Dobat 1985), we would expect that this situation is similar in the fruits that are produced by these flowers. We were unable to find any quantification of the proportion of petiolate fruits in the diet of bats relative to the diet of other major seed-disperser groups, but if the conjecture above is correct, we expect it to be higher.

5.9 Consumption of dry fruits

Many of the general trends described in this chapter also hold for dry fruits, but the sensory mechanisms likely differ. In dry fruits such as nuts, animals have fewer visual and olfactory cues to locate these fruits. However, seed-eating animals can use cues, like ethylene, that are related to plant growth, but there is a dearth of knowledge as to whether such cues are used to identify suitable food resources.

5.9.1 Sensory mechanisms

In general, intended communication of plants to attract animals to seeds and dry fruits that contain seeds (for example, for external dispersal – see Section 3.4) is less likely to occur than in fleshy fruits. Animals usually consume dry fruits to obtain nutrients from digesting their seeds, resulting in an overall detrimental effect on plant fitness. Yet, an important fraction of the seeds that are removed from the parent plant for consumption can be dispersed by these animals (see Section 3.3). Similar to the rules that govern the food choices of consumers of fleshy fruits, the consumption of dry fruits is in most cases explicable by optimal foraging theory. There is also no evidence that plants use communication, that is warning signals, to deter feeding on dry fruits. Rather, dry fruits are often defended by spines or a hard coat that is difficult to crack.

As expected by optimal foraging theory, there is ample evidence that seed predators such as wild boar (*Sus scofra*) and wood mice (*Apodemus sylvaticus*) preferentially consume larger seeds relative to smaller seeds (e.g., Gomez 2004). A notable exception are acorn woodpeckers (*Melanerpes formicivorus*), which store acorns of lower average mass than those available in their territories because the smaller acorns fit better in the available holes of their granaries (see Plate 6; Koenig and Benedict 2002). Presumably, seed predators can identify relative seed size visually, although further discrimination is probably based upon tactile information obtained during fruit handling. The preference for larger seed size can likely be explained because larger seeds yield more energy per investment of handling than smaller seeds. The consequences of this selective pressure acting early in the life cycle of a plant are interesting. After a seed is dispersed, large seeds are favoured during seedling establishment because they are superior in competing for space and can better resist herbivory, drought, and nutrient limitations (Leishman et al. 2000). Predation thus exerts conflicting selective pressures relative to those acting in the subsequent life stage of seedling establishment (Gómez 2004). The corollary is thus that predation on seeds impinges on the following life history stages and that selection on seed size will not be constant in space and time but depends on the importance of escaping predation relative to the probability of successful seed establishment.

Seed size is not only selected by the combined effects of seed predators and seed establishment but also by abiotic factors. Koenig and colleagues (2009) found a strong latitudinal decrease in the size of acorns of bur oak (*Quercus macrocarpa*) throughout North America. Bur oak populations in Texas had eight times higher dry masses per acorn than those collected in Minnesota. Given that bur oak is mainly dispersed by the same disperser throughout the range, it is unlikely that this pattern is explicable by variation in size preferences by seed dispersers. It is likely, however, that such pronounced differences in relative yield have consequences on the interactions among plants and animals. For example, acorn woodpeckers, which mostly consume acorns, more likely consume a larger proportion of seeds from populations with a relatively low dry mass, provided that the dry mass is not so low that exploitation is no longer profitable.

The general patterns of secondary compounds mediating the interactions between plants and fruit consumers also hold for dry fruits. For example, wood mice are the main seed predators of the buttercup (*Helleborus foetidus*). They prefer to consume ripe fruits over unripe fruits. However, similar to nutrient preferences in fruit pulp, such preferences are always context-dependent and not absolute. The preference for ripe fruits is lost in habitats where the predation risk is high, suggesting that mouse concern about predation can override the drawbacks of plant secondary compounds (Fedriani 2006).

Similar to fleshy fruits, there are competitive interactions among different consumers of dry fruits. Unlike fleshy fruits, these interactions are probably mediated to a lesser extent by biochemical compounds, mainly because of the lack of a medium where these compounds can become active. Consequently, the interactions are probably mainly characterized as a competitive exploitation of the fruit resource. Supporting this conjecture, feeding experiments demonstrated that grey squirrels (*Sciurus carolinensis*) did not discriminate between acorns that were infested by curculionid larvae and those that were not (Weckerly et al. 1989). Interestingly, however, the authors observed that squirrels made feeding decisions when handling fruits. They discarded acorns that were already deserted by curculionid larvae and that, consequently, lacked a nutritious endosperm. These observations suggest that squirrels use sensory information (touch) to consume all acorns that provide nutritional returns either in the form of an endosperm or in the form of curculionid larvae.

5.9.2 Mosaic co-evolution

The intricate interactions among different seed predators and seed dispersers are arguably best understood in the dispersal of North American pines. The lodgepole pine (*Pinus contorta*) is mainly dispersed by different jay species, whereas red squirrels (*Tamiasciurus hudsonicus*) and red crossbills (*Loxia curvirostra*) are important seed predators. Red squirrels are pre-emptive competitors that drive the antagonistic evolution of seed defences in pines (Benkman 1999). This is because they harvest more seeds than crossbills, and crossbills are uncommon in areas where red squirrels are present and common in areas where they are absent. Red squirrels preferred cones that were relatively narrow at their base because these were easier to bite through (at the base) and consequently easier to harvest. Red squirrels also preferred cones that had many full seeds and a high ratio of kernel mass relative to cone mass. In areas where red squirrels were present, these traits were counter-selected, but in areas where they were absent, the number of seeds and the kernel ratio increased, whereas the diameter of the cone base decreased (Benkman 1999). In the areas where red squirrels do not occur, crossbills exerted the primary selective pressure upon pine defences. Here, cone morphology also evolved in a direction that lowered the benefit: costs ratio for crossbills of feeding on cones (Figure 5.6). Asymmetrical competition between two seed predators thus led to a geographic mosaic of lodgepole pine populations that co-evolved with crossbills (in the absence of red squirrels) and those that did not (in the presence of red squirrels).

Co-evolution with pines in so-called co-evolutionary hotspots and the lack thereof in co-evolutionary cold spots creates a mosaic of divergent selective pressures upon crossbills that form a remarkably diverse complex of different forms in North America and elsewhere. The forms differ in bill morphology, which impinges on the vocalizations of these birds in similar manner as it does in Galapagos finches (Podos 2001, Christensen et al. 2006). In areas where different call types co-exist, birds pair assortatively, suggesting some form of reproductive isolation. Yet, hybrids are common, and the different forms of crossbills are genetically very similar (Parchman et al. 2006). The evolutionary significance of the interactions among pines, red squirrels, and crossbills can best be seen in areas like Newfoundland where red squirrels were introduced in 1963. Here, the local type of crossbill went extinct less than 30 years after the introduction of the red squirrels (Benkman et al. 2008).

The evolution of pine defences not only results in a selection mosaic upon other seed predators, it also entails differential selection upon seed dispersers.

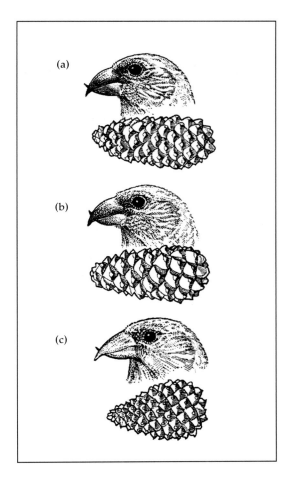

Figure 5.6 Red squirrels are pre-emptive competitors that drive the antagonistic evolution of seed defences in pines. The geographic selection mosaic between lodgepole pines (*Pinus contorta*), red crossbills (*Loxia curvirostris*), and red squirrels (*Tamiasciurus hudsonicus*) can thus be illustrated when comparing cone shape and bill shape of crossbills between sites where red squirrels are absent (a; and crossbills are the main selective factor), where red squirrels have recently been introduced (b), and where they were always present (c; and are thus the main selective pressure upon cone shape). Redrawn from Benkman 1999.

Increases in the seed defences of limber pine (*Pinus flexilis*) against red squirrels reduced the efficacy of the primary seed disperser, Clark's nutcracker (*Nucifraga columbiana*). At the same time, it increased the dispersal on the ground by scatter-hoarding rodents (*Peromyscus* spp.) (Siepielski and Benkman 2008). This example demonstrates that the evolution of seed defences in response to selection by seed predators can lead to a major change in the mode of dispersal (see Section 6.7 for a similar example of a change in the mode of pollination). Taken together these studies nicely illustrate the complex co-evolutionary dynamics of multi-species interactions.

5.10 The evolution of communication – unresolved research questions

To understand the evolution of communication in the complex interactions that characterize seed dispersal, an understanding of the *relative* selective pressures associated with communication to an intended receiver is required. This selective pressure needs to be evaluated by comparing non-communicative functions, which can be broadly related to plant physiology, to protection against abiotic factors and biotic agents, which interact with a given trait but do not respond to any information it provides. Given our current knowledge of plant-animal communication, it is easy to realize that we are only beginning to understand the complexities of such comparisons.

Many of the most fundamental questions about the evolution of communication remain. We have so far followed the traditional paradigm that plants communicate with animals, but, given the interactions among multiple species the first fundamental question will be to determine who communicates with whom. For example, do microbes such as yeast communicate with other fruit-eating consumers through the production of ethanol or is ethanol (only) a volatile by-product of competitive interactions among yeast species? Similarly, is the browning of fruit tissue by fruit rotting fungi only a necessary by-product of their feeding activity or does it occur because it conveys information to visually oriented frugivores? There will be many similar questions in other interactions, and the answers will often be difficult to determine. In many cases, information transfer will just be a necessary by-product of physiological activities, but unless we specifically test for a communicative role, we will continue to neglect the possible role of communication in ecological networks of multiple senders and multiple receivers of information.

If the principal question of who communicates with whom is resolved, the next question that needs to be addressed is which traits evolved mainly for their role in communication and which traits can be informative but did not evolve specifically for that function. Since it is often challenging to differentiate between signals (the former) and cues (the latter), the evolutionary significance of communication in plant-animal interactions remains to be evaluated. Despite these limitations, we predict some basic patterns. Hormones related to plant growth and senescence are olfactory cues that regulate plants' metabolism. Animals can exploit these cues because they provide reliable information on growth-related processes that are intimately linked to plant quality. Although they provide information, they are probably mainly selected for their regulatory function in the metabolism of plants. Similarly, the tissue softening during the ripening of fleshy fruits is probably a necessary process related to senescence and the increasing storage of water in pulp tissue. Yet, this process provides information on ripening stages to consumers that rely on tactile information when deciding which fruits to pick. At the same time, the ease with which a fruit can be detached from the plant can be a signal that reliably indicates the ripening stage of the fruit. Overall, and based upon the necessary link between certain traits and plant physiology, we predict that cues are common and widespread during fruit ontogeny. The use of cues thus provides a basic information transfer that might preclude the evolution of further communication by signals in many plant systems.

The next fundamental questions to ask are those that concern the reliability of information transfer. Cheating is ubiquitous and senders try to manipulate the behaviour of receivers to their own benefit, which usually does not coincide with benefits to the receiver. Yet, receivers are expected to respond only to those signals or cues that provide reliable information (at least on average). From an evolutionary perspective, the reliability of cues is less interesting than that of signals – because it is mainly determined by their non-communicative role. Hence, once a signalling function is established, the proximate mechanisms that are responsible for reliable information transfer are of particular interest. Given the number

of current limitations outlined above, it may seem premature to speculate on such mechanisms. However, the well-studied genetic underpinning to the ripening process of commercially valuable fruits provide an ideal tool to study the mechanisms underlying ripening and underlying the associated development of volatiles, aroma, or tactile cues or signals that could be used to communicate.

We propose that studies investigating the covariance among fruit quality and potentially informative fruit traits within populations are needed to assess the reliability of fruit traits. Such studies need to be complemented by information on food choices to analyse how animals select cues and whether selection driven by seed dispersers is consistent enough to translate into differential plant fitness. Furthermore, studies comparing the selective pressures acting on fruit traits between different populations are needed to analyse the patterns of geographically variable selection.

We finally hypothesize that the evolution of communication can be predicted by the different sensory abilities of interacting partners. In some plants, seed dispersers and seed predators will attend to the same cues because they are endowed with similar sensory abilities. In other plant systems, the sensory abilities will differ dramatically between animals that lower plant fitness and those that increase plant fitness. In these species, communication is more likely to evolve in traits that are can be better perceived by the interacting mutualist than in traits that are perceived equally by both. Note that this argument is one of relative differences in the sensory abilities of mutualists and antagonists; it does not require that a trait is solely perceived by the mutualist.

5.11 Conclusions

Throughout the chapter we have emphasized that fleshy and dry fruits as well as seeds represent evolutionary packages that evolve in reaction to the selective pressures of multiple species interacting with the plant. Some of the interacting animals are detrimental to plant fitness, whereas others increase it. This situation results in conflicting selective pressures that vary in space and time according to fluctuations in the relative abundance of the animals interacting with plants. We can therefore deduce that there is also substantial spatial and temporal variation in the selective pressures acting upon communication.

Plants mediate their interactions with animals mainly through biochemistry. Fruits offer an exceptional window into the proximate mechanisms resolving the interactions with different animals. This is because only in fruits do plants' interests change from defence to attraction of consumers during ripening. The various physiological and biochemical changes during ripening can thus be understood with reference to the change in the relative importance of defence and attraction. Importantly, not all of the changes during ripening evolved because they fulfil a communicative function. Toxic and many deterrent biochemical substances do not communicate, they act directly upon the receiver. However, because increased risk of fungal attack is associated with the partial consumption of a fruit, plants would clearly benefit from deterring unwanted fruit consumers by communicating to them. Yet, aposematic fruit signals are likely to be uncommon because plants likely experience lower fitness costs (compared to animals) associated with fruits being handled and ultimately rejected by an antagonist. An animal may loose its life when attacked, whereas a plant still survives when the fruits are attacked by antagonists even if it loses its fruit crop partially. The fitness costs of losing the fruit crop depend on the life history of plants, they are particularly high in annual plants and low in long-living trees.

We have described various morphological, physiological, olfactory, and tactile traits that change during ripening and that can thus be potentially informative to fruit consumers. Consistent with this conjecture of informative fruit traits, there is ample evidence that vertebrate and insect fruit consumers use sensory information to guide their feeding and egg-laying decisions. While it seems likely that some or even many of the traits that change during the ripening of fleshy fruits have evolved in order to communicate to seed dispers-

ers, there is a paucity of studies unambiguously establishing a signalling function. This dearth of studies contrasts to the commonly held view that many conspicuous fruit traits are signals to seed dispersers. This view often ignores the complexity of conflicting selective pressures associated with multi-species interactions as well as the multi-functionality of many traits. Finally, even if a signalling function is established in a species, it does not necessarily occur in all populations owing to the spatially variable selection upon the fruit traits that interact with multiple species.

The various volatiles and gaseous plant hormones whose concentrations change during ripening can be informative cues or even signals of fruit quality. As yet, however, the olfactory communication of fruits is not well known. Basic information on which volatiles attract which foragers and on the range that they are perceived is lacking. Given the high specificity of olfactory communication, it is possible that olfactory-guided seed dispersers and fruit predators use distinct volatiles to locate fruits. As such, olfactory communication might be much more specific than visual communication.

A fascinating aspect of the complexity of communication in multi-species interactions is that communication likely occurs also among different fruit consumers. Host marking behaviour is well known in many herbivorous insects, but communication among different types of fruit consumers is almost unknown. Communication could occur directly among the consumers or indirectly through specific compounds that are produced by the plant in defence against fruit consumption by fruit predators.

Having examined communication between plants that produce fleshy and dry fruits and the animals and microbes that consume these fruits, we now turn to flowers and examine communication in the context of pollination. In the next chapter we will discuss how floral traits such as morphology, symmetry, colour, and scent interact with animals and whether animals select floral traits in order to enhance these interactions.

CHAPTER 6

Floral communication and pollination

6.1 Introduction

Many of the core concepts on the evolutionary ecology of plant–animal interactions that feature prominently in our book have been developed and refined in the study of pollination (e.g., Harder and Barrett 2006). Moreover, decades of research have provided the functional basis for understanding communication in the context of pollination. In particular, the sensory ecology as well as the perceptual abilities of some pollinators are exceptionally well known (Chittka and Thomson 2001, Chittka and Brockmann 2005), as are the genetics underlying flower colouration and its ecological effects (e.g., Strauss and Whittall 2006, Rausher 2008). Owing to this unique constellation, it may thus seem paradoxical that communication theory has not featured prominently in the field of pollination biology (Schaefer et al. 2004, Raguso 2008). This is partly explicable by the difficulties of transferring a concept that has originally been developed in the context of intraspecific communication to the intricate interactions between different species where the selective pressures exerted by one species can be diluted by those of many other species.

In this chapter, we will illustrate how the fundamental concepts of communication apply to flowers, and where selective pressures upon floral traits deviate from those outlined in other chapters. Because studying the ecology and evolution of pollination has been a focus for naturalists at least since Sprengel (1793), we will fail miserably to cover all aspects related to the interactions between flowers and animals within one chapter. Moreover we have tried to minimize overlap with other works on flower – pollinator interactions (Chittka and Thomson 2001, Herrera and Pellmyr 2002, Waser and Campbell 2004,

Harder and Barrett 2006, Raguso 2008). Naturally, when comparing communication between flowers and the animals paying visits to them with other plant communication systems, we will draw most comparisons to fruits, because both fruits and flowers are reproductive organs that interact simultaneously with mutualists and antagonists. This comparison is even more obvious because the exterior of most fruits (their pericarp) develops from a single floral structure (the ovary wall). We can therefore postulate that factors influencing the development of the ovary wall in flowers may constrain adaptive evolution of the fruit pericarp through developmental linkage (Weberling 1989). Perhaps surprisingly, there have been very few quantitative tests of this idea. We also deem another comparison important. Flowers originate from vegetative structures, and studies in the last decade have emphasized direct and indirect interactions among herbivores and pollinators (see Section 1.6). As such, we will also compare floral traits to leaf traits. To facilitate the comparisons, we will structure this chapter according to the traits under consideration rather than according to the concepts (examples of floral communication ordered by concepts are given in Chapter 1). We will touch upon developmental or genetic constraints on the evolution of floral traits, but readers are referred to Conner (2006) and Ashman and Majetic (2006) for a thorough treatment of the genetics of floral evolution. In line with the focus of this book, we will mainly be concerned with communication.

We will start with a key question of communication; the relationship between the signal and its associated reward in flowers. We will then compare floral visitation and its associated selective pressures with those of seed dispersal and discuss differences in the extent of diversifying selection. Afterwards, we will turn to the suites of floral traits that likely play a role in communication: that is, morphology, colour, and odour. For each of these suites of traits we will discuss evidence that they are selected for a communicative function, that different pollinators differ in their selection upon these functions, and that their evolutionary trajectory is also influenced by other biotic or abiotic selective agents.

6.2 Relationship between signal and reward

Flowers can present very different rewards to animals visiting them. The most common rewards are nectar and pollen. However, given that ectothermic insects often need to raise their body temperature above the mean ambient temperature, flowers exceeding that temperature can present considerable metabolic rewards in terms of heat to insects (Seymour et al. 2003). Self-heating flowers evolved independently in six families with an estimated number of 900 species alone being pollinated by scarab beetles of the genus *Cyclocephala* (Seymour et al. 2003). This study showed that beetles resting inside the 3–4 °C warmer flowers had two- to five-fold lower metabolic rates than those outside the flowers. Thermogenesis is associated with elevated respiratory rates, which is why CO_2 emission can be a reliable indicator of heat production (Seymour et al. 2003). In the lab, bumblebees (*Bombus terrestris*) preferentially landed on warmer artificial flowers if the temperature difference between flowers exceeded 4 °C (Dyer et al. 2006). Insects may spend extended times basking on warmer flowers to heat up, and thereby increase the pollination rate of these flowers. This study also showed that bumblebees associated heat with floral colour rather than perceiving it remotely. Other types of floral rewards include resin and fragrances that are collected by specialized bees. For example, male orchid bees (Euglossini) collect fragrances from flowers and other resources and store them in specialized tibia pockets to use them later in courtship. Lastly, some flowers such as orchids provide overnight refuges to their insect pollinators. Again, overnight refuges can provide heat rewards, as shown by Sapir et al. (2006)

in a study on Iris spp. The iris flowers are dark and possibly gather heat by absorbing solar radiation. In these species, floral temperature is 2.5 °C higher than ambient temperature at sunrise, and the solitary male bees sleeping inside the flowers emerge earlier from them than from experimental shelters. Given the variety of rewards to animals, we hypothesize that the form of communicating with animals will depend on the type of reward offered by the plant. Since a phylogenetic component is to be expected both in the type of communication and in the reward offered, it would be necessary to analyse the conjecture that communication depends on reward type with phylogenetically informed methods.

The relationship between the visual or olfactory trait and the rewards it indicates is perhaps the most important difference that distinguishes floral communication from other plant–animal communication systems. In the moment a frugivore sees or smells a fruit, it knows that nutritional rewards are present. The particular composition of nutritional rewards may not be clear from receiving the fruit signal (but see Schaefer and Schmidt 2004, Schaefer and Braun 2009; Section 4.5), but the presence of some nutritional rewards is indisputable. The fruit is the reward and its quality may or may not be fully and accurately indicated by the visual or olfactory trait. Likewise, upon detecting volatiles or leaf colours herbivores recognize food resources whose quality may or may not be indicated by the stimuli from leaves (Chapter 9). The relationship between trait and reward is different in flowers because flowers are not the reward (except for florivores). Interestingly, it is a long-standing hypothesis developed by Pellmyr and Thien (1986) that communication to pollinators evolved as an exaptation from the ancestral defensive communication against florivores. This scenario is supported by the many defensive functions of plant pigments and volatiles (see Chapter 2). Floral communication to pollinators is different from that between fruits and frugivores because flowers usually indicate a hidden reward, except possibly for pollen and coloured nectar which may be visible before physical contact with the flower is made (Hansen et al. 2007). The relationship between floral signalling traits and reward is thus less direct than it is in many other plant–animal communication systems. As yet, this key difference between the communication systems has not attracted much consideration.

The indirect relationship between flowers and their reward entails distinct (compared to fruits) dynamics of communication. A central topic of communication theory is how the evolutionary stability of reliable communication is maintained (see Section 1.3). The concepts of communication theory of how evolutionary stability is maintained are less applicable to flowers than they are to fruits or leaves because no pollinator approaching a flower has an a priori knowledge that rewards are indeed present. Even in highly rewarding species that offer abundant nectar or pollen resources the flower may have been depleted by recent visits of competing pollinators (see Box 6.1). Pollinators thus need to be tolerant to variable nutritional returns linked to a given floral trait. This tolerance is important in evolutionary terms. It facilitates, for example, the repeated evolution of rewardless flowers precisely because tolerance leads to weaker selection against rewardlessness (see Section 8.4 for a discussion of mimicry). This tolerance likewise explains why handicaps that rely on cost-enforced signalling systems – and which are usually assumed to maintain honest signalling in intraspecific communication – are rare in floral communication. This conjecture is plausible because if pollinators are tolerant towards variable nutritional returns, they are unlikely to impose selection towards cost-enforced signalling. Based upon this argument we can then predict that the relationship between floral sensory traits and the quality or quantity of floral rewards will be generally more variable than it is in other communication systems. This prediction seems plausible but does not account for the paucity of cost-enforced signalling in other plant communication systems or, even more generally, among mutualists (Edwards and Yu 2007).

The tolerance of pollinators to variable nutritional returns also suggests that aversive learning of traits associated with non-rewarding flowers may be less pervasive than commonly assumed. A number of studies have shown that aversive learning can easily be shown in laboratory conditions and can lead to peak shift phenomena, where a preference can shift owing to a learned discrimination between dif-

ferently rewarding stimuli (Lynn et al. 2005). Explained briefly, peak shift phenomena occur if discrimination learning results in a stronger response towards new stimuli that are more differentiated from the unrewarding stimulus that the animal learned to avoid. While peak shift phenomena undoubtedly occur in the lab, it remains to be clarified how easily these results can be transferred to natural conditions, where flowers differ predictably in traits and in the rewards they offer, but also unpredictably depending on whether they have been recently depleted by other foragers. The consequences of this stochastic element have rarely been addressed for the evolution of communication.

If the logical argument above on the distinct dynamics of floral communication (compared to other plant communication systems) is correct, we need to address the fundamental question of how reliable floral communication is. Unfortunately, there is not enough information available to fully address this question. What is needed are intraspecific comparisons on the relationship between floral traits and rewards both within- and between-populations. Feasibly, the mean reliability of communication could vary across populations depending partly on the assemblage of competing species flowering at the same time and additionally on the consistency of the pollinator assemblage over years. In this scenario, reliability depends on the average reliability of flower species in a community and the pay-offs provided by each of these species. This scenario is thus one of market dynamics between competing flower species.

While the market metaphor has been used previously in the context of plant-pollinator interactions, market dynamics are not well explored on the scale of communities. Gumbert and colleagues (1999) found that the diversity of colours of common species within five plant communities did not deviate from chance, whereas the colours of rare plants did deviate from random expectations (see also McEwen & Vamosi 2010). A particularly interesting idea is that the pattern of common species is predominantly influenced by hub species which ground the local plant-pollinator networks (R. Raguso, pers. comm.). This pattern might be expected because rare plants can be under stronger selection pressure to secure pollination than common species. Unfortunately, no study has ever investigated whether the reliability in communicating with animal mutualists of common and rare species differs. In the study by Gumbert et al. (1999) there was no consistent trend in the form of the deviation in rare plants. In one community, rare plants were more similar to each other than expected by chance, whereas they were more divergent than explicable by chance in two other communities. It must be noted, however, that the innovative study by Gumbert et al. was conducted in communities that are strongly shaped by human land use. The anthropogenic impact, however, may greatly reduce the potential for market dynamics between compet-

ing species because species coexistence depends more heavily on external factors than on adaptive divergence in communication.

The conjecture by Gumbert et al. (1999) that rare species are under greater need for adaptive signalling predicts that negative frequency-dependent selection increases the reliability of communication. This conjecture could be extended to other species with enhanced need to attract pollinators, such as species that poorly establish in a given community. We know of no study that has tested this admittedly wild speculation, although experiments planting individuals of annuals into different backgrounds

---

Box 6.1 Pollinators add scent to flowers: adaptive marking or unavoidable consequence?

Many pollinators leave chemical footprints on flowers; these chemicals can play a role in causing pollinators to discriminate among flowers that have recently been visited and those that have not. Such chemical footprints consist of long-chained hydrocarbons (alkanes and alkenes). In general, these epicuticular lipids are found on the cuticula of insects and evolved presumably in order to prevent water loss in terrestrial habitats. Subsequently, a number of different functions evolved and some hydrocarbons are used in adhesion, while others play a role in nest-mate communication. Given that the chemical footprints on the flowers affect foraging decisions of subsequent foragers, they can be adaptive in that they help pollinators to avoid recently-depleted flowers. In this scenario the footprints possibly represent a communication system between pollinators that is imposed upon the flowers.

To address this potential communication system it is critical to distinguish between scent marks that are actively deposited on the flowers (signals) and those that are unavoidable consequences of the presence of insects (cues). Recent studies suggest that chemical footprints are a mere consequence of insect visitation and not an active flower-marking system that communicates to other pollinators. Workers of buff-tailed bumblebees (*Bombus terrestris*) left alkanes and alkenes in the same proportions on food and nests, as well as on neutral sites (Saleh et al. 2007). Similarly, bumblebees were equally repelled by footprints collected at feeders and those collected at neutral sites (Wilms and Eltz 2008). Both studies suggest that chemical footprints are involuntarily left on the flowers. Chemical footprints are retained on flowers in unchanged quantities for 24 hours at two ambient temperatures (Witjes and Eltz 2009). This study also applied synthetic footprints to flowers, which were retained on them for 48 hours. As such, chemical footprints might provide a relatively long-term record of flower visitation. If this holds generally, analyses of chemical footprints might be a relatively accessible way of determining floral visitation without the need for detailed flower observations. For example, Witjes and Eltz (2009) documented that the amount of footprint alkenes was correlated with the number of bumblebee visits during a day. The long-term record of chemical footprints means that they are less reliable cues for foragers than originally thought. This is because if the footprints remain on the plant for longer than the often circadian rhythm of nectar production, the footprint will not reliably indicate the reward status of a flower. Yet, if this situation is more widely the case, we would not expect flower visitors to attend to chemical footprints. As such, more work on the relationship between the reward actually present on a flower and the intensity of chemical footprints would be welcome to resolve how chemical footprints reflect the reward status of flowers.

of competing species and following the reliability of communication through generations would seem feasible.

Even though we currently lack knowledge on market dynamics, studies on the intraspecific reliability of floral communication should ideally be accompanied by analyses of the mean reliability of floral communication within a floral community. Another means to test our argument about the variability in the signal to reward relationship would be to use modelling to assess the likelihood of distinct evolutionary stable outcomes. As yet, we know of no model that formally compared the evolutionary stability of distinct signalling strategies among plant communication systems.

The variable relationship between signal and reward also entails behavioural modifications in pollinators that are not found in consumers of other plant parts. In particular, pollinators need to avoid flowers they or other pollinators have depleted recently. There is no comparable selective pressure upon frugivores or herbivores. Pollinators deal with this problem in two distinct ways. First, pollinators like bumblebees leave chemical footprints on the flowers they have visited. Smelling the footprints may enable the pollinator to assess the probable reward status of a flower prior to landing (see Box 6.1). Second, pollinators show remarkable cognitive abilities that allow them to subsist on floral resources which differ in their stimuli and profitability. Cognitive abilities include associative learning of the link between visual or olfactory stimulus and reward as well as long-lasting spatio-temporal memory on the availability, location, and relative reward of floral resources. For example, honey bees are thought to be remarkable in associative learning because they learn colours faster than vertebrates do (Pearce 2008). Given that the profitability of plant rewards such as nectar can differ among individuals but also within plant individuals throughout the day and season, colour preferences in honey bees shift during the day according to the circadian rhythm of the retrieval of colour memories. This circadian rhythm apparently tracks the circadian rhythm of nectar and pollen production in differently coloured flowers (Prabhu and Cheng 2008).

6.3   Diversifying selection

As outlined in Chapter 1, interactions with pollinators are on average more specialized than are those with seed dispersers (Blüthgen et al. 2007). The higher specialization is explicable because floral phenotypes (including morphology, scent, and colours) represent important filters structuring the interactions between flowers and floral visitors. First, the morphology of a flower can severely restrict the visitor spectrum attending to the flower (see Section 6.4.2). Second, floral morphology obviously influences not only the likelihood that pollen is transmitted by a floral visitor but also the sequence of events. Presenting pollen and stigmata that receive pollen so that pollinators touch them before they can exploit nutritional rewards (termed 'approach herkogamy') can be adaptive because pollinators provide the pollination service before plants deliver payment in nectar. This situation contrasts with that of seed dispersal where plants have no control of the movements of animal vectors once they leave the fruiting plant. Here, plants necessarily pay in advance – in terms of the energy provided by fleshy fruits – prior to seed dispersal event.

Most of the research on pollination has been guided by the concept of adaptive floral trait diversification that goes back to Sprengel (1793). It postulates that floral morphology is an adaptation to increase the likelihood of successful pollen transfer. There are numerous examples for this hypothesis, such as the presence of tactile and visual floral (or nectar) guides in flowers that help pollinators to orient on the flower and to find its nutritional rewards. Indeed, naive bumblebees spontaneously use visual floral guides prior to landing in their search for food (Lunau 1992). It is generally assumed that once the animal has landed these signals further help to channel the movements of the animals, and they can thereby increase the likelihood of successful pollen transfer. For example, mechanosensory input is crucial for foraging performance in *Manduca sexta* moths, and morphological structures such as grooves or ridges that guide an insect to the centre of the flower (see Plate 7) and can improve foraging efficiency (Goyret and Raguso 2006). There is also possibly an olfactory correspondent of nectar guides because flower parts of the tall buttercup (*Ranunculus acris*) differed in odour profiles (Bergström et al. 1995). Most remarkably, pollen odour differed strongly from all other plant parts, and emission from the petals differed among apical and basal petal regions thereby paralleling optical nectar guides. Floral characters that direct the movement of animals distinguish floral communication from the communication of other plant organs where no specific orientation of the animal is required. Consequently, floral communication involves a dimension that is not found in other plant–animal communication systems.

Arguably the most important difference between fruits and flowers is that plants need to entice animals to repeatedly visit conspecific flowers (termed flower constancy) in order to ensure pollen transfer. At the same time it might be beneficial for flowers to reduce heterospecific pollen transfer because it might clog the stigmas or lead to interbreeding. As such flowers are likely to be important in pre-mating isolation and speciation in plants – for thorough reviews of pollinator-driven speciation see Waser and Campbell (2004) and Johnson (2006) – unlike fruits and vegetative structures that have a less direct link to speciation.

However, it is important to note that mode of seed dispersal will influence the propensity to colonize new habitats and can therefore also indirectly influence speciation rates. In general, ethological isolation in flowers occurs when pollinators restrict their movements between flowers of different species and are more likely to visit flowers of the same species. The proximate basis for ethological isolation is the distinct sensory world of pollinators (Chapter 2). Ethological isolation can be an important pre-mating isolating barrier that has been well studied in sister species and hybrid zones. For example, a study on a hybrid zone in *Ipomopsis aggregata* and *I. tenuituba* demonstrated that ethological isolation mediated by difference in flower colour was more important than mechanical isolation owing to distinct floral morphologies. *I. aggregata* is characterized by red flowers with wide corolla tubes, and a high nectar production, whereas *I. tenuituba* has white to pink flowers with long and narrow corolla tubes and a low nectar production. Painting flowers of both species red reduced the percentage of conspecific seeds (Campbell and Aldridge 2006), demonstrating the importance of plant–animal communication in directing gene flow.

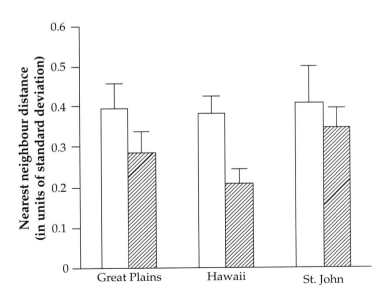

Figure 6.1 Flowers are more divergent from each other (in morphology and colour) than fruits of the same species in three North American florae. The results are consistent with stronger diversifying selection upon flowers compared to fruits. Redrawn from Whitney (2009).

Given the importance of flower constancy for the arrival of conspecific pollen, flowers should be under greater selective pressure to be memorizable for animals than fruits. We can thus predict that diversifying selection upon floral traits should be stronger than upon fruit traits. Indeed, animal-pollinated flowers from three North American florae were more divergent in morphology and colour than the fruits of the same plants (Figure 6.1; Whitney 2009). Animals' ability to identify and memorize flowers depends partly upon their visual and olfactory traits and on floral shape as well as on more general traits that are associated with plant identity such as plant height and leaf area. The latter traits are less likely to be under diversifying selection than the sensory traits of flowers (including floral size and shape).

Interestingly, adaptive divergence of floral traits is not the only hypothesis that can explain the higher divergence in flowers compared to fruits (Whitney 2009). The alternative neutral hypothesis is that flowers are more divergent in shape and colour because they represent multi-component structures, whereas the pericarp of fruits originates from the ovary wall. The multi-component origin of flowers might simply explain their higher developmental diversity. This conjecture can be tested by quantifying how much morphological divergence can be predicted by accounting for developmental structure. Furthermore, fruit shape is more constrained by physical laws than is flower shape. That is, most fruits consist primarily of water, which is best contained in a spherical form. Deviation from an ideal form is thus possibly more costly in fruits than it is in flowers. This hypothesis can easily be tested in an interspecific comparison. It predicts that there is an inverse relationship between the water content of fruits and deviation from a spherical form, whereas no causal relationship between form and content is to be expected in flowers.

Although the concept of diversifying selection exerted by pollinators upon floral traits is intuitively plausible, and although this process has likely played an important role in the diversification of angiosperms, it remains to some extent controversial. On one hand, interspecific comparisons consistently suggest that animal-pollinated clades are more species-rich than clades with other pollination

systems. This is consistent with the growing number of studies showing how pollinators can induce micro-evolutionary changes in flowers (e.g., Medel et al. 2003, Herrera et al. 2006). On the other hand, however, there is the obvious difficulty of linking micro-evolutionary processes to macro-evolutionary patterns, particularly in the context of multiple species (pollinators and flower predators) interacting with a given flower. As such, simple questions such as whether the diversity of syntopic flowers in terms of their morphology, colour and odour is adaptive – as expected under the market theory mentioned above – are still poorly understood. Unfortunately, patterns of floral scent are unexplored on a community level, and there are no studies to evaluate the relative importance of tactile, olfactory, and visual filters on a community scale (which obviously is a Herculean task). Given the considerable phylogenetic influence on the morphology of plant reproductive traits (see Jordano 1995a for fruits), the observed diversity in floral morphology might be selectively neutral, reflecting simply the fact that species of a more diverse phylogenetic background are more likely to co-exist by random or sorting processes than an identical number of closely related species. Phylogenetically-informed null models recently showed that co-flowering species tended to be more divergent in floral colour than expected by chance or by phylogeny (McEwen and Vamosi 2010).

Within a species, consistent geographical variation in pollinator assemblage can feasibly produce a pattern of diversifying selection (the geographical pollinator mosaic hypothesis), but there are few studies analysing intraspecific floral diversification in response to a geographical cline in pollinator-mediated selection. In the lavender (*Lavandula latifolia*), Herrera et al. (2006) found that clines in floral morphology were congruent with clines in pollinator-related selection, supporting the concept of adaptive diversification. These authors emphasize the need to move beyond correlational studies between the fit of plant and pollinator morphologies by demonstrating selection by pollinators upon floral traits as well as its variability across a geographical scale. These advances will be critical in order to distinguish between pollinator-mediated selection and ecological sorting which occurs if pollinators differ in their use of pre-existing flower traits.

Although diversifying selection may act more strongly on flowers than on fruits, there are many ecological and evolutionary similarities. While plant-pollinator interactions are on average more specialized than fruit–frugivore interactions, the selective pressures that specific animals exert upon flowers traits are usually similarly diluted into the overall selective pressures of multiple species interacting with flowers and including both mutualists and antagonists. This is because specific pairwise interactions between a single pollinator and a single plant are not the rule, at least not in temperate areas of the northern hemisphere. In flowers, such pairwise interactions are apparently more common in species-rich communities (Johnson and Steiner 2000), whereas pairwise interactions between fruits and seed dispersers if they occur are extremely rare and more likely to be found in species-poor communities. Another common theme in pollination and seed dispersal are pleiotropic effects on reproductive traits that can greatly influence their evolutionary trajectory (Delph et al. 2008, Armbruster et al. 2009). These pleiotropic effects are not well understood, but we will discuss some of them in this chapter.

We can conclude that diversifying selection exerted by animals should act more strongly on flowers than on other plant traits. Because ethological isolation can play an important role in premating isolation among flowers, diversifying selection is likely to act on floral traits that are important in communicating to animals. Quantifying the extent of adaptive diversification has often proved to be difficult, however, owing to ecological constraints – the dilution of the selective pressures of single species into the overall selective pressures of multiple species, as well as geographical variation in abiotic factors – and owing to the often unknown genetic and physiological constraints that may limit plants' responses to selection.

6.4 Floral morphology

Although we usually think first about colour when visualizing optical traits, form is an important property that might distinguish syntopic flowers. Floral morphology, in particular floral size, can show very high evolutionary rates (Barkman et al. 2008). Studies on pollinators found preferences for larger flowers in the field and demonstrated that these preferences can be innate in the lab (Mothershead and Marquis 2000, Martin 2004, Naug and Arathi 2007). The high evolutionary rates of changes in floral size may either be attributable to strong selection by pollinators or to few genetic or developmental constraints associated to size-related changes. One explanation for pollinators' preference for large flowers is that larger targets are easier to detect from afar than smaller ones. However, there are few studies documenting the selective pressures of pollinators on a micro-evolutionary scale for increasing floral size.

Some of the largest floral displays are found in the genus *Rafflesia*, where floral diameters approach 1 m in size. These species perform a deceptive strategy of carrion mimicry, attracting flies specializing on carrion that pollinate the flower. They are deceptive because the blossoms typically do not provide conditions for larval growth. Carrion mimicry has evolved independently in at least ten lineages and is always associated with large flower size (reviewed in Davis et al. 2008). Feasibly, large flowers produce more smell and thereby attract more flies than smaller flowers, and flies prefer visiting larger flowers (Kneidel 1984). This hypothesis requires a correlation between floral size and the intensity of the associated olfactory signal. Such a positive relationship was indeed shown in the cactus *Escallonia myrtoidea* (Valdivia and Niemeyer 2006). There is a further mechanism potentially selecting for large flowers in carrion-mimicking species. This is that insects strongly prefer larger carcasses because these provide better resources for their offspring. This is a compelling argument for strong directional selection upon floral size that could explain the unusually high evolutionary rates found in *Rafflesia* (Barkman et al. 2008).

Pollinators also prefer large flowers in the lab, where flowers are presumably easy to detect (Naug and Arathi 2007). This may be because at least in some species size-related traits are correlated with the quantity of nutritional rewards. In *Dalechampia ipomoeifolia* the strongest selection by resin-collecting bees acted upon the bract size of the flowers and not on the resin gland area (Armbruster et al. 2005). The authors argued that bract size might be the easier visible indicator of the quantity of rewards, given that differences in gland area are relatively small and therefore more difficult to judge from a distance. A further possible reason for pollinators to prefer large flowers is that flower size can be a state-dependent variable indicating the resource potential of the plant. For example, leaf and bud herbivory reduces flower diameter in *Oenothera macrocarpa* and *Alstroemeria exerens*, presumably because fewer resources are allocated to reproduction (Figure 6.2; Mothershead and Marquis 2000, Suárez et al. 2009). In both studies smaller flowers are less visited by pollinators and thus set fewer seeds. In the study by Suárez et al. (2009), herbivory also reduced the size of floral guides thereby potentially compromising pollen transfer because the orientation of pollinators on the flowers is hampered. Pollinators' preferences are also potentially explicable because herbivory can cause the induction of chemicals in nutritional rewards (Strauss et al. 2004). In this case, pollinators could use flower size as a cue to avoid plants affected by herbivory.

A specific morphological adaptation for communicating to pollinators is found in bat-pollinated plants. At least 528 plant species from 67 families are pollinated by bats (Fleming et al. 2009). Neotropical bats orientate and forage using their sonar system. As such, they perceive and select for conspicuousness very differently from other pollinators. Adaptation towards bats is found in the neotropical vine *Mucuna holtonii* whose flowers contain a small concave mirror that functions like an optical cat's eye, reflecting most of the energy of the bats echolocation calls back into the direction of incidence (Figure 6.3; von Helversen and von Helversen 1999). The floral structure is salient because a passing bat receives high-amplitude echoes for many calls, whereas the surrounding vegetation reflects few calls back. The small concave mirror – the vexillum measuring 19 mm across – is situated just above the

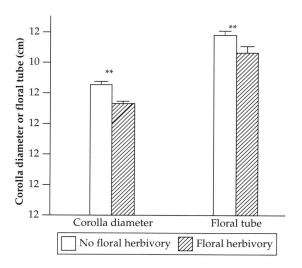

Figure 6.2 Flowers of *Oenothera macrocarpa* damaged by herbivores in the bud stages developed smaller corollas and shorter floral tubes than control flowers. Illustrated are means and S.E. Redrawn from Mothershead and Marquis 2000.

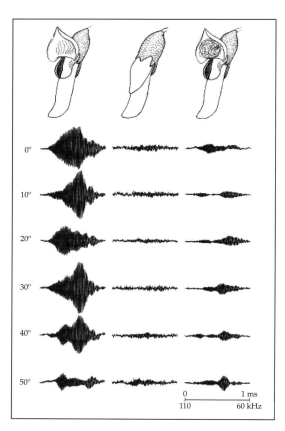

Figure 6.3 Mature flowers of the neotropical vine *Mucuna holtonii* sport a mirror-like vexillum (left) that reflects echoes emitted by foraging bats back to the bats. In various angles the echoes obtained from experimental signals are much stronger in the mature flowers, than in buds (centre) and flowers whose vexillum has been experimentally clogged (right). Redrawn from von Helversen and von Helversen 1999.

flower entrance and is raised upon floral maturation. Removal of the vexillum reduced bat visitation strongly so that bats only visited 21 per cent of manipulated flowers compared with 88 per cent of intact flowers. Finally, *Mucuna* spp. that are pollinated by small paleotropical Megachiroptera bats lacking the sonar system also lack the vexillum, suggesting that it evolved to increase conspicuousness to bats using echolocation. In a further study on that species, the authors showed that the nectar-feeding bat *Glossophaga commissarisi* discriminated among similar motionless objects based upon the spectral composition of their echoes (von Helversen and von Helversen 2003). In addition to differences in the spectral composition, differently-shaped objects also produce different sequential echoes (von Helversen et al. 2003). As such, nectar-feeding bats using echolocation can probably discriminate among differently-shaped flowers by integrating temporal and spectral echoes.

6.4.1 Accessory structures

Flowers are not the only structural parts that could present visual signals to animal mutualists. Like fruits, some flowers sport structures associated with the floral display that might fulfil a signalling role. Such structures can have functions unrelated to communication. For example, the sterile inflorescence of the South African *Babiana ringens* increases mating success by providing a perch for sunbirds and thereby directing pollen transfer (Anderson et al. 2005). However, as discussed previously, accessory structures can also be involved in communication. Bract size in *Dalechampia ipomoeifolia* may function as a reliable indicator of quantity of available rewards (Armbruster et al. 2005). It is probably the developmental correlation between bract and gland size that explains why insects respond to variation in bract size. Here, as in other species, the accessory structure exceeds the size of the rewarding structure and can therefore function as the most visible aspect of the display. For example, in the butterfly-pollinated *Mussaenda frondosa* the white bracts are much larger than the flowers. Removing the bracts reduced flower visitation by butterflies as well as subsequent seed set (Borges et al. 2003). This study suggests that bracts act as long

distance visual signals of the presence of floral rewards.

Although a communicative function occurs in the bracts of *D. ipomoeifolia*, it is obviously problematic to infer the origin from the current function of a trait. In an elegant analysis, Armbruster (1997) showed complex changes in function in the pollination of *Dalechampia* species. He suggested that floral resins evolved originally as floral defence. This feature served then as pre-adaptation that allowed the evolution of a specialized resin-reward pollination system. This pollination system is thus an exaptation of the original floral defence. When this pollination system evolved, floral resin had lost its original defensive function. Subsequently, there followed new defensive innovations. Bracts that originally evolved for their communicative functions to attract pollinators were co-opted for a defensive function that includes closure during the night to protect flowers from herbivory inflicted by nocturnal foragers. Interestingly, resin was later replaced by pollen and fragrance as pollinator reward five to six times, and in three to four of these the resiniferous bractlets were 'immediately' redeployed to protect flowers (Armbruster et al. 2009). These authors suggest that florivores exert strong selective pressures and that the easiest line of evolutionary response was to reactivate pre-existing developmental pathways.

Distinct selective pressures on accessory structures can also be seen in the dove tree *Davidia involucrata* where white conspicuous bracts form an umbrella over the flower head. Pollen-collecting bees preferred to visit flowers under white bracts (compared to green bracts) showing that the white colour functions as a signal to increase the conspicuousness of flowers (Figure 6.4; Su et al. 2008). The location of the bracts above the flowers shelters them from rain thereby reducing pollen mortality. It thus seems as though colour is a signal but the bracts and their position are most likely cues.

It might at first sight seem feasible that the large accessory structures of floral displays could function as handicap signals. The handicap hypothesis requires costs that could be associated with the production of large accessory structures. This conjecture was tested by Keasar et al. (2009), who analysed biomass allocation under distinct irrigation

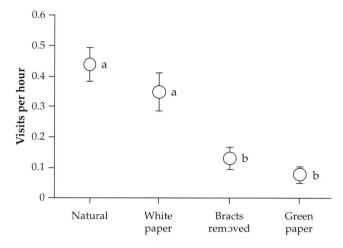

Figure 6.4 Conspicuous accessory structures can increase floral visitation. Bees visited more flowers with natural white bracts or experimental white paper bracts compared to flowers with experimental green bracts or without bracts in the dove tree *Davidia involucrata*. Illustrated are means and S.E., letters denote statistical differences between treatments. Redrawn from Su et al. 2008.

treatments in a greenhouse to assess the costs associated with flag production in *Salvia viridis*. Allocation to flags accounted for up to 0.5 per cent of plants' biomass and did not change between irrigation treatments, suggesting that it does not function as a condition-dependent handicap signal. In this species, the presence of flags influenced the number of floral visitors to a patch but had no influence on foraging decisions within a patch. Keasar et al. (2009) thus concluded that flags serve to increase the detection rate of flowers in isolated sites.

Finally, leaves can also function as accessory structures to attract animals. Weevils as pollinators of the dwarf palm (*Chamaerops humilis*) are attracted to the volatiles (a blend of mono- and sesquiterpenes) produced by the leaves and not the flowers (Dufaÿ et al. 2003). This is a very interesting study because it documents the transfer of a communicative function from reproductive tissue to vegetative tissue. This transfer is probably more widespread, being evident in the studies on the functions of accessory structures in pollination (discussed above) and seed dispersal (Schaefer and Braun 2009; Section 4.8). Because plants have usually more vegetative tissue than reproductive tissue, the transfer of communication to the leaves likely yields fitness benefits by increasing signal production and consequently the range of which a signal can be per-

ceived. Furthermore, the defensive properties of the mono- and sesquiterpenes (see Section 2.10) produced by palm leaves suggest that the attraction of weevils can be an exaptation of the original defensive function (Dufaÿ et al. 2003).

6.4.2 Symmetry

Compared to fruits or leaves, an important difference in floral morphology is that there can be selection by animals on floral symmetry. In general, flowers can be categorized into open radial symmetric flowers (actinomorphic) and closed zygomorphic types with bilateral symmetric flowers. Radial symmetry is the ancestral form, with zygomorphic flowers having evolved several times independently (Takhtajan 1991). The adaptive hypothesis on floral symmetry posits that zygomorphic flowers restrict the directionality of approach and movement within the flowers, leading to more efficient pollen transfer. The orientation of a flower is a further factor that can affect directionality of approach and thereby increase pollination efficiency (Ushimaru and Hyodo 2005) as it does in the sterile inflorescence of *Babiana ringens* (mentioned above). Actinomorphic flowers are not able to restrict pollinator movement. Because the movements of animals are more restricted in bilateral zygomorphic flowers, it has been suggested that zygomorphic flowers are visited by fewer and more specialized insects (Donoghue et al. 1998, Gong and Huang 2009). However, this pattern appears to be restricted to specific lineages: flower openness was not correlated with generalization in pollination across a large floral database (Olesen et al. 2007b). Independent of whether floral symmetry alters generalization patterns, if it alters the spectrum of floral visitors, it might be involved in pre-mating isolation. Examining this hypothesis throughout a phylogeny indeed yielded a correlation between zygomorphy and species richness (Sargent 2004). Furthermore, zygomorphic species were less variable in floral size than were actinomorphic species: suggesting stronger stabilizing selection on zygomorphic flowers (Figure 6.5; Gong and Huang 2009).

Symmetry and spatial orientation are two components that can be important in distinguishing flowers from other syntopic flowers, particularly because honey bees and birds can perceive and select for symmetry in the lab (Giurfa et al. 1996). However, studies failed to find that pollinators select against deviations from symmetry or from a given axis of orientation in the field. For example, in the monkey-flower *Mimulus luteus*, fluctuating asymmetry was unrelated to female fitness (Botto-Mahan et al. 2004). Evidence for selection upon floral symmetry would be particularly welcome because selection is likely to be more variable in the field depending partly on the density and profitability of other flowers. The lack of evidence is surprising because frequent mutants with radial symmetry are found in zygomorphic orchids and mints, making them ideal candidates to address this question (Rudall and Bateman 2003). In orchids particularly these authors discuss whether reversals in symmetry might lead to the formation of new species.

Theoretically, the axis of symmetry can be an important component shaping flower-animal communication. For example, in actinomorphic flowers pollen rewards are often visible upon approach to the flower. Thus, pollen-collecting or pollen-eating species can directly evaluate the quantity of reward present without needing to rely upon an associated signal or cue that indicates rewards. The situation is different in zygomorphic flowers, where the reproductive parts are often concealed and nectar is offered (Osche 1983, Lunau 2000). It is yet unresolved whether pollinators select for a higher visibility of rewards. On one hand, it is conceivable that such a selection pressure leads to more reliable communication; on the other hand, at least in insects with poor spatial vision, the visual assessment of rewards is necessarily restricted to close distances and may therefore not strongly increase foraging efficiency.

6.4.3 Floral integration

Berg (1960) was among the first to establish a quantitative criterion for examining adaptive flower morphology. She analysed the variance and covariance among reproductive and vegetative traits that might originate from developmental and genetic constraints. She found covariation within vegetative and within reproductive traits, but found that these morphologies were not strongly correlated among each other, a situation she termed correlation pleiades. More specifically, she found that those plants that are pollinated by specialized insects had stronger covariance among reproductive traits than other plants. She concluded that the adaptive significance of that pattern was to ensure the morphological fit between pollen deposition and pollinator so that the pollen is deposited on a definite part on the body of the insect.

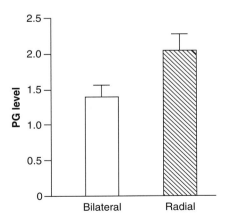

Figure 6.5 Bilateral symmetric flowers are visited by fewer pollinator guilds (PG = pollinator generalisation) than radial symmetric flowers in an alpine meadow in China. Redrawn from Gong and Huang 2009.

Her conclusion that effective pollination requires a fit between floral morphology and pollinator morphology and behaviour is intuitive and provides the basis for the evolution of mechanical isolation among flowers. The extent of mechanical isolation obviously depends on the extent of differences in pollinator morphology and their approach to the reproductive parts of the flower. For example, Muchhala (2007) showed that bats and hummingbirds select for different corolla widths. In an experiment, he showed that wide corollas guided bat snouts better, whereas narrow corollas guided hummingbirds' bills better. A poor fit between floral morphology and pollinator morphology resulted in variable entry angles and decreased pollen transfer of both pollinator types (Figure 6.6). Interestingly, generalization in plant traits was never favoured in this experiment because intermediate corollas always had lower pollen transfer.

Despite the expected selection towards an increased fit between floral traits and pollinator morphology, subsequent studies found that the relationship among the morphology of vegetative and floral parts as well as within floral parts were more complex than envisioned by Berg (Armbruster et al. 1999). Animal- and wind-pollinated species both showed lower variation in floral traits than in vegetative traits. Species with specialized pollina-

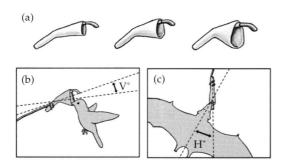

Figure 6.6 There is an adaptive trade-off in floral morphology in *Burmeistera sodiroana* flowers. Narrow corollas (a, left) guide bills of hummingbirds better, whereas wide corollas (a, right) guide the snouts of bats better. A mismatch in floral morphology and animal morphology results in variable entry angles that reduce pollination transfer. Hummingbirds have variable entry angles when visiting wide corolla (b), whereas bats have variable entry angles when visiting narrow corollas (c). The reduction in pollen transfer can be a functional explanation for the evolution of pollination syndromes.

tion had less variation in floral traits than those that were unspecialized or wind pollinated, as predicted by the correlation pleiades concept. Yet, the same was true for vegetative traits: showing that the patterns of phenotypic integration in plants are not solely explicable as adaptation to specialized pollinators (Armbruster et al. 1999). They can also be explicable by selection by antagonists such as herbivores and parasites (R. Raguso, pers. comm.). That floral integration can be unrelated to pollinators was also shown in an intraspecific comparison. Nine populations of the bumblebee-pollinated herb *Helleborus foetidus* differed both in the pattern and magnitude of covariation among floral parts (Herrera et al. 2002). Yet differences in integration were not related to variation in the composition and morphology of pollinators, nor were they distance-dependent. Instead, these authors argue that rather than adaptations towards the selective pressures of pollinators, genetic drift in the small and ephemeral populations of this species is the most likely cause for the observed differences in the covariance of floral traits.

Overall, the patterns of morphological floral integration were relatively low (e.g., 20 per cent Ordano et al. 2008) and lower than that of fleshy fruits (38 per cent, Valido et al. 2011). Consistent with this low degree of floral integration, recent genetic studies on *Petunia* flowers have shown that the genetic linkage between different morphological traits can be broken in a few generations (Venail et al. 2010). A study on *Schizanthus* species showed that pollinators do influence floral integration patterns and that patterns of floral integration were not strongly determined by phylogeny (Pérez et al. 2007). In pollinator-dependent species, the overall integration was lower in lineages where some floral traits had lost their function in pollination compared to those that retained trait functionality. The authors concluded that this pattern is a consequence of the decoupling among functional and non-functional traits thereby indirectly supporting the role of floral integration for pollination in *Schizanthus* species.

Taken together, these papers leave us in the uncomfortable situation that it appears difficult to judge the degree of floral adaptation towards pollinators. Floral adaptation undoubtedly occurs, for example to increase physical contact with pollinators (Temeles and Kress 2003) or to position pollen on specific body parts of the pollinator to reduce interspecific pollen transfer (Nilsson et al. 1987). However, few studies can quantify the extent of adaptations. There are several explanations. First, pollinators do not select for overall floral integration but rather for the combination of specific traits: that is, intrafloral integration (Pérez et al. 2007, Ordano et al. 2008). In other words, it is not the entire flower that is the target of pollinator-mediated selection but rather only specific combinations of floral traits. Whether these floral traits respond to pollinator mediated selection depends on their genetic architecture. Lack of additive genetic variation for particular traits and genetic correlations among floral traits can greatly constrain the potential for adaptations of specific traits towards pollinator-induced selection. In the ivy-leaf morning glory (*Ipomea hederacea*) morphological adaptations to reduce heterospecific pollen flow are constrained by unequal genetic variance of floral traits that can greatly deviate (up to 73°) the response to selection from that expected in the absence of variability in genetic variance (Smith and Rausher 2008). Thus, while genetic correlations form one of the foundations of phenotypic integration, this study highlights the necessity to also consider differences in genetic variance between traits in order to predict responses to pollinator-mediated selection.

The next possible explanation is that the multiple species interacting with plant traits exert conflicting selective pressures resulting in trade-offs between, for example, attractive and defensive functions. Galen and Cuba (2001) showed in an elegant experiment that ants predating skypilot flowers (*Polemonium viscosum*) select for more tubular flowers, which receive less pollen and set fewer seeds in bumblebee-pollinated populations. Selection on flower morphology thus depended on the relative costs and benefits of predation by ants vs. pollination by bumblebees. Spatial variation in pollinator and florivore abundance is thus likely to alter the resultant trade-off between attractive and defensive functions in *Polemonium viscosum*. Experiments on that species further showed that large flowers exact water-loss costs under dry conditions, reducing the survival probabilities of individuals with large

flowers and heavy seed set (Galen 2000). Yet, bumblebee pollinators, but not fly pollinators, enforce a positive relationship between corolla size and fecundity (Galen 1989). Taken together, corolla size in skypilot exacerbates a demographic cost of reproduction under drought conditions in bumblebee-pollinated populations.

These studies again testify to the importance of accounting for variation in the biotic and abiotic factors interacting with plants. Variation in biotic agents includes differences in species composition as well as differences in the relative abundance of species that interact with a given plant. Finally, there is a possible interaction between the effects of abiotic and biotic selective agents which can make it difficult to predict selection regimes for one population based upon results obtained from another population. Variation among plant populations in the number and relative importance of interacting species form the basis of the theory of the mosaic of co-evolution that predicts that populations vary in the extent of co-evolution with mutualists or antagonists (see Section 1.4 and Section 5.9.2).

Few studies have examined the patterns of integration among different types of traits. Frey (2007) found that floral colour evolution proceeded relatively independently from phenotypic integration of morphological traits. This situation is paralleled in fruits where colour traits show considerably lower values of phenotypic integration than morphological or chemical fruits traits (Valido et al. 2011). Interestingly, although floral colour was not consistently associated with other floral traits, the study by Frey (2007) documents various associations between colour traits and floral morphology. Pink flowers of the spring beauty *Claytonia virginica* were larger and had higher pollen viability compared to other colour morphs of the same species. These relationships emphasize that there are many covariances and pleiotropic effects among traits that are often considered in isolation. On one hand, these covariances can greatly constrain the responses to selection upon a given trait. On the other hand, non-random covariances among traits form the basis of pollination and seed dispersal syndromes (see Section 1.4) because they can indicate correlated selection induced by animals. Clearly, there is a need for further studies that analyse genetic variances and covariances in an ecological setting of pollinator-mediated selection to distinguish between the two scenarios.

6.5 Visual traits associated with flowers

6.5.1 Colour patterns and association to rewards

Floral colours differ from the colours of other plant parts in that patterns can play a major role in communication. Patterns can be adaptive because they guide pollinators to the food resources, and because they can facilitate species identification and might thus be potentially under diversifying selection (see Section 6.3). Analysing flight behaviour in insects showed that they often orient along pattern contours. Floral patterns can have various shapes. Commonly three distinct types of patterns are distinguished (Dafni and Giurfa 1998, Lunau 2000). First, lines, dots, and dash markings that guide pollinators can make the flowers more attractive. The adaptive value of these markings can thus be that they help pollinators to orient on the flower and help to identify the flower. These floral guides often have spectrally purer (more chromatic) colours than the surrounding peripheral flower colours, which may facilitate orientation on the flower (Lunau et al. 1996). The second type of pattern includes usually dark blotches and spots that mimic mating partners such as in sexually-deceptive *Ophrys* spp. The adaptive value of this kind of floral mimicry is also clear and discussed in Chapter 8. The third group includes any markings sufficiently similar to the colour of pollen and/or anthers and stamens to mimic these potential rewards. In the following we focus on this last group because these potential signals are likely to be taxonomically widespread.

Pollen colour is a good example of a direct link between communication and the type of reward offered by flowers. Pollen is predominantly (> 80 per cent of species) characterized by an absorbance of UV and blue light coupled with a high reflectance in the yellow part of the spectrum. The colour reflected by pollen is a conspicuous and reliable indicator of the pollen rewards offered in those primitive, radial symmetric, open flowers that are often pollinated by pollen-eating beetle, flies, and bees (van der Pijl 1960, Osche 1983). Because wind-pollinated gymnosperms also produce yellow pollen, the pigments imparting pollen colour (flavonoids and carotenoids) likely evolved to protect pollen against excess light and UV radiation and not in order to be easily detectable by pollinators (Osche 1983). Thus pollen colour qualifies probably as an example for an exaptation where a defensive function predates the evolution of a possible communicative function.

Because the association between pollen reflectance and the presence of pollen is consistent and necessary and already occurs in wind-pollinated species, it represents not a signalling system adapted to communication but an index of pollen rewards. This tight association between the presence of pollen and its colour, however, can be exploited by plants to manipulate pollinators. Hence, although pollen per se is not a signal, exploiting pollinators by mimicking larger pollen rewards than actually present would be a signal in floral communication (Vogel 1978, Osche 1979). In order to attract pollinators, plants can present a signal that is larger than their pollen resources by mimicking pollen colour through the androecium (which includes all stamens). Osche suggested that adaptations to mimic pollen are particularly important in derived zygomorphic species where the visual cue of pollen rewards is hidden within the flower and no longer visible to the approaching pollinators (see Figure 6.7). Like the pollen, the anthers are also pigmented by flavonoids and therefore appear yellow. There are various examples where flowers show morphological adaptations to present a stronger or supernormal signal to pollen-collecting insects (Lunau 2000). These examples include modifying other flower structures such as the inner corolla (*Narcissus pseudonarcissus*) or lip (*Mimulus guttatus*) so that they resemble extremely large androeciums or single anthers (Figure 6.7). Alternatively to these purely morphological adaptations, pollen rewards can also be feinted by colour patches that match pollen colour without any morphological change.

There is correlative evidence supporting the adaptive function of pollen mimicry in a large number of species. Analysing the colouration of 162 species, Heuschen et al. (2005) found (i) that the

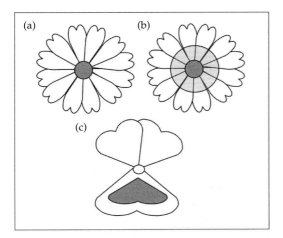

Figure 6.7 The visual cue of pollen (which is yellow with little UV reflectance) can be enhanced through pollen mimicry. In a radial symmetric flower pollen (grey area) is visible before pollinators contact the flower (a). Pollen mimicry can occur if the inner corolla has similar colour to pollen, thereby enhancing the pollen signal (b). In zygomorphic flowers spots on the lip can mimic pollen rewards (c). Redrawn and modified from Heuschen et al. 2005.

inner colour of actinomorphic and zygomorphic flowers appears less diverse than the peripheral colour, (ii) that the inner colour in these flowers is very similar to pollen colour according to a model of bee vision, (iii) that bumblebees prefer two-coloured flowers over uni-coloured flowers, and (iv) that they exhibit preferences for larger central areas and for centres that resemble pollen colour. The preferences for larger areas might be explicable because these represent a larger stimulus. As such, the two-coloured ray florets of many Asteraceae that visibly enlarge the central area might be an adaptation towards increasing stimulation of the receiver (Heuschen et al. 2005). While a differently coloured centre might simply help the pollinator to orient towards the nutritional rewards, this hypothesis does not explain the colour similarity to the colour of pollen. Moreover, Heuschen et al. showed that bumblebees preferred, among different colour combinations, a UV-absorbing yellow central spot that resembles pollen.

Although pollen mimicry may represent some of the clearest examples of adaptive floral colouration and although it is likely widespread, this topic has received considerably less attention than other types of floral mimicry. There is a dearth of studies on the micro-evolutionary processes that could highlight the selective pressures associated with communication. The pollen mimicry hypothesis makes two important predictions. First, that pollinators should select for colour similarity to pollen and second that pollinators should select for a larger stimulus (in terms of patch size). Hence, and in contrast to other flower parts, we do not expect selection towards increasing conspicuousness in the colouration of pollen-mimicking structures. We would therefore particularly welcome studies that scrutinize these predictions in order to evaluate whether variation in pollen mimicry translates into fitness benefits.

In contrast to the relatively high frequency of polymorphism in flower colour, there are few studies documenting pollen colour polymorphisms and its demographic consequences (but see studies by James Thomson, e.g., Thomson and Thomson 1989). In the annual herb *Nigella degenii* individuals produce either yellow or dark pollen, with dark colour being the putatively derived form. In this species, plants with dark pollen grains were more often found on less sun-exposed locations, suggesting a pleiotropic effect on pollen colour dimorphism (Jorgensen and Andersson 2005). Experiments by these authors supported this conjecture, because plants with dark pollen showed a higher mortality in drought and nutrient-poor conditions. Interestingly, the population differentiation for pollen colour exceeded that of other morphological traits and also that of AFLP markers of presumably neutral genetic divergence (Jorgensen et al. 2006). These results suggest that diversifying selection and local adaptation shape differences in pollen colouration. It is therefore intriguing that populations vary strongly in the frequencies of the two pollen colours morphs, and that the dark morph has not reached fixation in any of the populations studied. While pollinators sometimes responded differently to the pollen morphs, there is no clear evidence that pollinator-mediated selection maintains the colour polymorphism. Long-distance gene flow, pleiotropic effects as adaptation to micro-habitats, and temporally fluctuating selection are possible explanations for the maintenance of the pollen colour polymorphism. Since it seems relatively easy to measure the arrival of differently coloured pollen upon stigmata of these plants, we suggest this system is suitable for evaluation of the relative importance of pollinator-mediated selection vs. post-mating selection owing to pleiotropic effects.

In most species, nectar is clear and less strongly coloured than pollen. Yet, coloured nectar was known to the ancient Greeks and is currently known from at least 67 taxa (Hansen et al. 2007). This review showed that coloured nectar has evolved independently at least 15 times at the level of families. It occurs particularly on islands and in species occurring at high altitudes. The colours range from red (25 per cent) to yellow (22 per cent), brown (21 per cent), black (13 per cent), to amber-orange (9 per cent), and blue (3 per cent). Coloured nectar is often presented in such a way that it is visible before contacting the flower, and it occurs particularly in vertebrate-pollinated taxa (mostly birds and, in Madagascar, geckos). In some taxa, the chemistry involved in colouration is known. In *Aloe vryheidensis*, the brown colouration is produced by oxidation of phenolic compounds (Johnson et al. 2006), whereas in *Nesocodon mauritianus* the red nectar is produced by an aurone (Olesen et al. 1998). In the nectar of the aloe, phenols filter flower visitors because unspecialized birds are not repelled by the nectar, whereas specialized bird pollinators (sunbirds) and honey bees are apparently repelled by it (Johnson et al. 2006). The authors suggest four possible explanations for the presence of coloured nectar in this species. Coloured nectar might evolve as a visual warning signal towards unwanted floral visitors (sunbirds and honey bees), it might evolve to increase pollination efficiency by guiding pollinators to replenished flowers, it might evolve in response to colour preferences of pollinators or as a by-product of selection for bitter nectar. Of all these explanations, the one concerning innate colour preferences is probably the weakest because these are often labile and transient in animals foraging on plant rewards. The hypothesis of a warning signal is not supported by observations of sunbirds repeatedly probing flowers with dark nectar (Johnson et al. 2006). If the pigments are the chemicals that directly deter unwanted floral visitors or if the production of pigments is highly correlated to that of defensive compounds, it seems to us that by-

product selection is the most parsimonious explanation. Correlated selection upon nectar bitterness would represent an example of an index where nectar colouration necessarily indicates phenolic defences.

Although filtering the visitor spectrum of flowers might explain to a large extent the occurrence of coloured nectar, several studies have suggested that it serves as a signal to pollinators. An endemic gecko of Madagascar, *Phelsuma ornata*, for example, prefers coloured over clear nectar in artificial flowers (Figure 6.8; Hansen et al. 2006). The authors suggested that coloured nectar can be an honest, cost-enforced signal whose costs do not (necessarily) lie in the production of coloured nectar but in the absence of it. This occurs since coloured nectar is often clearly visible prior to contact with the flower (at least to vertebrates with good spatial acuity), the flower may not receive further visits until nectar rewards are replenished. Interestingly, the same argument was made for scent associated with nectar production (Raguso 2004). These are extremely interesting suggestions because the costs associated with the handicap would be shifted from the production of the signal to its maintenance in spite of pollinator visitation. Given the general tolerance of pollinators towards variable nutritional returns

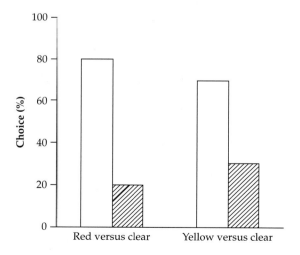

Figure 6.8 *Phelsuma ornate* geckos prefer red- and yellow-coloured nectar over clear nectar independent of the colours of the corolla. Redrawn from Hansen et al. 2006.

of floral cues, one might ask whether producing coloured nectar represents an evolutionary stable strategy. In analogy with the pollen-mimicry discussed above, it might seem feasible that a population of plants producing coloured nectar can be invaded by individuals producing floral traits that mimic the presence of coloured nectar. At low frequencies within the population, these plants might secure pollination success. Detailed observations of the behaviour of pollinators in combination with manipulations of the signal would be needed in order to scrutinize the handicap hypothesis of coloured nectar. Such data would then provide the basis for consideration of the evolutionary stability of alternative signal strategies.

6.5.2   Flower colour

The diversity of floral colours produced by the colouration of petals and sepals intrigued mankind long before modern scientific thought. Although the diversity may seem endless at a cursory look on an Alpine meadow, more than 1,000 flower colour spectra could be categorized into 10 colour categories (Chittka et al. 1994). Interestingly, there are almost no floral colours that absorb in the red part of the spectrum and there are no colours that absorb less in the UV than in other visible wavelengths. This situation contrasts with that of fruits, where black fruits that absorb in the red part of the spectrum are common, and fruits covered by epicuticular waxes that reflect strongly in the UV (as in many plums and grape varieties) have higher UV reflectance than reflectance in other parts of the visible spectrum (400–700 nm, Figure 2.4). Obviously, the limited diversity of floral colours can be understood by the limited combination of pigments and surface structures available to impart colours in petals and sepals. This hypothesis of physiological constraints does not explain, however, why black is a rare colour in petals and why UV-reflecting epicuticular waxes are likewise rare in flowers. A possible explanation is that neither black colour in fruits nor epicuticular waxes function in plant–animal communication but rather in defence against fruit predators (fungi) and abiotic factors (see Section 4.6 and 4.7). We speculate that this type of defence is more important in nutritious fruits than

it is in less nutritious flower structures such as petals.

Differences in floral colour within polymorphic species, among sister species, or among hybrids and their parental species provide some of the most compelling evidence of adaptive floral diversification (Hodges and Derieg 2009). If the sensory differences among pollinators are pronounced, floral colour can lead to considerable ethological isolation among syntopic morphs and species. For example, hummingbirds and insects differ proportionally in their visits to distinct floral colours of several plant species. Two species of *Ipomopsis*, *I. aggregata* and *I. tenuituba*, provide an illustrative example. *I. aggregata* has red flowers and is mainly pollinated by hummingbirds, whereas *I. tenuituba* has white to pale-pink flowers and is mainly visited by hawkmoths. Painting flowers of both species red reduced reproductive isolation by 20 per cent (Campbell 2004). This study thus demonstrates that floral colour can contribute to reproductive isolation, while also highlighting that other floral traits are required to achieve significant reproductive isolation.

More dramatic changes in the relative visitation rate of insects and birds were found in a remarkable study on two *Mimulus* species where floral colour alleles were interchanged between the red-flowering *M. cardinalis* and its yellow-flowering sister species *M. lewisii*. Substituting the *M. cardinalis* genes with those of *M. lewisii* produced dark-pink flowers leading to a 74-fold increase in bumblebee visitation, whereas substituting *M. lewisii* genes for those of *M. cardinalis* produced yellow-orange flowers resulting in a 68-fold increase in visits by hummingbirds (Bradshaw and Schemske 2003). This study and a similar one by Hoballah et al. (2007) on a transcription factor regulating anthocyanin expression in *Petunia* highlight that single genes can have strong effects on floral colours, resulting in pronounced shifts in the relative visitation rate of pollinators with distinct sensory abilities. Although these studies are consistent with pollinator-driven selection on floral colours, a comprehensive review on evolutionary transitions in floral colour documented that there is considerable evidence for pleiotropic effects of floral colour genes and selection by agents other than pollinators (Rausher 2008). Rausher also documented that mutations causing loss of function are more common in the anthocyanin biosynthetic pathway than those causing gain of function.

What are the proximate mechanisms underlying the role of floral colour in producing ethological isolation? Previously, it was assumed that colour preferences can be strong and consistent enough to produce diverging selection by distinct pollinators. Yet, as in seed dispersers, colour preferences in pollinators are often labile and transient. As such, the underlying proximate mechanisms of why, for example, red flowers generally belong to a bird-pollination syndrome have been termed a mystery (Rodríguez-Girones and Santamaría 2004). In particular, even though many insects are not particularly sensitive to red colours, red flowers are obviously not invisible to them. Indeed, insects such as bees do visit red flowers which they perceive as being mainly achromatic – that is, grey (Chittka and Waser 1997). It is important to note here that syndromes are not necessarily exclusive, and that they are not exhaustive. However, even a small difference of a few per cent in the likelihood of flower visitation of distinct pollinators – if consistent throughout evolutionary time – can have strong effects on the evolution of floral colour. As such, the mystery of red flowers can be understood in terms of the distinct visual world of hummingbirds and most insects. For birds, red is a particularly contrasting colour against a predominantly green background (Schmidt and Schaefer 2004, Schaefer et al. 2006), whereas it is not a particularly contrasting colour for bees. This may explain why the size of red nectar guides in the monkey flower *Mimulus luteus* is positively correlated to the proportion of hummingbirds in the pollinator assemblage (Medel et al. 2007). Thus, everything else being equal, the relative likelihood of visits to red flowers will differ among hummingbirds and bees.

Conceivably, the different visual abilities of pollinators could exert balancing selection upon flower colour polymorphisms if they consistently differ in their choice of different colour morphs. Eckart et al. (2006) found that the most common pollinator, a Melittidae bee, of the gunsight Clarkia (*Clarkia xantiana*), exerted positive frequency-dependent selection, while two less common pollinator species (bees from the Apidae and Halictidae) exerted negative frequency-dependent selection preferentially

visiting the rarer colour morphs. It is unclear in this study whether differences in the sensory ecology of the bees are strong enough so that differential flower visitation could be explicable by differences in sensory ecology. Alternatively, such differential selection patterns might be driven by direct competition between the commoner pollinator species and the two rarer ones.

Communication systems that are shaped by receivers with distinct sensory abilities are often termed private communication channels. The metaphor is useful to indicate that an intended receiver is more likely to respond to a specific signal than an unintended receiver. Yet, as in syndromes, it is not useful to think of exclusive use of this private communication channel by the intended receiver. In animals, private communication channels are often assumed to involve UV reflectance, because some groups (such as most mammals or avian predators) are less sensitive to ultraviolet light than their prey (Cummings et al. 2003, Håstad et al. 2005, but see Stevens and Cuthill 2007). In visual plant–animal communication, UV reflectance is perhaps less likely to feature in private communication channels, because most pollinators perceive ultraviolet light well (except for some mammals).

Although differences in floral colour can affect ethological isolation, many studies showed a high variance in the extent of ethological isolation even if disparate pollinators such as insects and hummingbirds are involved in visiting flowers. This is particularly true if divergent floral morphs or two flowering species are growing in strict sympatry where pollinators rarely restrict their visits to only one flower type. Often, however, reproductive isolation also involves a habitat component, with distinct micro-habitats favoured by distinct species. If, however, the differences in visual system between pollinators are not pronounced, it is unlikely that floral colour plays an important role in ethological isolation.

Although differences in floral colour are among the most obvious targets of pollinator-mediated selection, it is important to remember that floral colour also presents some of the most compelling evidence for pleiotropic effects, correlated selection on linked genes, and conflicting selection between pollinators and herbivores (see Section 1.7). Given that floral pigments such as anthocyanins play an important role in plants' stress responses in vegetative tissue (Section 7.3), it is not surprising that they are also involved in floral defence. If species are polymorphic in floral colour, it is usually the anthocyanin-pigmented morph that is avoided by herbivores or that is more resistant to drought, high irradiance, and heat stress (Strauss and Whittall 2006). Likewise, floral colour can be a by-product of local adaptation to soil, as discussed in Section 6.7 (Schemske and Bierzychudek 2007). The evolution of floral colour is thus an iconic example of how communication can be channelled by pleiotropic effects. However, this issue could also be phrased differently, in that it is not at all clear how much (if any) adaptation towards communication is occurring in floral colour evolution (Rausher 2008).

There are thus different scenarios of flower colour evolution. The null hypothesis is ecological fitting, where floral colour evolution is predominantly influenced by selective agents other than pollinators, and pollinators simply visit those flowers that they can easily exploit and/or that are particularly salient in their colour and scent. In this scenario, pollinators may still exert a selective pressure on the plant, but any response by the plant will be constrained by other ecological functions or developmental or genetic constraints. An adaptive hypothesis is that colours evolved in order to communicate to animals (both antagonists and mutualists). The third scenario is that colours evolved first as by-products and that their communicative function evolved secondarily as an exaptation.

A major shortcoming in our ability to discriminate among these scenarios is that pollinator-driven selection is not measured, but inferred. Many studies have documented discrimination among distinct floral morphs, but few studies demonstrated that such a discrimination resulted in consistent selective pressures. This argument likewise applies to studies analysing pleiotropic effects or selection mediated by herbivores where colours evolve as by-products. While damage is often reported, it is far less clear how this damage translates into plant fitness given that plants can employ distinct strategies ranging on a continuum from resistance to tolerance. We highlight these issues not to be critical of the work of others but rather to more clearly lay out

future research lines. One possible approach is obviously to compare population divergence at colour loci with population divergence at neutral loci to test for selection upon colour genes. While this approach would be supportive of local adaptation it would not identify the agents selecting for local adaptation and would thus not distinguish between pollinator-driven selection and selection by other agents (see Section 6.7). Thus, there is no substitute for empirical research that measures pollinator-driven selection directly in the field.

6.6 Floral scent

6.6.1 Comparison to vision

In general, olfactory communication by flowers is not as well known as visual communication. This reflects partly our own bias as visual animals, partly the more complex nature of olfactory signals (compared to visual ones), with more than 1,700 substances involved in floral scents currently identified (Knudsen et al. 2006), and partly the difficulty of modelling the sensory space of animals, which is defined by many more types of sensory receptors. In spite of these methodological challenges, there has been tremendous progress in studies of floral scent in the last decades, and many of the issues that we discussed in visual communication are likewise important in olfactory communication (Raguso 2008).

Yet, there are also striking differences between visual and olfactory communication. Odour travels slowly through the environment (being influenced by wind speed), whereas colour traits travel with the speed of light. At the same time, the directionality of odour plumes is low and depends on prevailing wind conditions, whereas the directionality of visual traits is immediate upon receiving it. Likewise, the intensity of odour plumes might vary strongly with internal factors of the flowers (such as nectar availability) and extrinsic factors, such as wind conditions. This variability might explain why some pollinators such as *Manduca sexta* reacted equally to odour mixtures over a 1,000-fold range in concentration (Riffell et al. 2009). In dense habitats, however, it might be adaptive to use olfactory traits instead of visual traits because the former do not require a clear line of sight. For a small herbaceous plant submerged in a matrix of tall grass it might thus be adaptive to use primarily olfactory communication. We know of no study that has investigated the possible link between habitat density and complexity (which may also vary seasonally) and the prevalence of olfactory communication.

Another central difference between olfactory and visual communication is that the composition of odours changes over time, whereas temporal variability in floral colour is less pronounced (except for flowers that change colouration after pollination has occurred; see Section 1.1). It is to be expected that the emission of compounds that attract pollinators should be tailored to their temporal activity and be less variable within that time window than the emission of compounds that fulfil other functions (see Section 6.7). There are indeed data supporting this conjecture (e.g., Raguso et al. 2003, Waelti et al. 2008). The study by Raguso and colleagues showed that odour chemistry from nine hawkmoth-pollinated species of tobacco *Nicotinia* spp. is species specific during the night but less distinct during the day. Obviously, the temporal variability of odour composition is also linked to the fundamental question of which chemical compounds constitute an olfactory signal. Volatiles are produced by different tissues, the petals, stigma, the anthers, and even pollen grains (Martin et al. 2009). The relative contributions of the different tissues to scent production varies between species and possibly also within species on a temporal scale.

Another methodological difference is that studies on scent first have to establish which odour compounds are perceived by a given animal. Doing so requires behavioural tests in wind tunnels, flight arenas, immediate Y-tubes; or at least measuring the antennal response of the focal animal. This is a much more laborious task than modelling the spectral sensitivity of photoreceptors because most olfactory communication systems in plants as in animals consist not of a single chemical compound but of blends of compounds. For example, only 9 of the more than 60 volatile compounds of the flowers of sacred datura (*Datura wrightii*) elicit a behavioural response in its pollinator, the moth *Manduca sexta* (Figure 6.9; Riffell et al. 2009). Owing to the complexity of volatile blends, it may often be the ratio of compounds rather than a single substance that determines the effectiveness of odours. For example, species can often be identified by their unique ratio of common volatiles (Hossaert-McKey et al. 2010). The complexity of blends thus requires a more complex approach to behavioural tests and

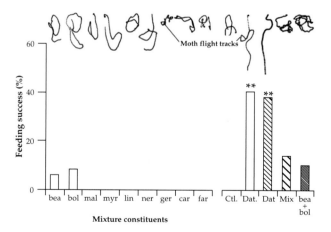

Figure 6.9 Only the entire bouquet of the flowers of scared Datura (*Datura wrightii*) or a blend of the nine most active compounds but not the single components elicit tracking behaviour of *Manduca sexta* moths as evidenced by the flight paths on top. Scents are abbreviated as follows: benzaldehyde (bea), benzyl alcohol (bol), methyl salicylate (mal), b-myrcene (myr), linalool (lin), nerol (ner), geraniol (ger), caryophyllene (car), and a-farnesene (far), control (Ctl), entire bouquet (Dat flower), blend of nine compounds (Dat mimic), a random mixture (Mix), and a combination of benzaldehyde and benzyl alcohol (bea + bol). Redrawn from Riffell et al. 2009.

statistical analyses. For example, compositional analyses of volatile compounds would be an adequate tool to assess the relative intensity of single compounds, but it has not yet been applied to odour analyses.

In contrast to colour, odour can represent a direct reward to specialized pollinators. Male euglossine bees collect fragrances from orchids that do not provide other rewards to them (as well as from other sources such as decaying wood) and presumably use the fragrances in their courtship. This specialized communication system is fascinating because it elucidates some of the consequences of plant–animal communication. Males of two morphotypes of *Euglossa viridissima* differ in their perfume collection in that only males of one type collect a specific compound (2-hydroxy-6-nona-1,3-dienylbenzyldaldehyde; Figure 6.10). Electroantennogram tests revealed that sensitivity to this compound differs between the two morphotypes, suggesting a proximate mechanism why related species collect different odour blends (Eltz et al. 2008). Hence, differential perfume perception translates into differential perfume collection; a finding that is consistent with a saltational olfactory-driven mode of perfume evolution in Euglossini (see Box 6.2 for general discus-

sion of how communication between plants and animals may affect between-animal communication). The authors went one step further and showed that males of 15 sympatric Euglossini species collect distinct blends of fragrances with chemical disparity among species being greater than expected by a null model throughout the reconstructed evolutionary history (Zimmermann et al. 2009). Interestingly, chemical disparity was greater within recently diverged lineages than among them, suggesting that divergent selection on perfume preferences acts quickly, likely because it conveys pre-mating isolation.

In contrast to visual characters, odours can also stick to a forager, and foragers can add odour to flowers (see Box 6.1). Odour that sticks to social insects can be transported back to a nest where it can be used to communicate good floral resources to nest mates. It has been shown that nest mates learn chemical cues associated with food resources. Social insects could collect scents actively in order to improve communication with their nest mates or passively owing to specific surface properties of the cuticle. If scent transported back to the colony increases visits of nest mates to the source of the scent one could envision selection towards odour components that are more likely to stick to the insect's hairs or cuticula. We know of no study that has investigated this admittedly speculative idea.

6.6.2 Additive effects of odours and colours

So far, we have emphasized differences between olfactory and visual communication. This poses the risk that both communication systems are viewed in isolation rather than in conjunction. Yet, most pollinators use both visual and olfactory traits to locate and identify floral resources. Indeed, honey bees learn to associate scent with rewards faster than visual cues (Menzel 1985). A combination of multi-modal traits – such as visual and olfactory ones – results in a higher discrimination ability of pollinators (compared to unimodal traits) and probably also increases their capacity to memorize flowers. For example, floral discrimination by colour cues is enhanced by the presence of scent (Kunze and Gumbert 2001). As such, it is not surprising that flower constancy – pollinators'

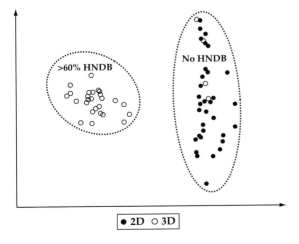

Figure 6.10 Differences in the perfumes collected by males of two morphotypes of the orchid bee *Euglossa viridissima* as revealed by multi-dimensional scaling analyses. Only males with three mandibular teeth (3D) collected 2-hydroxy-6-nona-1,3-dienylbenzyldaldehyde (HNDB). Males with two mandibular teeth (2D) could thus be differentiated in multidimensional scaling of odour profiles. Redrawn from Eltz et al. 2008.

Box 6.2 Pre-existing biases can drive animal communication systems

Perfume collection is an exceptionally tight communication system between plants and Euglossini bees because the very substances that are collected from plants are used by the pollinator during courtship. As such, it is reasonable to expect that differentiation in scent production by plants translates into differentiation in courtship and then pre-mating isolation among Euglossini bees. We would encourage studies that investigate whether olfactory or visual stimuli that pollinators (and other animals responding to plant traits) use to locate food resources translate more widely into the communication systems used by these animals (see Box 4.2 for a possible example on fruit colours and sexual signalling in primates). Species whose sensory system is hardwired to respond positively to a certain stimulus (e.g., by approaching it for foraging) show pre-adaptations that could later be exploited in a different context (e.g., mating). This kind of exploitation of biases that arose in a foraging context during mating has been documented in fish (Smith et al. 2004, Garcia and Ramirez 2005). Perhaps surprisingly, there is a dearth of investigations on this topic in plant–animal interactions in general, and in pollinators specifically. This might be partly explained by the often inconspicuous colours of many pollinators such as diptera or many lepidoptera. Generally, we would not expect that biases from foraging are exploited commonly in other contexts in generalist species that pollinate many flowers and thereby respond to a variety of different stimuli. However, in more specialized tropical species sensory exploitation could be more important than hitherto acknowledged.

tendency to restrict visits to a specific species during a time interval – is significantly improved in the presence of multi-modal signals (Gegear and Laverty 2005). It would thus be interesting to compare how divergent co-flowering species are in the visual and olfactory receptor space of pollinators. Given that the receptor spaces obviously differ in their dimensions and metrics, they cannot be compared directly. However, support for adaptive divergence would be given if species that resemble each other visually are more distinct in an olfactory measure and vice versa.

A combination of olfactory and visual traits is not only important because multi-modal signals allow a better identification of food resources (see Box 11.1), they are also important because both are used to locate flowers. Analysing the flight paths of pollinators often shows that they approach flowers indirectly (making many detours) and only fly directly towards the flower from a short distance. This is possibly explained because pollinators use the less spatially precise olfactory cue for long-distance or patch localization and visual traits for short-distance identification of the target. The relative importance of visual and olfactory cues for long-distance and short-distance

localization is most likely variable across pollinators, depending on the spatial acuity of colour vision, tracking abilities of volatile compounds, and the specific physical and meteorological properties of the environment at the time of foraging. For example, there are pronounced differences in the range over which birds and insects perceive visual cues (see Section 2.1). Overall, the literature now supports much sensory flexibility in flower finding by honey bees, bumblebees, moths, and other flower visitors under differing photic conditions.

Tracking abilities also depend on the chemical structure of plants' volatiles, which is yet another difference to visual communication. In the air, longer-chained compounds are less likely to dissipate through the environment than shorter-chained compounds, but may be more reliably linked to quality aspects of the sender (Brennan and Zufall 2006). Because size is the major determinant of volatility, at least some of the diversity of floral volatiles can be adaptive from the perspective of communication because different compounds might fulfil different functions. For example, some compounds might simply function as long-range attractants that merely indicate the location of a plant. Other, longer-chained volatiles might function at close range, revealing species identity and specific information on nutritional rewards. Because animals are assumed to maintain stimulus identity even in spite of changing concentrations of volatiles, such differential qualities of volatiles might encode distinct information. Interestingly, the differential qualities of volatiles would not be apparent in behavioural tests that are conducted over short distances such as in flight arenas. It is thus a long-standing idea that variation in molecular weight can explain which compounds function in short-range and long-range attraction. Fig wasps use different volatile compounds to locate fig trees with receptive figs, and once they are on the fig, to locate the opening (ostiole) through which they enter the fig (Hossaert-McKey et al. 1994).

6.6.3   Scent and pollinator attraction

As alluded to in Section 1.3, pollinators use volatiles to locate flowers, and there is variability in floral scents within and across populations. As such, there is the raw material for pollinator-mediated selection to act upon floral bouquets. Indirect support for the adaptive concept of olfactory signalling comes from analyses of the variability of scent in *Ophrys sphegodes* and O. *arachnitiformes* complex. In these species, stabilizing selection acts apparently on alkenes, the substances that elicit an electrophysiological response in the pollinator bee, *Colletes cunicularius*, because these showed less geographic variation than other scent components (Mant et al. 2005, see Dötterl et al. 2005 for a similar example in *Silene latifolia*). There is also direct support. Augmenting floral odour extracts in the damask violet (*Hesperis matronalis*) leads to a higher visitation rate of insects (Majetic et al. 2009). Even more importantly, the authors show that, in wild populations of damask violet, floral scent emission rates during the night are positively associated with seed production (Figure 6.11); likely because more pollinators visit flowers with high scent production. This study is notable because it links variation in scent production – a measure of olfactory conspicuousness – to a proxy of plant fitness, while accounting for variation in floral colour morphs which did not contribute significantly to variation in floral visitation and, consequently, seed production.

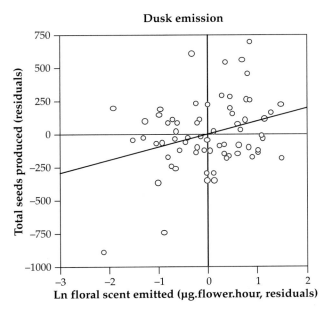

Dusk emission

Figure 6.11 There is a strong positive association between scent emission at dusk and seed set in *Hesperis matronalis* ($r2 = 0.77$, $p < 0.0001$). Redrawn from Majetic et al. 2009.

How important, then, is floral scent in producing pollinator specificity and, finally, ethological isolation? There are few studies that address all the necessary parts of the puzzle, but some convincing answers are available. For example, the campion wildflowers *Silene latifolia* and *S. dioica* differ in floral morphology, colour, and pollinator spectrum. *S. latifolia* has white petals and is pollinated by noctuid and sphingid moths; whereas *S. dioica* has red flowers and is visited by diurnal pollinators such as bumblebees, syrphids and butterflies. Unlike the pronounced qualitative differences in the odour profiles of fruits with diurnal compared to nocturnal dispersers (see Lomáscolo et al. 2008, 2010), only 4 of the 28 compounds identified were species-specific (Waelti et al. 2008). However, quantitative differences in odour emission were more pronounced and both species could be differentiated according to their odour profiles. Surprisingly, both *Silene* species showed similar circadian rhythms in odour profiles with monoterpenoids being mainly emitted during the night. In total, four compounds showed the same rhythm in both species. Applying phenylacetaldehyde to flowers of both species made their odour more similar and doubled the transfer of dye as a pollen analogue between species, suggesting that differences in the odour profile were important in maintaining ethological isolation.

Arguably the best evidence of scent driving pollinator specificity stems from sexually deceptive orchids that are thought to represent one of the most specialized pollination systems mediated by olfactory mimicry (see Plate 8; for a fuller treatment of floral mimicry, see Section 8.4). The sexually deceptive orchid *Chiloglottis trapeziformis* attracts males of its sole pollinator species, the thynnine wasp *Neozeleboria cryptoides*, by emitting a unique volatile compound, which is also produced by female wasps as a male-attracting sex pheromone (Schiestl et al. 2003). Two other co-flowering *Chiloglottis* species differ in their pollinators and floral odour bouquets with each pollinator being attracted by a single chemical compound. No cross-attraction between the two compounds and the two pollinator species is known, suggesting that ethological isolation is mediated by only two chemical substances (Whitehead and Peakall 2009). Subsequent chloroplast DNA analyses revealed substantial genetic differentiation between the two chemically distinct taxa (Ebert et al. 2009).

The communication system between the *Chiloglottis* orchids and their pollinators can be characterized as a private communication channel. Given that the dimensionality of an olfactory perception space of an animal is much higher than that of a visual space, private communication channels are more likely to arise in olfactory communication systems than they are in visual communication. The relatively high number of animals' olfactory receptors is matched by the discrete odour compounds that make up a blend. Hence, adding rare compounds to a blend can yield species-specific bouquets.

The link between scent and ethological isolation can be complex. *Ophrys* orchids primarily mimic female pollinators by scent, with visual resemblance between the flower and female pollinators being only approximate. Solvia and Widmer (2003) showed that in this genus gene flow occurs across species boundaries, suggesting that pollination by sexual deceit is not as species specific as often assumed. Similarly, although divergent selection upon floral scent imposed by pollinators explained the stronger population differentiation in behaviourally active compounds in Sardinian *Ophrys* spp., there was little genetic variation in neutral markers (Figure 6.12) – suggesting ongoing gene flow between the species (Mant et al. 2005). An alternative explanation is that orchids are under strong selection, evolving isolating mechanisms more rapidly than accumulating neutral genetic variation. This situation has been found in two closely related perennial wildflowers of *Aquilegia* spp. that are either pollinated by hummingbirds or by hawkmoths. These species differ morphologically (including floral colour) and ecologically, but not in genetic data – suggesting that rapid genetic diversification is limited to a few genome areas (Cooper et al. 2010).

Two other Sardinian species, *O. iricolor* and *O. lupercalis*, overlap somewhat in their scent profiles, although they attract different pollinators: the bees *Andrena morio* and *A. nigroaenea*, respectively. There was considerable hybridization with about 20 per cent of plant individuals being visited by the 'wrong' pollinator, and one of the parental species

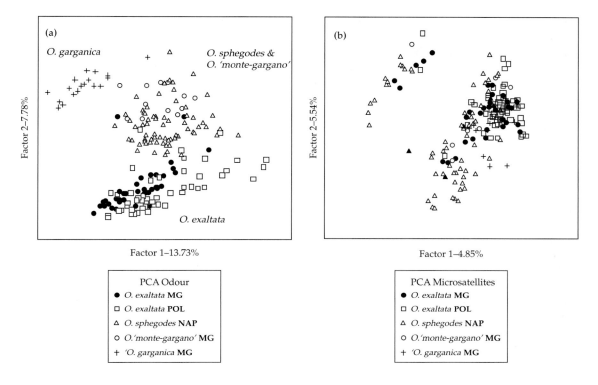

Figure 6.12 Differences in odour profiles exceed differences in micro-satellites among populations of three orchid species of the genus Ophrys. Populations were sampled at three sites, Napoli (abbreviated NAP), Monte Gargano (MG), and Pollino (POL). Sympatric populations of different species are relatively well differentiated in odour profiles (a) but less differentiated in microsatellites (b). are redrawn from Mant et al. 2005.

might become replaced by hybrids (Stökl et al. 2008). The situation is similar in three other *Ophrys* species that occur on another Mediterranean island, Majorca. All three species are pollinated by different *Andrena* species. They all use the same odour compounds for attracting their pollinators, albeit in different proportions. Consequently, the three species maintain distinct clusters in their odour profiles, but only two species are separated in a genetic space based upon AFLP markers (Stökl et al. 2009). This is an extremely interesting study as it suggests that phenotypic differences in odour profile strongly exceeds those of neutral genetic markers in one of the species pairs. In other words, there is much less introgression on loci coding for odour profiles than there is in neutral markers.

In contrast to sexually deceptive orchids, European food-deceptive orchids that do not provide any nutritional reward but also do not mimic female pollinators are often scentless (Jersáková et al. 2006). Because floral scent facilitates flower identification and memorization of unrewarding food sources, and bees discriminate 10–20 times faster among scents than they do among colours (reviewed in Wright and Schiestl 2009), it might be adaptive for these species to thwart identification by not presenting scent. Alternatively, if they produce scent, there is often a high interindividual variation in scent production among rewardless orchids; a characteristic commonly explained as adaptation towards deceiving pollinators and/or disrupting associative learning processes (Moya and Ackerman 1993; see Gigord et al. 2001 for a similar argument on floral colour in orchids). Yet, a study on the epiphytic orchid *Tolumnia variegata* showed that variance in fragrance within and among populations was unrelated to reproductive success (Ackerman et al. 1997). Taken together, the lack of scent in rewardless orchids presents some indirect evidence that one role of floral scent – apart from attraction – is to aid floral identification.

Apart from orchids and their often specialized pollination mode, there is unfortunately a dearth of studies investigating the extent that variation in scent structures pollinator guilds. There is clearly the potential for such structuring, given that the attraction of some animal vectors is known to rely on only a few chemical compounds. Yet, there is often not enough information available on the sensory abilities of pollinators on the one hand and on the chemical profile of floral scent in vegetation communities on the other hand to model the extent to which variation in floral scent explains variation in floral visitors. As such, the gap between studies on single species and studies on the ecological dynamics of vegetation communities is even more pronounced than it is in respect to floral colouration. This gap represents an unexplored frontier in the communication between plants and animals.

Scent-mediated interactions do not only occur between plants and animals, but also between fungi and animals as dispersers. *Botanophila* flies transfer gametes of grass-infecting *Epichloë* fungi with flies benefiting from the fungi because their larvae develop and feed on the fertilized fungal fruiting bodies. A close association between both partners should thus increase the fitness of fungi and flies. The flies are attracted by blends of only two compounds (Steinebrunner et al. 2008). Consistent with the conjecture that animals maintain stimulus identity in spite of changing volatile concentrations, variation in the ratio of these compounds among *Epichloë* fungi spp. had no effect on pollinator attraction. This communication system is, however, less specialized than the communication of the orchids discussed previously because specificity occurs only at the genus level. The three investigated species of *Botanophila* flies did not discriminate among the odours of distinct *Epichloë* fungi. Thus, these authors note that travel costs could explain the lack of specificity in this mutualism. Grass clumps infected by *Epichloë* are usually separated by considerable distances. The increased reproductive output caused by oviposition on specific *Epichloë* species might thus not outweigh the search costs

required for finding infections by a specific *Epichloë* species.

Is there evidence for honest signalling in olfactory communication? Obviously, there is no variation in the reliability of indices that are necessarily linked to the presence of rewards – like, for example, CO_2 emission and thermogenesis (Seymour et al. 2003). In the herbaceous perennial *Datura wrightii*, floral CO_2 emission coincides with peak nectar availability (Guerenstein et al. 2004). If this association is unavoidable, it is best seen as an index, but whether floral CO_2 emission is generally an index of nectar rewards is not well known (Raguso 2008). Experiments have shown that hawkmoths approach CO_2 upwind, suggesting that it is a redundant floral cue for attracting pollinators in addition to other floral odours and to colours (Goyr et al. 2008). However, hawkmoths prioritize visual over olfactory cues (Goyr et al. 2007). In carrion-mimicking flowers, CO_2 emission can be a deceptive signal imitating microbial presence. This interesting hypothesis needs to be evaluated in the wider context of carrion mimicry. Carrion mimicry includes the production of CO_2 and oligosulphides, as well as thermogenesis, and possibly visual mimicry of flesh and tactile mimicry of hairs by trichomes (Stensmyr et al. 2002, Angioy et al. 2004, Jürgens et al. 2006, Raguso 2008; see also Section 8.4.4 for further discussion on carrion mimicry). CO_2 emission could consequently be a signal mimicking microbial presence or be a necessary by-product of thermogenesis.

Unlike CO_2 emissions, floral scent is not a direct consequence of the presence of rewards, thus fulfilling all requirements of a true signal. Although there are convincing arguments to expect honest olfactory signals (Wright and Schiestl 2009), the honesty of floral scent is poorly investigated, representing another major gap in current knowledge of communication between flowers and pollinators. Based on logical reasoning, we expect that olfactory signals – because they change more quickly over the course of a day – can be more intricately linked to signal honesty in flowers than visual signals, which are always unreliable in recently-depleted flowers. This suggestion awaits empirical testing. The question of signal honesty in scent is intriguing because the temporal and spatial variability in odour composition and production could feasibly imply that individuals and populations differ in which compounds convey information to pollinators. While some studies have investigated temporal variation in scent production (Theis et al. 2007, Waelti et al. 2008), we know of no study that has addressed how this temporal variation relates to temporal variation in the rewards offered.

Ideally, studies should couple pollinator-driven selection upon floral bouquets with analyses of the heritability of the quantity and quality of scent production to track whether micro-evolutionary changes in floral scent are consistent with an adaptive framework of communication. The increasing use of paternity analyses based upon micro-satellites makes such studies feasible. Another promising research avenue is to manipulate the genes involved in volatile and reward production as it has been done in the tobacco plant *Nicotiana tabacum* (Kessler et al. 2006; see Section 1.7). While this research direction is particularly suitable for assessing pollinator-driven selection upon plant genotypes, a major limitation is that the molecular mechanisms involved in volatile production are still poorly known.

6.7 Pleiotropy and floral traits

While many studies are consistent with pollinator-driven selection on floral scent and floral colour, studies in the last decade have convincingly shown that other agents also influence both olfactory and visual traits. Indeed, it is a long-standing hypothesis that floral scent and colours are an exaptation of originally-defensive volatile compounds that deterred herbivores (Pellmyr and Thien 1986). This hypothesis is particularly interesting as it provides a potential route towards explaining the complexity of floral odours that is otherwise difficult to reconcile with pollinator-driven selection alone. In other words, if odour compounds fulfil different functions (some repel herbivores, others attract pollinators), and the compounds differ at the same time in their volatility, the complexity of odours can partly be explained by their multiple functions (defence vs. attraction, long-range vs. short-range signals). Obviously the interactions with multiple species and the resultant conflicting selective pressures imposed by them also explain the evolution of other flower traits such as toxic nectar that imposes costs by decreasing pollen flow to neighbouring plants (Adler and Irwin 2005).

Pleiotropic effects have recently been studied in flower colour polymorphisms. In an influential study, Schemske and Bierzychudek (2007) found that the mosaic of a colour polymorphism in the flowers of the North American wildflower *Linanthus parryae* was explicable by subtle differences in the environment rather than random genetic drift or differential selection by pollinators. In this species, plants produced either white or blue flowers; and the frequency of the two colour morphs differed strongly on a regional and local scale. Reciprocal transplant experiments showed that natural selection favours the resident morph because it produced more seeds than the other morph (Figure 6.13). Pleiotropic effects can also occur because petal colour morphs differ in their induced responses to herbivory. In the radish (*Raphanus sativus*), yellow and white floral morphs were preferred both by bee pollinators and by a suite of different herbivores. These colour morphs had lower inducible defences than the less-preferred anthocyanin-containing pink and bronze colour morphs (Strauss et al. 2004). Balancing selection by the combined effects of pollinators and herbivores may thus contribute to the maintenance of the colour polymorphism in this species, as it apparently does in the fruit colour polymorphism of *Acacia ligulata* (Whitney and Stanton 2004).

Although pleiotropy and correlated selection are topics that are increasingly studied in plant–animal interactions (for examples, see Chapter 1), their relative importance in shaping floral traits are is still not well known. In Australian Hakeas bird-pollinated species (but not insect-pollinated ones) have red colours and are better defended by cyanide suggesting that pre-adaptations to florivores can drive the evolution of floral syndromes (Hanley et al. 2009). Importantly, pleiotropy and correlated selection can occur directly because different animals interact with a given floral traits are indirectly because of systemic responses of plants. There are good examples of such indirect interactions where activity by herbivores on leaves alters the behaviour of pollinators (see Section 1.7 for thorough discussion). Few of the available studies on this topic have identified the proximate mechanisms of such interactions. A study on the wild

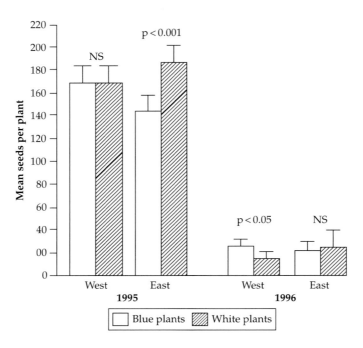

Figure 6.13 Reciprocal transplant experiments showed that the resident morph (blue in the west, white in the east) in *Linanthus parryae* flowers produces more seeds than the other morph. Thus, although allozyme markers revealed no spatial differentiation, the resident morph is favoured and flower colour pleiotropically linked to this selective advantage. Redrawn from Schemske and Bierzychudek 2007.

tomato (*Solanum peruvianum*) is remarkable because it shows that bumblebees rarely land on flowers from damaged plants (and if they do so, they stay less than two seconds), suggesting that avoidance is mediated by volatiles (because no change in colouration was detected; Kessler and Halitschke 2009). This study further showed that leaves and flowers of the wild tomato are largely congruent in the chemical composition of their bouquets, and that some of the compounds that were induced by herbivory were induced in both flowers and leaves. At the same time, flowers and leaves also produced distinct compounds in response to herbivory, again suggesting the presence of a wound signal that is transmitted throughout the plant. This study is particularly interesting, as it provides a mechanistic basis – based on the well-known communication system of induced plant volatiles (see Chapter 9) – as to why herbivory and pollination are interconnected.

The study by Kessler and Halitschke (2009) suggests that reduced pollination can contribute to the ecological costs of induced plant defences. It suggests that it is non-adaptive for flowers to change their emission profile in response to herbivory. However, since pollinators obviously also perceive the volatiles emitted from vegetative tissue, they could use leaf volatiles instead of floral scent to assess the damage incurred by herbivory if that damage lowers the quantity or quality of plant rewards. As such, the change in floral emission might be relatively unimportant compared to the overall volatile emission by the plant. It is also conceivable that there are ecological effects such that a higher abundance of herbivores can increase the predation risk of pollinators but this idea remains speculative so far. Overall, there is currently not enough evidence to disentangle the relative importance of induced floral vs. leaf volatile emission for the foraging behaviour of pollinators on one hand and the reproductive success of plants on the other hand.

Further work of Kessler revealed that herbivory can affect pollination even more intricately. The tobacco plant *Nicotiana attenuata* flowers usually at night and is pollinated by *Manduca sexta* moths. The moth is both a pollinator and a herbivore of the tobacco plant. When attacked by moth larvae, the flowering phenology changes, and plants produce flowers that open in the morning and are mainly pollinated by hummingbirds (Figure 6.14; Kessler et al. 2010). Given that foraging can often be correlated to oviposition, the authors propose that the change in the pollination system to hummingbirds

is adaptive to minimize herbivore-related fitness costs of the plant.

The question of why the odour profile of flowers changes after damage occurred on vegetative parts leads to the general debate over the extent to which the presence of plant secondary compounds is adaptive in reproductive organs or simply an unavoidable result of their presence in other plant tissue (e.g., Ehrlen and Eriksson 1993). This topic

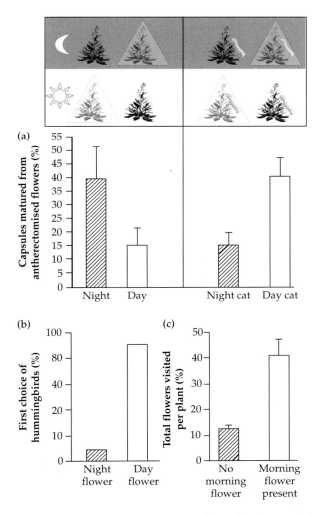

Figure 6.14 Herbivory by the larvae of the main pollinator *Manduca sexta* changed the pollination system in the tobacco plant *Nicotiana attenuata*. When no herbivores where present *Manduca sexta* is the most important pollinator because plants set more seeds that were accessible only at night (to *Manduca sexta*) compared to those that were accessible to pollinators (hummingbirds) only during the day (a; left). This pattern was reversed when herbivorous larvae were present on the plants (a, right; cat treatment). The shift to hummingbird pollination is explicable because plants change to opening flowers in the morning when infested by herbivorous larvae, and hummingbirds prefer to visit flowers that open in the morning (b) and plants that have flowers in the morning (c). Redrawn from Kessler et al. 2010.

is controversial, likely also because there is no single explanation that accounts for all secondary compounds. Thus explanations are probably specific to taxon and/or compounds and not easily generalized.

There is some evidence that the presence of defensive volatiles can be adaptive, reducing the visitation rate of flower predators. Ants, which often steal nectar rewards without pollinating the flowers, can not only select for floral morphology (see Section 6.4.3) but also for defensive odour components. Experiments in the lab revealed that nectar-consuming ants were significantly repelled by a high proportion of floral scent from different species. In particular, the terpenoids geraniol, linalool, and α-pinene all repelled ants, suggesting that this repellent function might explain their widespread occurrence in floral bouquets (Junker and Blüthgen 2008).

Compared to visual traits, there is considerably less evidence on how abiotic factors influence odour bouquets. While geographic variation in fragrance composition has been noted in relatively many species (e.g., Ackerman et al. 1997, Dötterl et al. 2005, Chess et al. 2008), the causes of this variation are far from understood. There are several possible explanations for this phenomenon. First, geographic variation could simply reflect random variation caused by drift. Second, it could also be selectively neutral if certain compounds are more easily produced in one environment than another: for example, owing to a precursor that is obtained from the environment. Finally, such geographic variation could represent adaptations to the prevailing abiotic factors or to geographic variability in the composition of the biotic agents such as pollinators interacting with the plant. One possible research avenue would be to analyse floral bouquets of different species along well-pronounced environmental gradients, such as altitudinal gradients or gradients of increasing humidity or sun exposure. Ideally, such analyses would be run concomitantly on both flower bouquets and leaf bouquets, to analyse the extent of correlation among them. If the correlation among the odour profiles of leaves and flowers is not strong, this would already provide compelling evidence for the independence of odour composition in flowers.

6.8 Perspective

There is considerable evidence that plant traits, in particular morphology, colour, and odour, influence pollinator behaviour. There is further evidence that plant diversification is consistent with expectations based upon pollinator-driven selection. Yet, evidence for pollinator-driven selection is often elusive, and it remains to be documented that it overrides the selection of other agents (e.g., herbivores, abiotic factors). We suggest that studying how selection by other agents channels the evolution of floral communication is a particularly important research topic that is likely to change our concept of communication. Another important challenge is to define the information conveyed by visual and olfactory traits. It is undisputed that flowers and fruits convey messages like 'come over here', 'keep out', 'remember me', and 'you can leave now' (Schaefer et al. 2004, Raguso 2009, Wright and Schiestl 2009). However, it is currently unknown whether information is more specific than attracting animal vectors to a given site. If plants transmit more specific information, the question of signal honesty arises. We proposed that signal honesty in flowers might be conveyed by olfactory signals rather than visual ones because odour composition changes more quickly. Other neglected topics are whether the diversity of floral colours and odours is adaptive on an ecological scale, and whether the spatial distribution of floral resources affects the form and reliability of communication. While adaptive divergence has been supported by some studies, few studies were able to disentangle the relative importance of different selective pressures. In summary, the multiple functions of many plant traits including odours and colours, the multiple species that most plants interact with, and the local background (co-flowering plants, spatial distribution of food rewards) of these interactions all call for considering more explicitly the ecological factors that influence the evolution of communication. Communication theory has been predominantly influenced by evolutionary biologists that laid the foundation of how communication evolves and how it is maintained. These theoretical foundations are extremely valuable, but they do not yet encompass the complexity found in the communication between plants and animals.

(a)

(b)

Plate 1 The eyes of vertebrates perceive a strong colour contrast between the yellow flowers of a bush and its background from a distance (a). The compound eye of insects has lower spatial acuity of colour vision and can only perceive the colour differences between yellow flowers and their background from a short distance (b). Photos © H. Martin Schaefer

Plate 2 The conspicuousness of reproductive organs can be achieved through strong colour contrasts (see Plate 4) but also through phenological adaptations. Blooming before leaves have developed makes flower displays conspicuous. Photo © H. Martin Schaefer

Plate 3 Fleshy fruits within plant communities can be very diverse as seen in this image of fruits from Barro Colorado Island, Panama. Adaptive diversification has been proposed in flowers, but is relatively unknown in fleshy fruits. Concepts such as market theory can be tested against null models to analyse whether adaptive divergence occurs. © Christian Ziegler

Plate 4 Red accessory structures are common in fruit displays. They increase the contrast of the displays for avian vision. Photo © Eliana Cazetta

Plate 5 The fruit display of *Xylopia langsdorffiana* shows differential investment in advertising relative to providing fruit rewards. The red inside of the fruit capsule is only an advertisement, as it does not provide nutritional rewards. The nutritional rewards are found in the white fruit pulp, while seeds are black and contrasting to the white pulp and to the red fruit aril. Photo © Eliana Cazetta

Plate 6 Acorn woodpecker at a granary. This species lives in groups; birds in the groups store the acorns they collect for later use in so-called granaries. Photo © Martha Marks / iStockphoto

Plate 7 The flower of the nectarless orchid *Chloraea bletioides* has morphological flower guides that direct pollinators towards the centre of the flower. By directing the movement of pollinators, chances of successful pollination are increased. Photo © Rodrigo Medel

Plate 8 Orchids of the genus Ophrys are sexually deceptive in that they mimic the scent (and less accurately) the form of female Hymenoptera and thereby attract males attempting to copulate with the flower and thereby pollinating it. Photo © Stéphane Bidouze / iStockphoto

Plate 9 Red leaf colouration in autumn can communicate either as signal or as cue relating to plant defences. Alternatively, it may be a phenomenon that is mainly associated with plant protection against adverse abiotic conditions. Photo © H. Martin Schaefer

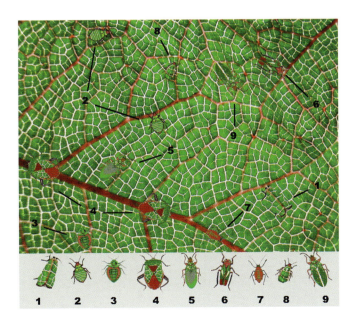

Plate 10 Leaf variegation has been suggested to either function in camouflage because it makes leaves more difficult to identify by shape, or to function in reducing camouflage of herbivores because these are ill-matched on at least one of the colours. However, different colours in a leaf may complicate the visual search of predators of herbivores and different colours may also provide more opportunities for the camouflage of differently-coloured insects – as illustrated here with a range of artificial insects. © Gregor Rolshausen and H. Martin Schaefer

Plate 11 Seeds although usually dull can be intensively coloured. The red colouration of the Acer spp. seeds can be a pleiotropic effect associated to the red colouration in vegetative tissue (which is visible in the stems of the leaves and developing in the leaves) or it could be associated with senescent tissue in the seeds. There is currently no experimental evidence that it serves as warning signal to repel predators from the seeds. Photo © H. Martin Schaefer

Plate 12 Crab spiders often ambush pollinating insects on flowers. Interestingly, some crab spiders contrast strongly to the flower background (as seen by bees), and some bees preferentially approach the contrasting floral display where such crab spiders are present. Crab spiders thus exploit innate sensory biases of pollinating insects to approach contrasting floral displays. Photo © Ludmila Yilmaz / iStockphoto

Plate 13 The colours of thorns have been proposed to form a warning signal to deter herbivores and to avoid that plants are trampled by an animal. The intensity of light (brightness) reflected from thorns can differ strongly from that of the remaining vegetative and wooden parts of the plant as seen in this picture. Brightness contrasts are easily perceived by large mammals which are dichromats. Photo © H. Martin Schaefer

Plate 14 Fungi mimicking flowers: the fruiting bodies of some fungi form what are called 'pseudo-flowers', and are thought to gain gene transport by luring pollinating insects. Bee on a pseudo-flower caused by the rust fungus *Puccinia monoica* infecting *Boechera holboellii* (synonym=Arabis). Photo © Barbara A Roy

Plate 15 It has been suggested that the dark coloured spots on *Xanthium stomarum* may mimic ants and thereby repel herbivores. Photo © Simcha Lev-Yadun

Plate 17 Transparent windows in the pitchers of carnivorous plants may be a form of sensory deception by the plant because they may confuse prey that orient towards these windows rather than to the aperture of the pitcher. This conjecture awaits experimental testing. Photo © Tamara Kulikova / iStockphoto

Plate 16 Carnivorous plants often provide nutritional rewards to insects they prey upon. Nectaries are often located on the entrance of the pitcher. Here a bumblebee feeds on the nectar provided by the carnivorous plant. Photo © red_moon_rise / iStockphoto

CHAPTER 7

The potential for leaf colouration to communicate to animals

7.1 Introduction

Leaves are the main photosynthetic organs of most plants. They are typically optimized to catch sunlight, being flat with a large surface exposed to the sun. Photosynthesis uses light energy to convert CO_2 into sugars and therefore requires light-absorbing green pigments, chlorophyll (see Section 2.6.1). Owing to their primary role as photosynthetic organs, leaves do not have such an obvious signalling function as fruits or flowers. While most leaves are green in order to photosynthesize, there is pronounced interspecific and even intraspecific variation in leaf colouration and some of this variation is used in remote sensing to characterize species composition (e.g., Gamon & Surfus 1999, Ribeiro da Luz 2006). Examples of extreme interspecific variation are the leaves of some plants that can be red, brown, or purple throughout their entire lifespan (some Japanese maples provide good examples of this). Others are green on the surface exposed to sunlight but can be whitish, yellow, or red underneath. Such non-green colouration appears to be more common in tropical plants and in understory plants, although patterns can be quite variable (Dominy et al. 2002). Some species feature leaves that are variegated: with the background green colouration being overlaid with red, yellow, or white patterning. Even among completely green leaves, there can be considerable variation between species in colouration, from silvery greens, through pale and lime greens to the darkest greens you can imagine.

While most of the colour variation outlined above concerns interspecific differences, a leaf can vary in colouration over its lifetime. It is quite common for young leaves to be predominantly red, only turning green as they fully open. Further, it is common for leaves of deciduous plants to change colour (sometimes to spectacular yellows and reds) prior to abscission. Those of us living in temperate areas are very familiar with the dramatic change to the appearance of the natural world driven by such autumn leaf colour changes; although this phenomenon is also known from tropical areas where it is less synchronized. Even some evergreens have been reported to show seasonal variation in colouration (Hughes et al. 2005).

Thus, there is considerable variation in leaf colouration both intra- and interspecifically, and it is the aim of this chapter to explore whether any of this variation is exploited by animals, and whether any such exploitation acts as a selective force on the plants. This has been a particularly active field of enquiry recently; in particular, the communicative functions of autumn leaf colouration have been intensively debated. However, in order to evaluate such communication-based hypotheses for the evolution of leaf colouration, we must also consider alternative hypotheses. In fact, there are several hypotheses offering potential explanations for autumn leaf colouration (summarized in Table 7.1). While posing alternative hypotheses is part of the general scientific endeavour, it is particularly important in the context of leaf colouration because most leaves primarily fulfil a physiological function (photosynthesis) throughout their lives. Moreover, since alternatives to the signalling hypothesis are generally simpler to explain, we

Table 7.1 Summary of the different major hypotheses for leaves changing colour in autumn

Purely physiological hypotheses	
Chlorophyll breakdown	As the leaves are about to be shed, their green chlorophyll breaks down to colourless compounds that unmask colourful molecules that existed in the leaves all along. This hypothesis applies only to carotenoids and not to anthocyanins.
Increased production of anthocyanins	Anthocyanin levels in leaves may rise in autumn to extend the effectiveness of photosynthesis in low temperatures and perhaps to act as an anti-oxidant.
Hypotheses involving animals cuing on leaf colouration	
Nutrient translocation hypothesis	Leaves with yellow or red colouration (due to carotenoids or anthocyanins respectively) are well protected from photo-inhibition (see above) and so can extract nutrients very effectively from leaves in autumn, to the benefit of the plant (and any herbivores on it) in spring.
Defence indication hypothesis	Red colouration may be a reliable cue of plant defences for herbivores, since defensive compounds of the phenylpropanoid pathway and anthocyanins share a common biosynthesis pathway.
Hypotheses involving co-evolution of animal responses and leaf appearance (i.e. signalling)	
Reproductive insurance hypothesis	Colour change is a signal that warns insect herbivores that nutrients will shortly be diverted to reproductive tissues and so will be diverted away from leaves, thus leaf-attacking insects would do better to target non-reproductive trees.
Signalling to frugivores	A means of long-distance signalling to frugivores to enhance discovery of fruit and thus transport of seeds.
Leaf signal theory (also called co-evolutionary theory)	Bright leaf colours in autumn are a signal revealing the level of defensive commitment of the individual plant, and comparison of the signals between individuals of a species allows insects that migrate onto plants in autumn to preferentially select less-defended plants. The co-evolution of colour preference in insects and bright colours in plants would allow well-defended plants to reduce their parasitic insect load, as well as allowing the insects to preferentially settle on the most profitable (least defended) plants.
Ant attraction	Leaf colouration signals plant quality to those aphids that are tended by ants. After aphids colonize a plant, ants are attracted to that plant and defend it from other herbivores.
Indirect defence	Terpenes are used by plants to attract the enemies of herbivorous insects to the plant. They are also used as direct defence against herbivores. Green colouration in autumn is required to prolong photosynthesis in order to fuel the production of terpenes. Thus green plants late in autumn are visually or olfactorally aposematic and flag that they are effectively indirectly defended in comparison to coloured plants. This conjecture needs to be tested against the direct defence provided by terpenes.

will consider these first. We will devote much of the chapter to autumn colouration, not because this is necessarily the most important phenomenon, but because it has attracted significant attention recently and is a good example of the dynamics of leaf colour changes.

7.2   Characteristics of autumn leaf colouration

In the broadest sense, changes in leaf phenotype during autumn involve the sometimes-spectacular colour change that occurs in the leaves of many deciduous species prior to leaf drop (see Plate 9).

However, this general statement hides a great deal of variation. Firstly, although autumn colour change is much more obvious in temperate regions, colourful senescent leaves are still shed by many tropical species. Changes in abiotic factors, particularly variation in temperature, are less pronounced in the tropics, resulting in a less synchronized leaf shedding compared to temperate regions. Even in temperate regions, not all deciduous plants change leaf colour prior to abscission, and at least some evergreen plants show seasonal change in colouration (Hughes et al. 2005). Leaf colouration in general (Castro-Esau et al. 2006) and autumn colour changes in particular can be very variable between and even within species, with green leaves becoming yellow, brown or red when senescent (e.g., Schaberg et al. 2003, Rolshausen & Schaefer 2007). Interspecific variation in leaf colouration results from differences in pigmentation; in the physical structure of the leaf such as waxy epicuticular layers and light scattering on air-cell interfaces; from adaptations to sun and shade exposure; and water content (Vogelmann 1993, Logan et al. 1996, Richardson et al. 2003). Interspecific differences in autumn leaf colour changes are usually attributed to differences in plant physiology, particularly the extent of nutrient resorption (Niinemets & Tamm 2005). Intraspecific differences in leaf colouration between populations can be linked to differences in latitude, altitude, prevailing weather conditions, soil, light conditions, and other aspects of micro-habitat (Richardson et al. 2003). At least some of this variation is explainable by variation in abiotic stressors. For example, autumn leaf colour changes typically occur first at higher altitudes and slowly spread towards lower altitudes. Interestingly, however, there is also intraspecific variation in leaf colouration and the timing and onset of ontogenetic leaf colour changes even within the same locality. Some of this variation might still be explicable by differences in micro-habitat use (e.g., soil quality and water availability) or mechanical damage, but some of this variation is apparently linked to biotic interactions such as herbivore damage. It is precisely this intraspecific variation at a given site that has attracted theorists to postulate a signalling function. Given this complexity of colour variation, we should not expect to find one mechanism that 'explains' leaf colour or autumn colour change. Rather, we would expect there to be several

contributor mechanisms, the relative importance of which will vary from case to case. To disentangle these processes, we consider some recently cited mechanisms in order of increasing complexity. We do not claim to be exhaustive, but rather focus on those mechanisms that we feel are most important (see Archetti 2009a for other mechanisms).

7.3 Mechanisms not involving communication with animals that might underlie autumn leaf colouration

7.3.1 Loss of chlorophyll and photosynthetic function

Until recent years, if textbooks mentioned autumn colour change at all, then this would be the explanation given. Green leaves are green because they are filled with the chlorophyll required to drive photosynthesis. Leaf senescence is a tightly regulated process that starts in the leaves before any change in leaf colouration is visible (Garcia-Plazaola et al. 2003). It is characterized by two processes, pigment degradation and protein remobilization, occurring from late summer on. In general, leaf senescence is a highly plastic trait that is influenced by environmental conditions such as irradiance, temperature nutrient supply, CO_2 concentrations, as well as other abiotic and biotic stressors (Wingler et al. 2006). Autumn is characterized by changes in some of these environmental factors, most notably dropping temperatures and still-high but declining light levels, both of which may inhibit photosynthesis, particularly in combination with each other. Eventually there comes a point where the weather is sufficiently poor that the chlorophyll can no longer perform its photosynthetic function in deciduous plants. Hence, the costs of maintaining leaves exceed the benefits of keeping them. At this point chlorophyll breakdown is accelerated and chlorophyll is split into simpler compounds to enable the plant to recover proteins from the plastids of the (soon to be shed) leaves. Proteins and important nutrients such as phosphorus are then stored in the permanent part of the plant for re-use the following spring. An important aspect to this is that the breakdown products of chlorophyll are colourless. Thus as the green colour of chlorophyll disappears, pigments are unmasked that were always in the leaves but whose colour was hitherto hidden by the dominant effect of chlorophyll. A good example for this unmasking effect of the breakdown of chlorophyll is the resulting appearance of carotenoids that impart yellow and orange leaf colour in autumn (Lee et al. 2003, Ougham et al. 2005). This unmasking effect is visible in many studies. For example, the percentages of green and yellow leaves are directly inversely correlated in sugar maple (*Acer saccharum*) (Schaberg et al. 2003). Degradation of chlorophyll generally results in an increase in leaf reflectance in the spectral range from 550 to 740 nm, while the reflectance in the range between 400 and 550 nm remains unaltered due to the presence of carotenoids (Merzlyak et al. 1999). Carotenoids have been present in the leaves throughout their life, and only the breakdown of chlorophyll allows the yellow and orange colours to become more apparent. It would be simplifying, however, to view carotenoid levels in leaves as generally being fixed and unaltered. In evergreens, carotenoids are synthesized when the risk of *photoinhibition* increases (Verhoeven et al. 1999) and there are seasonal changes related to cold acclimatization (Han et al. 2004). The unmasking of chlorophyll does not account for the bright red colours that some autumn leaves show, because anthocyanins, the red pigments, are not present in most mature leaves (Lee 2002, Gould & Lee 2002, Gould et al. 2009). Hence, anthocyanins are actively produced in leaves prior to their abscission.

An important question to ask is whether the observed intraspecific variability in leaf colouration is consistent with a purely physiological view of autumn colour changes. Regrettably, there are not a lot of data on this topic. However, recent studies on Arabidopsis show that naturally occurring variation in the regulation of plant senescence has a genetic basis, which strongly influences the onset and the development of senescence (Balazadeh et al. 2008).

Although very simple, this mechanism has a number of satisfying properties. It predicts that senescent leaf colour change would be primarily a feature of nutrient resorption because both deciduous and evergreen plants recover half of the nitrogen and phosphorus pool from their leaves prior to dropping them (Aerts 1996) and why it occurs in autumn in temperate regions (when seasonal weather changes reduce the photosynthetic productivity of leaves).

7.3.2 Physiological functioning of anthocyanins

At first sight, it would seem surprising that plants invest in the production of compounds for leaves that are soon to be shed. However, anthocyanins seem to be effective in two roles associated with photoprotection. Firstly, they act as a sunscreen: absorbing photons. In autumn the number of photons can exceed the requirement for photosynthesis (e.g., Gould et al. 2010). Secondly, they act directly as anti-oxidants, scavenging free radicals (Lee and Gould 2002, Hatier and Gould 2009; but see Manetas 2006). These functions may be particularly valuable in temperate autumn conditions that are characterized by a combination of low temperatures and high but variable light levels. In such circumstances, risk of oxidative damage is high because the photosynthetic apparatus does not function efficiently at low temperatures. Therefore, photons entering the plastid are not absorbed in photosynthesis, but may result in the formation of reactive oxygen species, particularly singlet-excited oxygen ($^1O_2$) and superoxide (O_2^-). These reactive oxygens are potentially harmful as they might oxidize molecules in the chloroplasts. Light-absorbing pigments, such as anthocyanins, intercept some of the photons that would otherwise reach the chloroplasts and thereby reduce the risk of photo-oxidative damage indirectly (Feild et al. 2001, Hoch et al. 2003). For example, exposure to strong white light reduced photosynthetic efficiency by 60 per cent in leaves containing anthocyanins compared to almost 100 per cent in leaves void of anthocyanins (2005). Such protective effects are not limited to deciduous leaves, as the study of Hughes et al. (2005) on the evergreen understory herb *Galax ureolata* documents. When exposed to cold temperatures and high light, *Galax* leaves produce anthocyanins causing the leaves to turn red. These anthocyanins dissipate in spring, and leaves can subsequently persist for up to three additional growing seasons. The authors suggest that the anthocyanins function as anti-oxidants and light attenuators.

Consistent with a photoprotective function of carotenoids and anthocyanins, leaf senescence occurs more rapidly in sun-lit leaves compared to shaded leaves (Garcia-Plazaola et al. 2003). For example,

Furuta (1990) observed that in the Japanese maple *Acer palmatum* leaves in shade turn yellow in autumn, whereas those in direct sunlight turned red (through the production of anthocyanins). This would suggest that the benefits of the investment in anthocyanins are linked to illumination, thereby corroborating the sunscreen hypothesis. Likewise, although anthocyanins shade chloroplasts, anthocyanin-synthesizing individuals maintained similar or higher photosynthetic efficiencies in stressful environments of high light exposure coupled with low temperatures than phenotypes that were unable to produce anthocyanins (Pietrini et al. 2002, Hoch et al. 2003, but see Manetas 2006; Nikiforou et al. 2010). Importantly, the stressful environment resulted in irreversible photoinhibition in anthocyanin-deficient mutants of three species (Hoch et al. 2003) and the resorption efficiencies were lower in phenotypes that did not express anthocyanins (Hoch et al. 2003). Further, there may be a need to prolong the season of effective photosynthesis in deciduous plants so that late-season photosynthesis can be used to provide the energy to transport nutrients out of the leaves into storage in the perennial parts of the plant. If anthocyanins can stop premature, irreversible photo-inhibition and allow more compounds to be salvaged from the leaves, then this function alone can make their synthesis a worthwhile investment for the plant.

The second role of anthocyanins in reducing oxidative stress is that they can quench free radicals directly. Anthocyanins are among the most potent anti-oxidants *in vitro*. Their role as anti-oxidants *in vivo* is not as well studied as their photoprotective role. Physical wounding in *Pseudowinteria colorata* resulted in the formation of the reactive H_2O_2 molecule (Gould et al. 2002). H_2O_2 induced chalcone synthase expression, a key enzyme in anthocyanin biosynthesis; and anthocyanins accumulated in response to wounding in *P. colorata* (Richard et al. 2000). Importantly, areas containing anthocyanins recovered more quickly from H_2O_2 bursts than green parts of the leaf (Gould et al. 2002). Similarly, leaves containing anthocyanins in the mesophyll cells displayed higher resistance to damage caused by reactive oxygen species than green leaves (Kytridis & Manetas 2006). However, the anti-oxidant scavenging function is contentious (van den Berg & Perkins 2007), and more research is needed to assess interspecific variability in the function and mechanisms of quenching free radicals as well as intraspecific variability caused by differences in the location and concentration of anthocyanin pigmentation.

7.3.3 Other colour-related physiological functions

Maximal anthocyanin pigmentation often occurs when night temperatures are just above freezing coupled with mild daytime temperatures (Steyn et al. 2002). It has been suggested that temperature-induced anthocyanin synthesis may act as an anti-freeze, and that these molecules are produced simply to lower the freezing point of tissues in the leaves, to avoid structural damage through freezing in autumn when there may still be strong enough daylight to make photosynthesis viable and/or before the plant has completed removing valuable compounds from the leaves (Chalker-Scott 1999). This mechanism is plausible for some high-latitude and/or high-altitude species (e.g., sugar maple: Schaberg et al. 2003), but a direct role of anthocyanins in acquiring cold hardiness has been questioned (Leyva et al. 1995). Further, extensive change on leaf colouration occurs in many temperate environments before sub-zero temperatures occur. Indeed in some areas leaves will change colour and subsequently be shed before freezing temperatures are experienced. It has also been suggested that colouration may affect radiative heat transfer in a way that raises leaf temperatures and thus enhances photosynthetic energy production; however, empirical evidence in support of this is lacking (Hoch et al. 2001, Wilkinson et al. 2002). Lastly, it has been suggested that the production of anthocyanins in the leaves prior to abscission benefits the plants after the leaves are shed, if these leaves condition the soil below the plant in a way that makes it harder for competitor plants to flourish (Ougham et al. 2005); but again we know of no empirical support for this idea.

7.3.4 Summary of non-communication functions

We can see that the selective pressure on anthocyanin production will be dependent on both the costs of their production, which are largely unknown (but see review of Gould (2004), and the benefits that can accrue from a reduced risk of photo-inhibition and increased nutrient translocation to perennial parts of the plant. Both these costs and benefits will be affected by climatic, geographical, physical, and other environmental factors as well as by plant physiology. Thus, the two major explanations for autumn leaf colour that are unrelated to communication may be very important mechanisms that between them can explain much of the variation in the transient appearance of red leaf colouration, particularly the autumn leaf colouration discussed in Section 7.2. This is not to say that communication-based theories are unimportant, but it is worth emphasizing that the recent growth in interest in communication-based explanations of autumn leaf colour has not been driven by obviously unsatisfactory aspects to non-communication-based explanations. With that important caveat, let us turn to these communication mechanisms, again starting with the simplest.

7.4 Autumn leaf colouration

The key question for us to consider is whether any given variation between plants in leaf colouration can be used by herbivores or frugivores to select the most suitable plant for them. This seems plausible for autumn leaf colouration, because at least some of the insect enemies of plants select their host in the autumn. For example, aphids are a geographically widespread pest group, and some species have a mobile stage in autumn that colonizes new hosts (see references in Hamilton and Brown 2001). It has also been demonstrated that such aphids can show non-random colour choice on which artificial targets they land on, landing on red targets less often than on green or yellow targets (Döring et al. 2009). As yet it has still to be demonstrated that such non-random colour choice is important when aphids choose to settle on plants, rather than correlated cues such as plant volatiles (Holopainen 2008, Holopainen et al. 2010) or the chemical composition of phloem governing host choices. Although some correlative studies have demonstrated a relation between plant colour and selection by autumn-flying insects (e.g., Hagen et al. 2003, Hagen et al. 2004, Archetti & Leather 2005; Archtti 2009b), these studies do not demonstrate that selection is made on the basis of colour. The one experimental study that avoided

potential confounders (Schaefer and Rolshausen 2007) found no difference in aphid numbers attracted to leaves of mountain ash (*Sorbus aucuparia*) that were artificially coloured red or green. There was also no difference in attraction towards naturally green over artificially green (or artificially red over naturally red) colours. Aphids did show preference, however, preferentially attacking trees with fruit (see Figure 7.1), possibly because fruit-ripening trees dispose of more nutrients that are transported to the fruits (see the nutrient translocation hypothesis, Section 7.4.1 below). Reflectance in this study was manipulated so that it corresponded to naturally occurring variation in leaf colours, which is important since the colour vision of aphids and other ecologically relevant autumn-flying herbivores differs from that of humans (see Chittka & Döring 2007, Sinkkonen 2009).

Even if a preference based on colour is demonstrated, this does not indicate that the colour itself is a signal. That is, demonstration of a colour preference by herbivores does not imply that the herbivores must exert the selection pressure that led to non-green leaf colouration in autumn. Alternatively, it may be that the colouration is a consequence of the types of physiological changes described in Section 7.3, and this colouration is a cue providing useful information to autumn-flying herbivores,

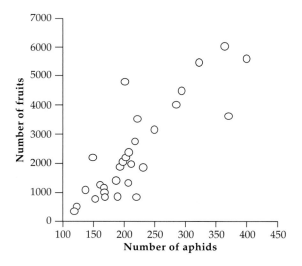

Figure 7.1 For mountain ash trees with a higher number of fruits on a tree had more aphids. Redrawn from Schaefer and Rolshausen (2007).

but the change in behaviour of the herbivores caused by this information does not itself induce directional selection on leaf colouration. We thus examine the evidence that leaf colour is an informative cue rather than a signal that evolved to transmit information.

How can non-directional selection by herbivores be explained if they exert colour preferences? As noted throughout this book, it is essential to account for all selective pressures acting on a phenotype. As such, selective pressures exerted by herbivores might be overridden by selective pressures owing to plant physiology. If selective pressures from herbivores are weak relative to other selective pressures, then colouration would more appropriately be described as a cue rather than a signal (see Section 7.4.2). It is important to note in this regard that although there are convincing examples of how differential selection by autumn-flying herbivores affects proxies of plant fitness (Hagen et al. 2003), the effects on plant fitness have yet to be fully substantiated. Such a demonstration will, admittedly, be a difficult task; particularly when studying long-lived trees that may produce offspring each year or may be dependent on masting events.

Another possibility is that colouration does not provide any useful information to herbivores on the relative suitability of different plants, but that herbivores still select based on colour because they detect some colours more easily than others. If colouration affects the ease with which plants are detected by herbivores, leaf colouration will be subject to herbivore-driven selection pressure, but the strength of this pressure relative to others can vary. To complicate things further, there is no reason why colouration could not both affect detection of the plant and provide inadvertent information about intrinsic plant quality. Thus demonstration of fitness consequences of differential section by autumn-flying insects does not of itself demonstrate that colouration acts as a signal of intrinsic plant quality, unless differential detectability can be accounted for.

With these cautions in mind, let us consider some of the potential explanations for autumn leaf colouration that involve sensory systems of frugivores or herbivorous insects.

7.4.1 A cue to nutrient movement out of the leaves

Holopainen and Peltonen (2002) proposed the *nutrient translocation hypothesis*. This suggests that leaves with yellow or red colouration (due to carotenoids or anthocyanins respectively) are well protected from photo-inhibition (see Section 7.3.2) and so can extract nutrients very effectively from leaves in autumn, to the benefit of the plant (and any herbivores on it) in spring. Thus, autumn-flying insects might be expected to preferentially target such plants for two reasons. On the proximate basis, colourful leaves with less defensive volatiles are a cue for high nutritional food sources because nutrients are remobilized and the phloem is nutrient-rich (see White 2009, Holopainen et al. 2010). On the ultimate basis, the offspring of herbivores might do better on such a plant the following spring, when higher levels of stored nutrients are again available. Thus, again, insect herbivores should actually preferentially select plants with yellow or red leaves. While many insects are known to prefer yellow colours (many human-build insect traps are yellow), there is very little suggestion that this occurs for red leaves (see Chittka and Döring 2007 for a review). Indeed Döring et al. (2009) speculate that the function of red pigments in autumn leaves may be to mask the yellow pigments that would otherwise provide an attractive cue to herbivorous insects.

Archetti (2007) presented a mathematical model of this mechanism and argued that an evolved signal can be evolutionarily stable, although under quite restrictive circumstances. However, there is currently no strong empirical evidence in support of it. While Hagen and co-workers (2003) show that herbivore damage in spring was inversely correlated with the intensity of leaf colouration of the host plant in the previous autumn, other authors found no such evidence (see Ramirez et al. 2008). Further, this theory assumes that autumn leaf colouration is not positively correlated with investment in defensive compounds, and Schaefer and Rolshausen (2006) argue that such a correlation is likely for red colouration (see Section 7.4.2). As an explanation of autumn leaf colour change, this mechanism has the strength of being naturally associated with senescence, but the weakness that it does not naturally predict within-plant variation in the colour of different leaves.

The idea that autumn leaf colouration may be a useful cue as to the attractiveness of plants the following spring is plausible. Furuta (1990) observed that in the Japanese maple *Acer palmatum* leaves in shade turn yellow in autumn, whereas those in direct sunlight turn red (through the production of anthocyanins). Plants whose leaves were predominantly in the shade one year tended to produce leaves earlier the next year. This author reports that autumn-flying aphids landed preferentially on yellow (shaded) leaves and not on red ones. Thus they may be using yellow colouration as a means of identifying the trees on which they can minimize their overwinter period of dormancy. Alternatively, their preference is explicable because yellow leaves are simply more conspicuous to aphids than red leaves (Chittka and Döring 2007, Döring et al. 2009 Chittka and Döring 2007), These studies testify to the difficulties of distinguishing between the various hypotheses based only upon observational studies.

7.4.2 A cue of investment in defences by the plant

Schaefer and Rolshausen (2006) suggested that red colouration may be a reliable cue of plant attractiveness to herbivorous insects without insects necessarily having been an evolutionary driver of red colouration. That is, they suggest that red colouration could act as an informative cue rather than as a signal. Their argument is that anthocyanins and defensive compounds originate from the same phenylpropanoid pathway (see Section 2.6.3). Accordingly, red colouration is pleiotropically linked to the production of defensive phenolic compounds. Activation of this pathway (for example, to produce anthocyanins) automatically leads also to quicker production of defensive compounds, because the precursors for both pigments and defensive compounds are already present. In this scenario, anthocyanins that accumulate in autumn owing to their photoprotective role thus also indicate the quick activation of defensive phenolic compounds. A twist to the hypothesis is that it also predicts that if defensive phenolic compounds are produced, red pigments are more

quickly produced. They term this the *defence indication hypothesis*. The advantage of this hypothesis is that it is not restricted to red autumn colours, but applies more generally to red leaf colouration, for an example involving young expanding leaves (see Karageorgou and Manetas 2006, and Section 7.6). One drawback to the explanatory power of this hypothesis is that it does not explain why different leaves on the same plant can show different colours. Also, it rests on the assumption that there are no allocation conflicts between pigmentation and defence. This physiological link between the production of defensive molecules and pigments could explain the observed negative correlation of aphid numbers and the intensity of leaf colouration through (at least) two quite different mechanisms. Firstly, colour could be used by mobile insects as a cue to help avoid heavily defended plants. Alternatively, insects could entirely ignore the colour and sample trees at random with respect to leaf colouration and preferentially depart from heavily defended trees (which happen also to be brightly coloured). The sampling method of the study of Rolshausen and Schaefer (2007) (involving coating leafed branches with glue) should measure only the initial decision-making of insects (since they should be unable to leave the first plant they land on) and so should allow empirical separation of these two mechanisms. They did not find evidence for colour-based selection of plants.

Rolshausen and Schaefer (2007) found no temperature-related effect on the leaf colouration of two mountain ash (*Sorbus aucuparia*) populations in environments with contrasting autumn temperatures. They did find a correlation between daytime temperature (rather than nighttime temperature) and photoperiod against the extent of colouration change. This correlation supports a photoprotective role for the pigments involved in colour change. These authors did, however, find a response of insects to autumn colouration (or something correlated with it) such that trees with a strong colour change in autumn had fewer aphids. It may make sense for aphids to avoid strongly coloured plants if strong colour change is indicative of a particularly nutrient-stressed individual that must invest heavily in anthocyanin synthesis to enhance nutrient recovery from leaves prior to abscission (as Rolshausen and Schaefer suggest).

As articulated by the defence indication hypothesis of Schaefer and Rolshausen (2007) leaf colour would be a cue not a signal. That is, the herbivores profit from information gained from leaf colouration, and the change in their behaviour as a result of this information may affect plant fitness, but this effect is likely to be overridden by other, mostly abiotic factors that influence leaf colouration. This is plausible if autumn-flying insects have a negligible effect on plant fitness, and/or if selection pressures on leaf colouration due to purely physiological mechanisms (see Section 7.3) were much stronger. At the moment, we lack good empirical exploration of the relative strength of the different selective pressures, a fact that complicates the assessment of all the various hypotheses. In the terminology of Maynard Smith and Harper (2003), red colouration is likely to be a *cue*, rather than an *index*. An index is a signal that is honest because it is impossible to fake. This would be the case if it was impossible to produce any defensive compounds without producing pigments or vice versa; but this is not the case, particularly because there is no physiological and biochemical link to those defensive compounds that are produced by pathways other than the phenylpropanoid pathway but that might still be a deterrent to herbivores. Production of defensive compounds acts to prime the metabolic pathways in a way that makes production of pigments easier (and vice versa), but their production is not an obligate consequence of production of chemical defences. Ougham et al. (2005) suggest that colour could be an unfakable index, if it was inextricably linked to low nutrient levels. For example, low nitrogen levels are known to induce leaf senescence (e.g., Ono et al. 1996). However, such a link between nutrients and pigmentation, if it exists more generally across species, is certainly variable depending on which nutrients are considered. There is ample evidence that gene expression in anthocyanin synthesis is strongly upregulated by sucrose (e.g., Takeuchi et al. 1994, Solfanelli et al. 2006), which suggests that a general link between sucrose and anthocyanin pigmentation is likely. Furthermore, anthocyanin

biosynthesis has been linked to an imbalance between high sugar concentrations and low nitrogen levels in plants, conditions that generally induce senescence (Wingler et al. 2006). This is consistent with their photoprotective function, because the light interception of anthocyanins may reduce the rate of photosynthesis and thereby reduce the sugar–carbon imbalance. In sum, for the time being, it is probably prudent to regard anthocyanin pigmentation as cue, rather than as an unfakable index. In the evolution of communication it is quite likely that what was once a cue evolves into a signal. Such signalling of the quality of plants is considered fully in Section 7.5. Next we consider some other potential signalling mechanisms.

7.4.3 Signalling that leaves are about to be shed

Lev-Yadun and Gould (2007) suggest that autumn leaves signal that they are about to be shed. Obviously, autumn leaf colouration is a cue that leaves will be shed in the near future, but these authors suggest that it may also be a signal. Such a signal may be biologically plausible because most insects live on leaves, but does not seem a promising candidate for explaining variation within and between species in the extent of autumn colour change. Observed variation in the relative timing of colour change and abscission also seems problematic to this theory. One would also have to think carefully about what would keep such a signal honest. If such a signal discourages pests from settling on the leaves, then there would be a selective pressure on the plant to change colour earlier and earlier (relative to abscission), and there would need to be some balancing cost to constrain this tendency. Such a cost may exist if the signal can only be produced at a cost of reducing photosynthesis. However, we then have the logical problem of identifying the benefit to the plant of this signal. That is, if the signalling leaf is no longer very active in photosynthesizing and is shortly to be shed, then a signal that discourages insects from attacking the leaf could be of little benefit to the plant. It might benefit the plant if insects initially alight on the leaves but then shortly move onto other parts of the plant. However, in this case, where insects only spend a short period on the leaf, it is hard to see why insects would pay heed to a signal that the leaf is shortly to be shed. Empirically, such a signal would only be selected for if ecologically important pests both select host plants in autumn and attack leaves (see White 2009 for further consideration of such insects). for further consideration of such insects). Considerably more work needs to be done to build theoretical and empirical support for this recent hypothesis before it can be considered an important driver of leaf colour change.

7.4.4 Signalling investment in reproductive tissues

Another co-evolutionary theory was introduced by Sinkkonen (2006a). This *reproductive insurance hypothesis* builds on the observation in Sinkkonen (2006b) that the extent of autumn colour change was correlated with the extent of seed set in the mountain birch (*Betula pubescens* sp. *czerepanovii*). This hypothesis suggests that colour change is a signal that warns insect herbivores that nutrients will shortly be diverted to reproductive tissues and so will be diverted away from leaves, thus leaf-attacking insects would do better to target non-reproductive trees. Trees would benefit from the reduced damage to their particularly important reproductive tissues. At the moment, this theory lacks empirical evidence that insects would actually benefit from responding to such as signal, and also lacks identification of a mechanism that would keep such a signal honest. Moreover, at least for phloem-sucking insects, the information that nutrients are mobilized and will be transported from the leaves to reproductive tissues may actually induce decisions to land on the plant rather than to stay away from it (see nutrient translocation hypothesis, Section 7.4.1; Holopainen and Peltonen 2002, White 2009). Sinkkonen suggests that both the trees and their specialist aphid species have a shared interest in the successful reproduction of the tree. Whilst shared interests can facilitate signal honesty, there must be a shared interest in avoiding some outcomes of the current interaction (Maynard Smith and Harper 2003), not the longer-term linking of ecologies described by Sinkkonen. This mechanism also does not seem to offer an explanation for between-leaf variation in colour change within a single plant. The correlation reported in Sinkkonen (2006b) could have

a physiological basis in diversion of nutrients that has nothing to do with signalling. Rolshausen and Schaefer (2007) found more fruit on trees with less dramatic autumn colouration in their study of mountain ash (*Sorbus aucuparia*). They interpret this as nutrient-stressed individuals restricting their fruit production and also investing heavily in anthocyanin to enhance nutrient recovery from leaves prior to abscission. Although one could argue that stressed individuals could not afford to invest in anthocyanins, autumn colouration does seem to be associated with physiological stress (amongst other things). However, in the same species, Schaefer and Rolshausen (2007) found more aphids on trees with more fruit, which they speculate was triggered by a trade-off for the plant between investment in reproduction and investment in defensive compounds. Since the aphids were collected with glue, this study suggests that aphids are selective in where they land (rather than simply moving on from poor trees). Since Schaefer and Rolshausen present evidence that leaf colour is not used in this decision, it may be that the appearance or smell of the fruit itself is used or some other volatile of the plant (see Holopainen 2008). A high-quality micro-habitat might lead to high investment in fruit and (quite separately) to high levels of nutrients in the phloem. Aphids might be attracted to such a high-quality phloem by cuing of features of the plant itself or the micro-habitat. Such cuing seems logically possible, but arguing that leaf colouration is a signal requires a (currently missing) explanation for why non-reproductive trees would not cheat by signalling. The first challenge for future empirical studies is to convince that correlations between reproductive investment and aphid numbers are not driven by some third factor (such as soil quality).

7.4.5 Signalling to frugivores

Stiles (1982) suggested that leaf colour change in autumn functioned to alert frugivores to ripe fruit present on the plant. Whilst we have already discussed evidence that reproduction does seem to be one of the factors associated with autumn leaf colour change, a strong link with the timing of fruit ripening has yet to be demonstrated. Stiles himself hints at a limitation in the value of this signal if it still remains after the fruit has been eaten, and becomes diluted through the autumn as more and more species change colour. Stiles does not consider what would keep such a signal honest, although plants may have little incentive to signal that they have ripe fruit when they do not. That is, a plant has little to gain from signalling that it has fruit when it does not. This may be a rare example of a signal that does not need costs to keep it honest: because both parties (the plants and the frugivores) want the same thing – frugivores to eat fruit. The most direct test of this theory comes from Facelli (1993), who attached colourful plastic ovals to infructescences of *Rhus glabra*. As well as an unmanipulated control, there was a factorial design of colour (red or yellow) and size (large or small) of attached plastic discs. Only the treatment with large red flags caused increased fruit removal compared to the control (see Figure 7.2): this despite the fact that yellow flags contrasted more strongly with the red fruits of this species. The differential effects of large red and yellow flags suggests that the mechanism underlying any effect cannot simple be mechanical (e.g., increasing shaking to the infructescences in the wind), but must be at least in part visual (see Section 6.4.1 for a similar study on accessory structures in floral displays). It may be, however, that in this study frugivores were drawn by large red disks because they could be mistaken for large concentrations of fruit.

Stiles' (1982) theory was also evaluated by Willson and Hoppes (1986). In an extensive cross- species comparison on North America deciduous tree species, they found no evidence for Stiles' predictions that autumn leaf colour would be particularly prominent in species with inconspicuous fruiting displays, or in species with oily fruits (that are likely to rot if not discovered quickly). They also found no evidence of synchrony of fruit ripening and leaf colour change. In an experiment, they added either red or green plastic leaves to grape vines, with portions of the vines left as unmanipulated controls. They found no effect of their manipulations. In interpreting this experiment the authors acknowledge that the plastic leaves were a poor match in shape and size for the plant species concerned, and so frugivores may not have reacted to them as they would have to natural foliage of this plant. None the less, Willson and Hoppes (1986) do not provide any empirical support for the fruit-flagging theory. In the light of this, and the logical problems highlighted above, we do not see this as likely to be an important underlying mechanism of autumn leaf colour change. The use of non-ripe fruits as flags to attract frugivores to nearby ripe fruit and the patterns of fruit ripening are discussed in Section 4.8 and Section 4.2, respectively.

The reverse of fruit flagging is suggested by Groom et al. (1994). They observed that the Australian plant species *Hakea trifurcata* has fruit that remains green when ripe and has two distinct leaf types (broad and needle). They suggest that the fruits are confused for the broad leaves by granivores. In support of this hypothesis, they demonstrate that a major natural granivore of this species (the white-tailed black cockatoo, *Calyptorhynchus funereus latirostris*) removed fewer fruits from branches from which broad leaves had been experimentally removed compared to unmanipulated controls. However, an alterative or complementary explanation to the 'self-crypsis' favoured by the authors is that the broad leaves simply shield the

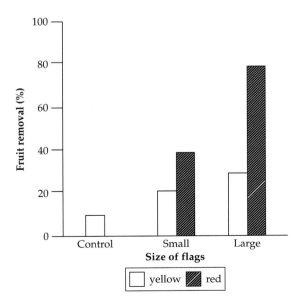

Figure 7.2 The percentage of fruit removed from panicles attached to plastic foliar flags of different colours and sizes and from unmanipulated controls. Redrawn from Facelli (1993).

fruits from view and/or make it more physically awkward to access the fruit. Hence, although plausible, the sensory explanation suggested by the authors remains without strong support.

7.5 Signalling to herbivores about intrinsic plant quality

The late Bill Hamilton was certainly one of the most important evolutionary theorists of the twentieth century, and is renowned for the breadth and originality of his ideas. One of his last ideas to be published was the *co-evolutionary theory* of autumn colour: the idea behind this theory is that the bright leaf colours in autumn are a signal revealing the level of defensive commitment of the individual plant, and comparison of the signals between individuals of a species allows insects that migrate onto plants in the autumn to preferentially select less-defended plants. The co-evolution of colour preference in insects and bright colours in plants would allow well-defended plants to reduce their parasitic insect load, as well as allowing the insects to preferentially settle on the most profitable (least-defended) plants. This theory was introduced in Archetti (2000) and Hamilton and Brown (2001), being later defended against criticism in Archetti and Brown (2004). Although it was originally called the *co-evolutionary theory* we prefer the name *leaf signal theory* suggested by Schaefer and Wilkinson (2004), since there are several alternative theories that also involve co-evolution between plants and animals (see Section 7.4). We will devote considerable space to this theory, because of its subtleties and because it has generated more research interest than any of the alternative theories. We will leave consideration of whether this bias is justified until the end of the chapter.

7.5.1 Assumptions of the theory

The theory rests on the assumption that at least some of the insect enemies of plants select their host in the autumn, and that these insects exhibit a preference between alternative plants based on difference in colouration. These assumptions have been discussed earlier in the chapter. It is certainly true

that some winged insect herbivores select host plants in autumn, although evidence that leaf colouration and not correlated factors influence plant selection is currently lacking.

In order for communication between plants and herbivores to have been co-evolutionary, it seems reasonable to assume that both sides benefit. Thus a further assumption underlying this theory is that variation in leaf colour leads insects to exhibit preferences based on (or correlated with) colour that in turn leads them to preferentially settle on less-defended (or otherwise more advantageous) plants, and that this preference triggers further evolutionary changes in the plant.

There has been some confusion in the literature over what property of host plants the signal should provide information on. Archetti (2000) suggests that it is plant vigour or quality, with high-quality individuals being able both to invest strongly in defences and in signalling strongly. However, as Hamilton and Brown (2001) point out, defensive commitment need not be positively correlated with vigour. They suggest that increasing signal strength should indicate increasing investment in defences (this is also the stance taken by Archetti & Brown 2004). However, we feel that the suitability of a host to parasites will be a function of both the level of defensive investment by the plant, and the value of the resources potentially available if these defences are overcome (see Section 5.5.1 on titration of rewards and toxins in fruit). Thus, it may sometimes be optimal for a pest to select a more strongly defended plant if the value of the resources on offer can more than compensate for the increased defences. In this case, we feel that the signal should be one of suitability: in order to be valuable to the intended receivers (herbivorous insects). Suitability only correlates with investment in defences if the rewards to an insect that overcomes these defences are the same across plants, and there are at least some indications that nutrient rewards are naturally variable between plants (Wingler et al. 2006). The relationship between suitability and plant vigour is likely to be complex and situation-dependent: because resource allocation may be dynamic throughout ontogeny (Boege and Marquis 2005), and because a vigorous plant may be able to invest heavily in defences but may also have intrinsically higher levels of resources available to herbivores. Thus, we feel that this is an important conceptual issue, and evaluation of empirical support for this theory needs to take careful account of the quality that is being signalled. Archetti (2009c) provides a useful generalization of this theory: showing the signalling is still theoretically evolutionarily stable if the assumption of a strong correlation between plant vigour and defensive investment underpinning the original theory is relaxed.

It seems highly plausible that leaf colouration could usefully inform host plant choice by herbivorous insects. There is good empirical evidence that some insects exhibit choice that is related to colour (e.g., Hagen et al. 2003, Hagen et al. 2004, Archetti and Leather 2005). Thus either exercising this choice benefits those insects, or we have to explain why they make choices that are to their detriment. This last possibility cannot be completely dismissed: it could be that differential settlement on plants of different leaf colour is the result of differential ability to detect leaves of different colour. That is, it may be that red leaves are difficult for herbivores to detect (in comparison with green leaves), and thus insects land more often on green leaves not necessarily because such leaves are better for them, but because such leaves are easier to find (Chittka & Döring 2007, Döring et al. 2009). The study of Ramirez et al. (2008) on the southern beech aphid (*Neuqueenaphis staryi*) found a preference for green leaves in autumn but no evidence that the offspring of aphids landing on redder plants in autumn did worse the following spring.

Thus, in contrast to the results of Ramirez et al., another assumption of this theory is that preferential settlement by the insects must lead to a benefit to those plants that signal most strongly. That is, that the behaviour of herbivores in response to variation in plant colouration influences plant fitness, such that plants benefit from emitting the signal. There is certainly some evidence that aphids, and particularly some species that are mobile in autumn, can have a markedly detrimental effect on hosts, and that strong variation in level of infestation occurs and can be linked to plant fitness (see references in Hamilton & Brown 2001). However, we know of no study that has demonstrated that any

correlation between plant health and vigour and extent of autumn leaf colour change is mediated by autumn-flying herbivorous insects. Admittedly such a study would be challenging in the wild, but should be possible in a greenhouse situation where herbivore numbers can be differentially controlled between treatments. If signalling to herbivores benefits plants, then we would expect to see a stronger correlation between plant fitness and strength of signalling in a treatment where herbivores were plentiful compared to a treatment where such herbivores were excluded.

Hagen et al. (2003) found a negative correlation between degree of colour change in autumn and insect damage the following spring in a between-individual plant comparison using mountain birch (*Betula pubescens*). They found that measures of stress (fluctuating asymmetry of leaves) or reproductive investment (catkin production) were positively correlated with insect damage the next spring, which is a similar finding to that reported by Rolshausen and Schaefer (2007) for mountain ash. Hagen et al. interpret this as stress or reproductive investment requiring a reduction in defensive investment, and insects using autumn colouration as a reliable signal of this investment in order to target less-defended trees. They also report that trees with high insect damage in spring showed less change in autumn colour, which they interpreted as such trees extending their period of photosynthesis to make up for losses due to insect attack. However, as these authors themselves admit, these correlations could also be explained via third factors. For example, there may be no direct connection between colouration and vulnerability to insects, but both may separately be linked to some measure of intrinsic plant quality (which might show up through the correlations these authors find between these factors and fluctuating asymmetry). They suggest that manipulation is really needed, and suggest adding fertilizer to some trees in order to manipulate the resources available to the trees. They also make another point that deserves to be followed up: that host plant selection based on leaf colour (or any cue or signal) is more likely for insect species that have low levels of larval or nymphal dispersal. For such species, effective host selection by winged adults will be more crucial to ultimate fitness. Unfortunately, the dominant insects in their study were two moth species that feature dispersal in the larval stage through ballooning (thus providing larvae with the ability to mitigate costs due to poor host plant selection by their mother).

The final important assumption underlying this theory is that the signal is honest: on average there is a correlation between level of defensive commitment (or more generally in suitability to pest insects) and variation in colour. In order for honesty of the signal to be maintained, we generally need signallers to be subject to a strategic cost of signalling that increases with signal intensity (Maynard Smith & Harper 2003, Hoch et al. 2003). One plausible source of such costs is biochemical costs associated with the production of anthocyanins. Wilkinson et al. (2002) argued that these costs are likely to often be very low, but see Gould (2004). In any case low costs are still non-zero; metabolic costs may be associated with enzyme production and activity, with the transport of anthocyanins into cell vacuoles and with the conjugation of cyanidin molecules to monosaccharides (Gould 2004). These costs may potentially be high enough to maintain a correlation between signal strength and the vitality of the individual (and so its ability to bear these costs). The variation in signal strength will be less if the costs are low, however, and so the challenge to the insect of detecting a difference between individuals will be greater. This in turn adds costs of signal discrimination, which increase the risk of dishonest signalling by the senders (Dawkins and Guilford 1991). One prediction of the formal model of this theory by Archetti (2000) is that such signalling would break down if there was low intrinsic variation between individuals. However, in many species there is considerable variation between individual plants in the intensity and timing of leaf colour changes (see Figure 7.3; and, e.g., Balazadeh et al. 2008) that has not be satisfactorily linked to variation in plant suitability to herbivores (or any other measure of 'quality'). Moreover, importantly, for signalling to be honest the key correlation is between signal intensity and unsuitability of the signaller to pest insects. It may be that this correlation comes about because the most healthy and vigorous individuals can bear the combined costs of investment in anti-pest defences and of investment in signalling. However, differential allocation between signalling

and defences is certainly possible (e.g., investing proportionately more in the signal than in defence) unless the two are inextricably linked at a physiological level (see Section 7.4.2). Further, the suitability of a plant to a pest will be a trade-off between the costs and benefits of that plant to the pest. It may be that a healthy plant can afford to increase the costs to the pest by increasing investment in defence, but the very vitality of the healthy plant may provide enhanced feeding opportunity to the pest also. Hence, at present, evidence for a physiological cost to signalling is not strong.

Hamilton and Brown (2001) suggest that signalling requires paying the cost of cessation of photosynthesis. However, while some studies support this view (see, for example, Manetas et al. 2006 and Kytridis et al. 2008 for young red expanding leaves), others are inconsistent with it (Pietrini et al. 2002, Hoch et al. 2003). Schaefer and Wilkinson et al. (2002) argue that the photoprotective role of anthocyanins may actually aid photosynthesis, by lowering the frequency of cold-induced photo-inhibition. There does seem ample evidence that many of the compounds that influence autumn leaf colour have the potential to provide physiological benefits to the plant, and thus early cessation of photosynthesis does not seem a likely route for the imposition of costs required for maintaining signal honesty.

It may be that the cost associated with signalling is ecological, analogous to the cost of aposematic colouration by defended animals to their predators. According to this conjecture, increasing investment

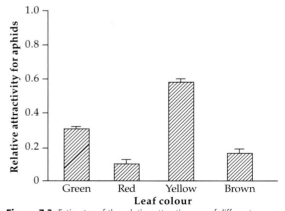

Figure 7.3 Estimates of the relative attractiveness of different-coloured tree leafs. Illustrated are means and S.E. Redrawn from Döring et al. (2009).

in signalling leads to increased conspicuousness to pest insects, and thus increased encounter rate with these pests. Only those plants that are unattractive to pests (perhaps through investment in defences) can accept such high contact rates because pests that discover them choose not to settle on them. However, this theory seems unlikely because red is certainly not a conspicuous colour to aphids that are commonly discussed as selective agents in autumn Chittka and Döring 2007, (Döring et al. 2009).

There are situations where signals can be cost-free and honest. The most commonly cited work in this regard is that of Lachmann et al. (2001). However, by this theory, the signal is only cost-free and honest if an individual that unilaterally switches from signaller or receiver strategy at equilibrium pays a cost for this. However, for the type of signal described here, there would seem to be no cost (and indeed obvious advantage) to a weakly defended plant individual switching to signalling strongly. Hence, the theory of Lachmann et al. (2001) does not seem to provide a strong foundation for expecting the signals under consideration here to be cost-free.

In summary, the signalling mechanism of leaf colouration is entirely logically consistent, and the elegant and careful modelling of Archetti has demonstrated convincingly that such signalling could be evolutionarily stable. They key question now is to explore how important this mechanism is likely to be relative to other potential mechanisms. This is essentially an empirical question, and we have devoted this section to exploration of the empirical support for the assumptions of the theory. Our conclusion is that many of the key assumptions required for this signal to be maintained are currently without strong empirical support. We next turn from exploring the assumptions of the theory to its predictions.

7.5.2 Predictions of the theory

Archetti and Brown (2004) suggest the following two predictions from the signalling hypothesis:

1. 'On an intraspecific level, individuals of signalling species vary in expression of this signal, with defensively committing individuals producing more intense displays, which prove aversive to the monophagous insects.' Thus we should expect to find intensity of signalling and intensity of defences to be correlated at the individual level, the more an individual invests in signalling the less that individual is settled upon by insect pests.

2. In a between-species comparison, 'species suffering greater insect attack invest more in defence and consequently more in defensive signalling than less troubled species'. Thus we should expect to see interspecific positive correlations between strength of signalling, strength of defences and intensity of herbivory from autumn moving insects.

Regarding the first prediction, there is certainly evidence for green leaves being preferentially settled on over red leaves (see Figure 7.3; and references in Chittka and Döring 2007, Döring et al. 2009). There is also evidence for a link between defensive compounds and compounds that strongly influence leaf colouration (Schaefer and Rolshausen 2006), but correlations between defence, colouration, and avoidance by herbivores is by no means a unique prediction of the co-evolutionary hypothesis. Thus, if a study found that those plants with the most dramatic leaf colour change were avoided by herbivorous insects and had higher levels of defensive compounds, then this would be consistent with the signalling hypothesis under discussion, but could also be consistent with the mechanism of colour as a cue (rather than a signal) of investment in defence (as discussed in Section 7.4.2).

We suggest that the logic behind the second (interspecific) prediction is not as solid as first appears. It rests on the assumption that if the intensity of herbivorous attack increases so the investment in anti-herbivore defences will increase. This assumption is a restatement of the arms race analogy that has been used in discussions of predator–prey co-evolution. However, if the anti-herbivore defences are costly (as seems likely), the best response to increased attack may not be increased investment in defence; it may be decreased investment if this allows greater investment to be made in (for example) growth or reproduction (that is, there may be complex interplay between the alternative strategies of tolerance and resistance to herbivores: see Mauricio 2000; Koricheva 2002; Leimu & Koricheva 2006; Nunez-Farfan et al. 2007). The optimal strategy will depend on the fine detail of the various functional relationships. A further weakness of the logic behind this prediction is that it assumes that if herbivore intensity increases and this produces increased investment in anti-herbivore defences in the plants, then this change by the plants has no effect on attack intensity. In contrast, one would expect that increased anti-herbivore defence by a given plant species might be expected to trigger a reduction in attack rate, at least initially, on that species through switching of herbivores to alternative species or individuals.

If we set these concerns with the logic of the prediction aside, Hamilton and Brown (2001) demonstrate a convincing correlation between the diversity of host-specialist aphids associated with a species of tree and the intensity of autumn colour (see Figure 7.4). Whilst this is interesting, we should note that diversity of specialist aphid species is a highly indirect measure of herbivore pressure, and the authors had no data to explore correlations with strength of investment in defence or suitability to aphids.

Schaberg et al. (2003) show that species with red leaves relocate more nutrients to perennial tissue than individuals with fewer red leaves. Schaefer and Wilkinson (2004) argue that those species with the brightest autumn colours are able to mobilize more nutrients from the leaves prior to abscission than less bright species. They suggest that only photosynthetic tissue that is well protected against photo-inhibition could supply the energy for such nutrient translocation. Since these salvaged resources can be put to further use by the plant, this leads to the photo-inhibition hypothesis predicting a positive correlation between plant vigour and strength of signalling. In contrast, as Ougham et al. (2005) point out, it is often the weak trees that are more strongly coloured, which is consistent with weak trees having the greatest need of protection and nutrient recovery. In this case, there would be a negative correlation between plant vigour and strength of signalling. This issue testifies to the difficulties of deriving robust predictions based upon our current knowledge on plant physiology. Also, it is important to remember that the patterns may differ between species, but we lack currently data to predict such interspecific differences. At any rate, the correlation, be it negative or positive, between plant vigour and plant colouration has (at least initially) nothing to do

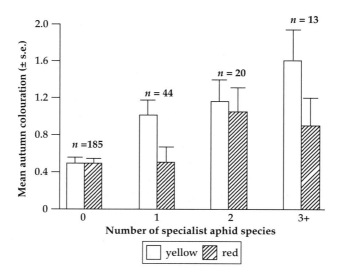

Figure 7.4 Autumn colour (evaluated from field guides and using a 0–4 scale of increasing intensity of colour), as a function of number of specialist aphid species mentioned in the same field guides as being associated with a particular tree. Illustrated are means and S.E. Redrawn from Hamilton and Brown (2001).

with insect herbivores – although these pests (particularly phloem-sucking insects) may then use the colour change as a cue of nutrient mobilization and translocation. Potentially, selection caused by this information use by herbivores could turn this cue into a signal. Consequently, the key issue here is that the second prediction of Archetti and Brown (2004) – like the first – is by no means exclusive to Hamilton and Brown's co-evolution theory.

Wilkinson et al. (2002) suggest geographic range as an important potential confounding factor in Hamilton and Brown's dataset. Those species with larger geographical ranges are likely to experience a wider range of specialist herbivores and also (since autumn leaf change is linked to latitude and climate) are likely to experience autumn colour change over at least part of their range. When Hamilton and Brown (2001) attempted to control for geographic range, the strength of their correlations was much reduced.

In conclusion the two key predictions of the co-evolutionary hypothesis (as identified by its main proponents) do not allow it to be separated empirically from more parsimonious explanations of autumn colour change. This is not to say that the co-evolutionary hypothesis can be disregarded. Rather, it suggests that further empirical work would be better used elsewhere than addressing the two predictions considered in this section.

7.5.3 A further theory

Yamazaki (2008) proposed the hypothesis that leaf colouration signals tree quality to those aphids that are tended by ants. After aphids colonize a tree, ants are attracted to that tree and defend it from other herbivores. The main assumption is thus that the colonization of myrmecophilous aphids and their tending ants leads to an overall reduction in herbivory. There is some evidence that this is indeed the case in several species (see references in Yamazaki 2008), although it has yet to be demonstrated in species with non-green leaf colouration. This hypothesis of tritrophic interactions can be distinguished from all other hypotheses – except for the nutrient translocation hypothesis – by its prediction that there should be more, not less, aphids colonizing strongly coloured non-green

plants. This hypothesis has the drawback that it is restricted to species that are colonized by myrmecophilous aphids. The maintenance of signal honesty is also interesting. It may be that signalling is honest at least in part because the interests of plants and aphids converge: i.e., ant protection. A plant of low quality (to aphids) would be unlikely to support sufficient aphids to be attractive to ants, thus if there is any cost to signalling it may be that signalling is only attractive to those plants that can allow aphids (and thus ants) to thrive. Such costs would occur naturally if other herbivores (that do not benefit from ant protection, and indeed are normally attacked by ants) are attracted to the signal, as well as ant-attracting aphids.

Yet another tritrophic hypothesis related to leaf colouration is the 'indirect defence hypothesis' of Holopainen (2008). This suggested that terpenes are used by plants to attract the enemies of herbivorous insects to the plant. He suggests that green colouration in autumn is required to prolong photosynthesis to fuel the production of terpenes. Thus 'green plants late in autumn are visually or olfactorally aposematic and flagging…that they are effectively indirectly defended among coloured trees.' Under this mechanism, yellow leaf colouration would be a cue that herbivorous insects could use to identify plants that had decreased their photosynthesis and were thus unable to continue attracting potential enemies of the herbivores. For plants, yellow colouration would still be favoured if the benefits from physiological mechanisms to do with salvaging of compounds (see Section 7.3.1) from leaves prior to abscission outweigh costs of increased herbivore attraction that non-green colour may trigger. One might expect the relative costs and benefits to vary between species and between micro-habitats. However, we currently lack evidence of the fundamental assumption of this theory that green colouration is linked with attraction of predatory insects in autumn.

7.6 The functioning of non-green leaf colouration in non-autumnal contexts

It is relatively common for young leaves to have a strong reddish appearance, turning to green as the leaf develops (a phenomenon often called 'delayed greening'). This appears to be both because anthocyanins are present and because chlorophyll input is delayed (Kursar and Coley 1992). As in autumn colouration, there are distinct physiological and ecological hypotheses to explain this phenomenon. The physiological hypothesis is related to photoprotection. This is because the young expanding leaf is under construction and the photosynthetic apparatus may not yet function completely, and there is a need to reduce incoming light that might otherwise result in the formation of free radicals. The ideas behind this hypothesis are thus identical to those related to leaf senescence.

The ecological hypotheses invoked are also the same as those involved in autumn leaf colouration. Red colouration in young leaves has been linked to reduced herbivory: either because such leaves are cryptic to herbivores, or because the red colouration is correlated with investment in defensive compounds and/or indicates that the young leaves are not yet of sufficient nutritive value to be attractive (see Dominy et al. 2002 for an overview). Evidence in support of these theories is restricted to a few species. Numata et al. (2004) compared eight sympatric congeneric *dipterocarp* tree species, some of which showed delayed greening (i.e., red leaves in early development). In a transplant study to control for variation in local environment, they found that leaf damage from insect herbivores was less for the delayed-greening species, despite there being no difference in measured growth parameters across species. However evaluation of leaf damage in naturally occurring seedlings failed to reproduce this difference. The authors did not speculate on the mechanism behind the difference picked up in their transplant study. Karageorgou and Manetas (2006) compared the young leaves of the evergreen tree *Quercus coccifera*. This species shows strong between-individual variation in the extent of delayed greening in young leaves. They found reduced herbivore damage on leaves with a more pronounced red colour and a positive correlation between the anthocyanins responsible for colouration and total phenolics (which are likely to act as defensive compounds). They speculated that the difference in herbivore damage may be due to a combination of aversiveness of the phenolics, red colouration making leaves difficult to detect by

herbivores (an idea that first appears in Stone (1979), who suggested that understory plants might have colouration that mimics that of dead leaves in the leaf litter) and red colouration making herbivores more easily seen by their enemies. This last idea is generally credited to Lev-Yadun et al. (2004) and discussed in the next section (Section 7.7). At the moment, the very small amount of work that has been done on red colouration in young leaves does not allow us to reliably distinguish between the various physiological or ecological hypotheses that may contribute to the phenomenon.

Lev-Yadun and Ne'eman (2006) suggest that often young prickles, spines, and thorns are aposematically coloured, but this colouration fades with age. They suggest that this change is adaptive because continued investment is required to maintain the colouration and older, larger, or otherwise better protected organs have less need of aposematic colouration. This interesting idea deserves rigorous testing using phylogenetically controlled comparisons of spine colour over time in comparison to the stem or other structure to which they are attached. At the moment it remains a plausible but untested hypothesis.

7.7   Leaf variegation

Variegation generally involves parts of the leaf where chlorophyll is either missing or screened by surface pigmentation. Either way, it is expected that variegation reduces photosynthetic performance of a leaf (Sadof and Raupp 2001, Yang and Sadof 1995, Sadof et al. 2003). One might thus conclude that variegation can only be maintained if there is some mitigating benefit. One proposed benefit that has been suggested is that variegation serves to reduce herbivore damage (see Smith 1986 for a full discussion of possible mechanisms). Smith (1986) found that variegated leaves of *Byttneria aculeate* suffered less herbivore damage than unvariegated leaves. However, different leaf morphs were associated with different micro-climates (with variegated leaves being more common in drought-stressed habitats with high sunlight levels; conditions where a reduction in photosynthesis may not be as costly as in shaded habitats). Thus, this study could not

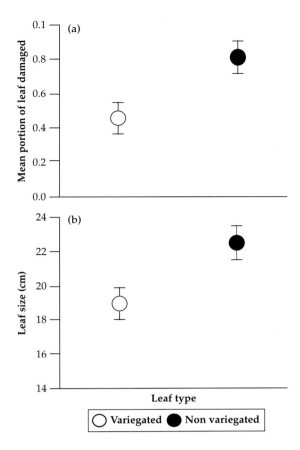

Figure 7.5 (a) Campitelli et al. (2008) found that variegated leaves received less herbivore damage than unvariegated leaves in the North America understory herb Virginia waterleaf (*Hydrophyllum virginianum*). (b) Variegated leaves were also larger. Illustrated are means and S.E. Redrawn from that publication.

rule out micro-habitat-specific changes in plant physiology and/or herbivore behaviour as possible confounding factors. Campitelli et al. (2008) also found that variegated leaves received less herbivore damage than unvariegated leaves in the North America understory herb Virginia waterleaf (*Hydrophyllum virginianum*). However, since variegated leaves were in general older and larger than unvariegated leaves in the same plant (see Figure 7.5), confounding factors cannot again be excluded, particularly because leaf biochemistry is known to vary with age. The fact that variegated leaves have been exposed to herbivores for longer but remain less damaged might seem to add support to the hypothesis that variegation provides defence from herbivores. However, before we can safely support this interpretation, we need to check that the plant does not invest more in defensive compounds in the first-produced (variegated) leaves. A further potential confounding factor may be position with respect to the ground, with early (variegated) leaves being nearer to the ground. That is, further work is required to demonstrate that variegation is not associated with some other trait affecting utility to herbivores (such as levels of defensive compounds), which could act as a confounding factor.

Soltau et al. (2009) found that variegated leaves of the tropical understory herb *Caladium steudnerifolium* has less damage from leaf miners than unvariegated leaves. The researchers followed up this observation with manipulative study. In one experiment leaves were either naturally uniformly green, naturally variegated, or naturally green but manipulated to appear to humans to mimic the variegated morph through the addition of typewriter correction fluid (Tipp-Ex). Seven out of 126 (5.6 per cent) of green leaves were mined compared to none of the 114 naturally variegated leaves and none of the 141 manipulated leaves. Whilst this is suggestive that the appearance of variegation acts to deter miners, it may be that the winged females were reacting to volatiles given off by the correction fluid. Thus, the authors repeated the experiment, now adding a fourth group where leaves experienced the same manipulation as with correction fluid but now using correction fluid thinner, which is colourless. The results are as shown below.

	Uniform	Variegated	White fluid	Clear fluid
Total leaf no.	164	174	169	153
No. attacked	18 (11%)	5 (3%)	11 (6%)	14 (9%)

To the authors, these results are suggestive that the appearance of variation may reduce the attractiveness of leaves to miners. We are more cautious. These data do suggest that naturally occurring variegation is associated with lower levels of miner infestation. However, this could be explained by third variables (see above). Moreover, if the data for their two experimentally manipulated treatments are analysed as a 2 × 2 contingency table, then there is no statistical difference in infestation between the white and colourless treatments, and thus no evidence that the appearance of variation affects infestation by miners. This is not to say that there is no sensory effect of variegation; this experiment suffers from low statistical power to detect an effect, because the overwhelming majority of leaves in all categories were not attacked by miners.

Recently, leaf mines, the often-conspicuous tunnels of leaf miners and the associated death of leaf tissue, have been suggested to function as visual signal or cue to deter other herbivores from consuming the leaf infested by the leaf miner (Yamazaki 2010). This hypothesis is based upon the interest of leaf miners to reduce feeding of herbivores on the leaf while they are still present in the leaf. There are several possible mechanisms to explain why herbivores may consume fewer leaves with leaf mines. These leaves may be more difficult to detect (analogue to the hypotheses on variegation above), leaf mines may be associated with *induced defences* of the plant (against the leaf miners) and thereby act as a reliable cue of such defences, or leaf mines may mimic fungi infestation. Because this hypothesis has only recently been proposed, experimental evidence for it is thus far lacking. Experiments like the one by Soltau et al. (2009) that use paint to manipulate leaf mines are a useful approach to investigate this hypothesis.

Lev-Yadan et al. (2004) suggest another herbivore-related function for variegation. They suggest that within-leaf variation in colour makes it harder for herbivores to be cryptic to their predators by background matching. Visually camouflaged herbivores cannot present a good colour match to all the different parts of a multi-coloured leaf and must thus either restrict their movements to those parts where they are well matched, or suffer higher predation when moving on non-matching parts of the leaf (both of which benefit the plant). This is an intriguing idea but currently lacks empirical testing. Also, it has to be kept in mind that multi-coloured leaves represent more complex visual backgrounds (see Plate 10), which are known to complicate and prolong search times of insectivorous birds even if prey items are not cryptic (Merilaita et al. 2001, Dimitrova et al. 2009). Thus, although the match between background and prey

can be reduced by multiple leaf colours, the effect that this may have on prey survival is not well resolved at present.

In the future, it would thus be interesting to study directly whether herbivores show a preference for non-variegated leaves and whether position of herbivore damage (or simply herbivore movement) on a leaf is linked to variegation. Finally, it would be interesting to explore the importance of background matching by herbivores as a defence against their predators, using translation experiments and/or dummy prey. It may also be that variegated plants simply provide a range of backgrounds that can be exploited by a range of different-coloured herbivores (as suggested by Schaefer & Rolshausen 2006, see Plate 10). Thus, comparison of herbivore diversity on variegated and non-variegated leaves would be instructive.

Givnish (1990) speculated that in forest herbs certain types of variegation featuring relatively large patches of colour (a condition he describes as 'mottled') might reduce detection by vertebrate herbivores in the sun-dappled forest understory. He thus predicts that mottling will be more common in low-growing forest herbs than in plants of other growth forms. He further predicts that mottling will be more common in herb species that experience generally high levels of irradiance, so that the costs of variegation in terms of light-gathering potential are lessened. These predictions were borne out in a comparative analysis of the flora of the north-eastern US. This study also found mottling to be common in evergreen herbs in temperate deciduous forests, which Givnish suggests may be due to high pressure from vertebrate herbivores when these herbs represent the only forage available during winter, and (because of low nutrient availability in such forests) damage by herbivores may be comparatively expensive to tolerate for these herbs. Whilst these observations are suggestive, to date the camouflage hypothesis has not been tested directly either by observation or manipulative experiment (despite calls by Givnish 1990 and by Allen & Knill 1991).

Cahn and Harper (1976) consider white clover (*Trifolium repens*), which can be characteristically polymorphic for white marks. They observe that marked leaves are prevalent in well-grazed pasture, which they suggest may indicate that these marks provide protection from vertebrate grazers, but again (despite calls by Givnish 1990 and Allen and Knill 1991) no direct evidence for this conjecture has been forthcoming. A manipulative approach is essential to eliminate confounding factors such as herbivores affecting the light environment by closely cropping grasses or correlated selection by herbivores on genotypes that are pleiotropically linked to the white marks.

In conclusion, there is currently suggestive evidence that variegation may influence plant choice by herbivores, but there has yet to be definitive evidence from a study that discounts confounding factors. Such evidence could be obtained from a manipulative experiment like that of Soltau et al. 2009 but with greater statistical power (through greater sample size or higher herbivore pressure). If such evidence is forthcoming, we might appropriately turn to consider the mechanism by which the appearance of variegation influences herbivores. It may be that variegated leaves are less easily detected than uniformly green ones. Alternatively or additionally the variegation may mimic previous herbivore damage (for reference to other plant mimicry see Chapter 8), causing the leaf to be avoided by herbivores in order to avoid competition from other herbivores or avoid plant defences that have been induced by previous damage. If, however, it appears that variegation does not influence herbivores, then we must look elsewhere for an explanation for its maintenance. Either way, the selective forces on leaf variegation very much warrant further investigation.

7.8 Gaps in current understanding

As Schaefer and Rolshausen (2006) discuss, it should be possible to differentiate their *defence indication theory* that red colouration acts as a cue from the *leaf signal theory* that it is a signal. An environment with a typically cold autumn should stress plants more and produce a more extensive leaf colour change under the *defence indication hypothesis*. In contrast, cold autumn weather should reduce pressure from autumn-flying insects and so lead to less benefit in producing a signal. For progress to be made in this field of complex physiological and ecological interactions among different biochemical compounds and among different species, it seems essential not only to account for different hypotheses but to more clearly direct research to those issues that allow the experimenters to contrast alternative hypotheses. This consideration is particularly important since a number of predictions that are usually regarded as specific to a certain hypothesis do not rule out alternative hypotheses.

As Schaefer and Wilkinson (2004) suggest, there should be value in comparison between plants that show autumn leaf colour change and genetic variants that do not. This would allow exploration of how the gene regulation of autumn colour change is linked (for example) to investment in defensive compounds, or how rates of photosynthesis are affected by adverse weather (such as low temperatures). Temperature could be manipulated in a greenhouse study to avoid the effects of confounding factors (such as light levels or herbivore numbers) that might correlate with temperature in the wild. Since reproduction also appears to be a significant factor in autumn leaf colouration in some species, fruit set too can be manipulated in greenhouse experiments in those plants that are insect-pollinated, by controlling the availability of pollinating insects.

We need more choice trials using appropriate herbivore species (that is, those that select their hosts in autumn). Archetti and Brown (2004) suggest that one advantage to laboratory-based studies is that incident light can be manipulated to explore whether herbivores respond to colour or to other (perhaps olfactory) cues that are linked to colouration. As well as choice trials, measurements of herbivore fitness after making different choices (related to autumn leaf colour) would be particularly valuable. For example, Archetti (2009b) demonstrated that aphids (*Dysaphis plantaginea*) showed a preference for settling on apple trees with fewer red leaves in autumn in a mixed-species orchard. The following spring, he artificially colonized trees and found that the success of aphid colonies was higher on trees of the varieties that had been preferentially selected by aphids the previous autumn.

No matter how we progress, it would be advisable to measure autumn colouration as it would be expected to appear to potential herbivores (through spectrophotometric measurements combined with an understanding of the physiology of the eye of relevant species). As Chittka and Döring (2007) point out, we very much need data on receptor spectral sensitivities as well as post-receptor neural processing from ecologically relevant species. We should clearly differentiate between different non-green colours and be slower than some previous authors to lump all non-green colours together. It is clear that, both physiologically to the plant and sensorily to the animal, yellow is very different from red. Indeed Döring et al. (2009) suggest that plants may produce red pigments in their leaves to mask yellow colouration. This idea stems from the demonstration in their study and others that flying insects seem preferentially attracted to yellow colouration over green and to green over red. Thus, if physiological processes (such as the loss of chlorophyll in senescing leaves) cause a yellow appearance, it may be selectively advantageous for the plant to invest in red pigments in order to reduce their rate of attack by flying insects. This is a plausible idea, worthy of further investigation.

In evaluating different hypotheses, we must also consider whether they satisfactorily explain why colour change occurs at the time it does (for most species this means autumn but not summer). For example, explanations based on prevention of cold inhibition of photosynthesis naturally involve a clear difference between summer and autumn. However, explanations based on signalling to insects in flight must explain why signalling occurs in autumn but not in summer (when generally more insects are on the wing). This issue is raised by Wilkinson et al. (2002). These authors also raise the issue that no theories so far naturally explain why senescence and resulting changes in leaf colouration can often be highly heterogeneous within a single plant (see Balazadeh et al. 2008).

More basic plant biochemistry is required if our understanding is to advance. It seems true that compounds important to pigmentation may be suitable as sunscreens and as anti-oxidants; but it may be the case that other compounds that are more abundant in leaves can do these jobs more effectively, or that non-pigmented compounds could be produced by the plant to perform these functions. Thus, where plant colour change involves production of pigments in autumn, we would welcome a stronger explanation for why plants produce anthocyanins.

7.9 Conclusion

Hamilton's insight in this field remains an important one. His ideas have stirred an immense amount of interest from ecologists and evolutionary biologists where previously there was very little. It is clear that there is a question to answer. At least some herbivorous insects that fly in autumn do seem to use leaf colour (and/or traits such as volatiles that correlate with colour) in order to select host plants. What is not clear is the mechanism or mechanisms underlying this selection, and the importance of the selective pressure that this choice generates relative to other concomitant selective pressures. It is important to remember that several mechanisms may contribute to the relation between leaf colouring and insect choice in one particular situation, and the relative importance of different mechanisms may vary between situations. Thus, it is of paramount importance to begin to assess the relative importance of different selective pressures. At the moment, Hamilton's suggestion of co-evolutionary signalling between plant and herbivore remains a plausible candidate mechanism, but we cannot be sure that it is an ecologically or evolutionarily important mechanism. Indeed, parsimony of required assumptions, would argue against this mechanism in comparison to plausible alternatives. The next few years will bring careful, critical testing between alternative mechanisms and considerably advance our understanding of autumn colour range. In so doing, this research is likely to kick-start greater interest in other examples of non-green colouration (such as in young leaves). We predict that this will be the chapter of our book that dates most quickly! We further predict, to echo the arguments of Richard Dawkins (1998), that a greater understanding of the mechanisms underlying the changing colours of leaves in autumn will not detract, and will generally enhance, the pleasure that humans gain from observing this phenomenon.

CHAPTER 8

Plant crypsis, aposematism, and mimicry

8.1 Introduction

Spectacular examples of visual traits in prey animals that provide protection from predators are staples of popular science books, zoology textbooks, and natural history TV programmes. Such traits have been a crucial testbed for evolutionary thinking right back to correspondence between Darwin and Wallace and continuing to this day (reviewed in Ruxton et al. 2004). One might expect that plants utilize similar mechanisms for protection from herbivores, and this chapter will explore this analogy.

In Section 8.2, we consider crypsis: traits that reduce the rate at which the bearer is detected by antagonists. In animals, species that have very potent defences against would-be predators often give up crypsis and advertise their defences by way of conspicuous signals. This combination of defence and signal is called aposematism. Many plants have potent defences against would-be herbivores, and by analogy we might expect aposematic signalling of those defences. The careful modelling of Augner (1994) and Augner and Bernays (1998) demonstrates the plausibility of such signalling and suggests that our understanding of the evolution and functioning of aposematism in animals should transfer without difficulty to plants. The empirical foundation of such an analogy is the subject of Section 8.3.

Visual mimicry of and by plants will be the main focus of this chapter. However, importantly, much of this mimicry will be quite different from that studied in animals. In animals, mimicry is mainly assumed to be driven by the selective pressures of potential predators that may avoid attacking a defenceless Batesian mimic that looks like a defended model, or that learn more quickly to avoid several defended species that share the same signal (Müllerian mimicry). In contrast, most consideration of plant-related mimicry has focused on manipulation of pollinators. In other words, studies on plant mimicry are mainly interested in the attraction of mutualists rather than repulsion of antagonists. That said, there may still be what might be considered Batesian and Müllerian mimicry between plants. For example, a rewardless flower that pollinators mistake for a rewarding flower might be thought of as a Batesian mimic. A group of co-occurring species that are sufficiently similar in appearance to be treated alike by pollinators, such that individual visitation rate is increased by the higher effective density of the species-complex, might be considered an example of Müllerian mimicry. Roy and Widmer (1999) provide fuller discussion on the use of the terms Batesian and Müllerian mimicries in plant biology; for commonality with the existing literature, we use the descriptions above as the basis for usage of these terms throughout this chapter.

The biggest conceptual challenge to progress is that mechanisms other than mimicry might be used by rewardless plants in order to encourage visits by pollinators, and mechanisms other than mimicry might lead to similarity of appearance between the flowers of co-occurring species. Thus, we consider the need for care in disentangling mechanisms when exploring such floral mimicry (Section 8.4.1), before considering the empirical evidence for floral mimicry (Section 8.4.2). Before leaving floral mimicry we consider mimicry of flowers by other plant parts and even by other types of organism as a way to dupe would-be pollinators into visiting (Section 8.4.3), and the reverse mimicry by flowers of non-plant entities in order to dupe animals into inadvertently pollinating (Section 8.4.4). There is the potential for mimicry among non-floral plant parts, and this is the focus of Section 8.5. Our final topic in mimicry is the potential for animals to mimic parts of plants (Section 8.6).

8.2 Crypsis in plants

Crypsis is one mechanism of camouflage (see Stevens and Merilaita 2009 for careful definitions of these terms) involving traits that reduce the rate at which the bearer is detected by antagonists. Crypsis is commonly described in predator–prey relationships, and it might seem reasonable to expect plants to similarly reduce their risk of detection by herbivores. However, putative examples of crypsis in plants are few, and experimental explorations of plant crypsis almost absent.

Although we know of no study that has demonstrated this, it is difficult to imagine any explanation for the extraordinary likeness of South African *Mesembryanthemacae* plants such as *Lithops* species to stones other than to provide protection via crypsis. However, we can think of few other strong cases of crypsis in plants. It has been suggested that variegation in leaves might reduce detection by herbivores (see Chapter 7). Wiens (1978) also suggests some cases of mimicry of dead grass by cacti and of dead sticks by succulents, but these suggestions do not seems to have been studied further.

The best evidence of plant crypsis comes from the experimental study of Klooster et al. (2009). Reproductive stems of the plant *Monotropis odorata* possess a dense covering of dead bracts that the study demonstrated provide a close match in appearance to the local leaf-litter substrate. Experimental removal of the bracts caused increased herbivore damage and lower fruit set (Figure 8.1). The authors also demonstrated that manipulated plants that escaped herbivory produced fruit as successfully as control plants that avoided herbivory, suggesting that herbivory (rather than any physical factor) was at the heart of the fitness difference found in their

study. They argue that herbivory can be reduced in this species without large cost in reduced pollination and seed transport since most herbivory occurs prior to floral anthesis (and production of floral fragrance) and both flowers and fruit provide strong volatile cues. This however seems a very specialist adaptation: *M. odorata* is the only species in its genus that retains dead bracts in this way, and this strategy may be possible only because this plant is non-photosynthetic: it is a mycoheterotroph that acquires carbon from associated mycorrihixal fungi.

Fadzly et al. (2009) postulate that selection by now-extinct herbivorous moas may explain colouration changes over leaf ontogeny in the New Zealand lancewood (*Pseudopanax crassifolius*) and possibly also some other species from New Zealand. Seedlings have narrow brownish leaves that are mottled in appearance, saplings produce larger leaves that have thorn-like dentitions along their margins associated with bright colouration, whereas adult plants have oblong, green leaves of relatively ordinary appearance. The authors postulate that seedlings may have obtained protection from moas through crypsis against the leaf litter, whereas the saplings may have been protected by spiny thorns. In support of their argument, they provide spectrometric analyses that suggest that birds would find differentiating the seedlings from leaf litter a challenging task. They also point out that the closely related *P. chathamicus* from the Chatham Islands off New Zealand (that never supported moas) does not show the same ontogenic change in appearance. The hypothesis of Fadzly et al. is credible. The relatively short time since moas have gone extinct may have not allowed sufficient time for adaptations to them to be lost. Manipulative enclosure experiments might be possible using extant ratite birds. Such experiments could tease apart whether brownish colouration in seedlings or their mottled appearance have any protective effect so that these are consumed at a lower rate than appropriate controls.

For seeds, there is evidence from both field and aviary that matching the substrate colour can reduce predation by birds (Saracino et al. 2004, Nystrand and Granstrom 1997). These studies focus on fire-associated seeds; association with fire may give the predictability of background colouration (grey ash) that makes crypsis more practical relative to more

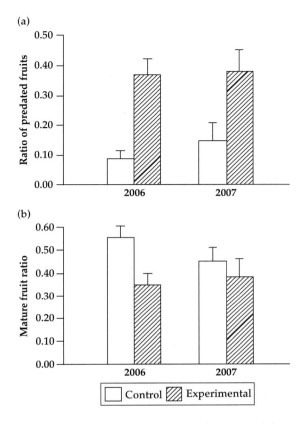

Figure 8.1 The (a) fraction of predated fruit and (b) proportional production for fruit for both control plants and those with bracts experimentally removed in the study of Klooster et al. (2009). Illustrated are means and S.E. Redrawn from that publication.

heterogeneous backgrounds. Saracino et al. (1997) noted that the Aleppo pine (*Pinus halepensis*) releases dark grey-brown seeds immediately after a fire, whereas later-dispersing seeds were yellowish. They interpreted this as crypsis first against ash and later against the sandy substrate that is exposed as the ash is removed by weather. Seeds offer perhaps the most tractable plant stage on which to explore crypsis further. Seeds of some plants sport more than one colour. It is currently unknown whether multiple colours can have a protective effect because the seeds are more difficult to identify. Many species have been reported to show between-seed variation in their seed colouration (Nystrand and Granstrom 1997), and the importance of visually-hunting granivores in maintaining this variation would be worth exploration. This might be driven by variability in background colouration and/or by

frequency-dependent selection (for example though the formation of search images for seeds of a particular colour) by the granivores. We would also be interested in cross-species studies of the relation (if any) between substrate colouration and seed colour. At its most basic level, this might explore whether plants on light substrates (sandy soils) produce lighter-coloured seeds than related plants that grow in darker (higher organic matter) soils (Capon and Brecht 1970 report such a correlation without providing quantitative data). The colouration of seeds can also be pleiotropically linked to colouration in vegetative parts (see Plate 11).

It is important to remember that some defensive adaptations may affect colouration but this may not be an important part of their defensive function. For example, high densities of tiny, hair-like structures called trichomes on a leaf generally give it a silvery appearance, and they generally reduce herbivore damage, but this reduction is considered to be primarily driven by physical, rather than sensory, mechanisms. Although sensory factors are not impossible (Lev-Yadun 2006), no experimental evidence for this currently exists.

It may be that over-reliance on human sensory systems is causing us to underestimate the frequency of plant (and indeed animal) crypsis in the natural world. Because humans are primarily visual hunters, there is a tendency for us to focus only on visual crypsis, however crypsis can function in other sensory modalities (see Ruxton 2009 for a review). Further, especially invertebrate herbivores may be heavily reliant on olfaction to detect suitable food resources. We finish this section with an example of the dangers of judging crypsis as perceived by humans rather than the relevant organisms. The crab spider *Thomisus spectabilis* commonly sits on the daisy *Chrysanthemum frutescens* (see Plate 12). To human eyes it is cryptic on its preferred flower, and different colour morphs of the spider select flower morphs that enhance their crypsis to humans (Heiling et al. 2005). However, the spider contrasts with the flower in the UV, and this contrast is readily detectable by relevant potential prey. Amazingly, prey preferentially land on spider-occupied flowers over empty ones (Heiling et al. 2006). Many pollinators have an inherent preference for flowers with strongly contrasting markings (Lunau et al. 1996) and the crab-spider exploits this sensory bias to lure prey.

8.3 Aposematism in plants

Aposematic signalling by plants has had an effective advocate in Simcha Lev-Yadun. In a series of papers (reviewed in Lev-Yadun 2009), Lev-Yadun has identified several candidate aposematic plants. However, as he acknowledges in his review, what is now needed is experimental investigation of such candidate systems to test if aposematism is actually occurring.

For example, in a series of papers, it has been argued that thorns, spines, and prickles are generally conspicuously coloured (Lev-Yadun 2001, 2003a, 2003b, Halpern et al. 2007, Lev-Yadun and Gould 2007, Lev-Yadun and Halpern 2007) (see Plate 13). Firstly, it would be useful to compare the colouration of such defensive structures with structural non-photosynthetic structures on the same plant (e.g., stalks and stems) to explore whether the colour of physically defensive structures is fundamentally different from non-defensive structures. Such comparison should be done objectively, evaluating appearance from the viewpoint of the visual systems of potential herbivores. The plants need to be surveyed in their natural habitat, rather than against the white page of a field guide (Lev-Yadun and Ne'eman 2004 argue convincingly that even green can be a conspicuous colour in a desert environment). Secondly, if the appearance of such structures is contrasting, then an aposematic function implies that the appearance influences the behaviour of herbivores such that their damage to the plant is decreased. We thus need manipulative field experiments where the colouration of defensive structures is altered and the consequences of this for herbivore behaviour and plant fitness measured. Lev-Yadun and Ne'eman (2006) argued that loss of conspicuous aposematic colour often occurs 'when the defended organs get less edible to large herbivores because of their increased size, mechanical rigidity or chemical defence, or when there is no need for such defence' (Lev-Yadun 2009: 178). Thus, we would welcome study of how the timing of colour change on defensive structures compares with any change in the appearance of other structures on the same plants. It would also be useful to explore whether colour change in defensive structures can be linked to changing attractiveness to herbivores and ultimately to plant fitness.

Midgley (2004) presented data from several African spiny plants that show a negative correlation between spine whiteness and hardness. This was interpreted as a trade-off between investment in the conspicuousness of the signal of the defence (whiteness) and investment in increasing the efficacy the defence itself (hardness). His argument is that if spines are cryptic then they must be hard enough to stand up to occasional inadvertent attack by vertebrate browsers. In contrast, if the spines are conspicuous then the spines only have to have sufficient strength to be unattractive for browsers to deliberately eat them (along with nearby plant parts). It would be worth analysing conspicuousness; that is, whether white spines are more contrasting to relevant browsers under natural lighting conditions than darker spines. Further, Midgely's argument suggests that a survey of naturally-browsed plants should find more damage to darker-coloured spines, and that artificially darkened spines should be damaged more than controls of the same strength, and artificially lightened spines should be damaged less than controls. It remains plausible that the correlation suggested by Midgley results from physiology – in order for spines to be stronger their physical structure might require a darker colouration (as is often the situation in animal colours, e.g., bird feathers) – rather than any effect on the sensory systems of browsers. However, both the sensory and structural mechanism may work in concert (especially if the structural mechanism leads to underlying reliability in the signal).

Lev-Yadun (2003b) suggests that many plants show thorn mimicry. That is, some plants have parts that look like thorns but are not stiff and sharp like thorns, or they have patterning on surfaces that give the appearance of spines. Some of the photographs that Lev-Yadun provides make such deception seem highly plausible, but again manipulative experiment is required. Firstly, we need to manipulate the appearance of such plants to explore whether their patterning affects herbivore behaviour in ways that benefit the plant. If such a benefit can be demonstrated, then we can turn to the question of whether this benefit occurs because the herbivore is duped into concluding that the plant is more heavily defended by spines than it actually is. Experiments to probe the mental state of the herbivore in such a way will be challenging, but progress might be made by comparing the effect of 'pseudothorns' on a diversity of herbivores that vary (perhaps through previous exposure within a species, or morphological differences between species) in how aversive they find thorns.

As both Lev-Yadun (2009) and Augner and Bernays (1998) argue, if there is aposematic signalling in plants, then both Batesian and Müllerian mimicry are logically plausible. Although Lev-Yadun (2009) suggests some plausible systems where such mimicry might occur in plants, we again emphasize that we must first confirm that there is actually an aposematic signal, then we can ask

whether herbivores generalize their experience of such signals across individuals of different species. Thorn mimicry is interpreted as Batesian mimicry by Lev-Yadun (2003b), however we suggest that there may be a generalized avoidance of thorn-like structures by the herbivore rather than the specific misidentification of one species for another that occurs with mimicry.

Lev-Yadun and Halpern (2007) suggest that some very toxic fungi that live on plants are aposematic and that this aposematism might have the potential to turn the fungus's relation to the plant from parasitism to a mutualism: with the fungus paying for the nutrients it takes by reducing risk of herbivory. The evolution of this is relatively easy to understand in theory since both plant and fungus might benefit if herbivory of fungus-infected plant parts is reduced and the fungus is not extremely costly to the plant. The benefit to the plant is less certain if the fungus is very costly. Under such circumstances, plants might gain from herbivory of fungus-infected leaves, if this stops spread of the fungus across the plant. Indeed, it is even possible that an infected plant might actively recruit herbivores in this instance (cf. Chapter 9). However, again we await evidence that the conspicuous appearance of such fungi actually have an effect on natural herbivores that provides a fitness benefit at least to the fungi.

Inbar et al. (2010) argue that some plant galls are aposematic. They argue that some gall-inducing insects manipulate plant development so as to load the walls of the gall with defensive chemicals and to advertise this defence with conspicuous colouration. We would welcome comparative study of whether conspicuousness (that is, contrast with the rest of the leaf surface) of galls varies with levels of defensive compounds. Even more pertinently, we would welcome demonstration that any potential antagonist of the gall-dwelling insect is influenced by gall colouration. Such studies are particularly valuable because an alternative hypothesis for the accumulation of the red colour is that many galls have is that anthocyanins are involved in the stress response of the plant to the gall-inducing organisms.

Sherratt et al. (2005) investigated aposematic signalling among mushrooms. In a study based on field guides they could find no evidence that plants considered toxic to humans were distinctively coloured, but toxins did seem to be associated with reports of strong odour (Table 8.1), suggesting the possibility of olfactory aposematism or biochemical pleiotropy (see Chapter 2). Again, an experimental approach is required to take this further, for example exploring whether potential herbivores find the odours of such plants aversive and what the molecular mechanisms of such a correlation between odour and toxins are. Dobson and Bergstrom (2000) suggested that the pollen of some wind-pollinated plants have volatile compounds that repel insects that might otherwise consume the pollen. Their description suggests to us that (pending further study) the volatiles might best be thought of as a cue (or even an index) rather than a signal, since the same compounds appear to be associated both with defence and volatile emission. However, again we await manipulative experiments to explore whether detection of such volatiles leads to changed behaviour of herbivores that then leads to improved plant fitness. Although it is not framed in those terms, Massei et al. (2007) provide an experimental test of olfactory aposematism. Rabbits (*Oryctolagus cuniculus*) show a strong preference for red clover (*Trifolium pratense*) over white clover (*T. repens*), something the authors linked to volatile emissions from white

clover. Rabbits were demonstrated to eat considerably less red clover that had been sprayed with white clover extract compared to a control group given a control spray (Figure 8.2). Although it is possible that the pairing of red clover appearance with white clover olfactory cues might have triggered a neophobic reaction because of its unfamiliarity, these results are suggestive that white clover benefits from olfactory aposematism. If technically possible, a follow-up study could involve removal of the volatiles from the white clover (perhaps by genetic manipulation), or removal of the ability to detect the odour in rabbits, to explore whether this causes an increase in herbivory.

8.4 Floral mimicry

Permanently unrewarding flowers have evolved many times in the Angiosperms (Renner 2005). Plants that do not offer a reward to pollinators are generally considered to obtain pollination by animals through deception. As we discuss at greater length in Schaefer and Ruxton 2009, this can be achieved by two related (but different) mechanisms: mimicry and *perceptual exploitation*. We begin this section by differentiating between these two mechanisms.

Table 8.1 The association between odour (four-point scale) and edibility (two-point) for mushrooms in North America and Europe. Redrawn from Sherratt et al. (2005).

	Edibility		
Odour	**Edible**	**Poisonous**	**Total**
North American data			
None	8	1	9
Not distinct	95	14	109
Smells	28	15	43
Unpleasant	6	10	6
Total	137	40	177
European data			
None	53	2	55
Not distinct	31	3	34
Smells	128	13	141
Unpleasant	6	13	19
Total	218	31	249

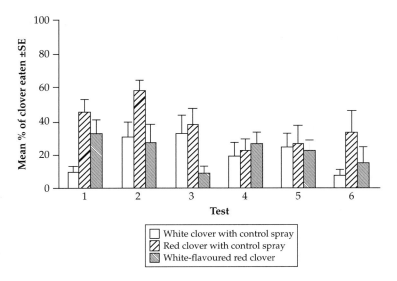

Figure 8.2 Results of six replicate experiment of Massei et al. (2007) on clover attractiveness to rabbits. Illustrated are means and S.E. Redrawn from Massei et al. (2007).

8.4.1 Batesian floral mimicry and perceptual exploitation

Batesian mimicry by unrewarding food-deceptive flowers has been a productive research topic, mainly because mimicry has been pitted against an alternative scenario, termed *general food deception*, where rewardless flowers do not mimic a particular model species, but rather exploit pollinators' tendency to react to large and/or conspicuously-coloured flowers (Nilsson 1992, Jersáková et al. 2006). The main difference between Batesian mimicry and general food deception is thus that there is a specific model only for mimicry. General food deception relies on *perceptual exploitation*, a term that we prefer to use because it illustrates the underlying mechanism.

Perceptual exploitation posits that receivers have pre-existing sensory or cognitive biases for particular traits and that selection therefore favours any sender that evolves a trait matching these biases (Ryan and Rand 1993). Because most flowers differ from their predominantly leafy green background in colour and shape, at least some pollinators have developed innate sensory biases to identify potential food resources against the background. Such sensory biases induce pollinators to preferentially respond to large and contrasting floral displays (Naug and Arathi 2007) or to particular colours

(Raine and Chittka 2007). Such biases can be induced by recent experience. For example, after visiting unrewarding purple flowers, pollinators preferentially visit another white floral colour morph. This effect persists even when white ping-pong balls are used (Dormont et al. 2009). Rewardless flowers may thus capitalize on these pre-existing biases to entice pollinators.

Assessing the relative frequency of Batesian mimicry and perceptual exploitation in rewardless flowers may allow us to evaluate the relative biological significance of each concept. Although flower species from 32 families are permanently rewardless (see Renner 2005), detailed data are available only for orchids, which are the prime example of deceptive flowers, with roughly a third of all species not offering any rewards to pollinators. To achieve pollination, the most commonly employed strategies of rewardless orchids are food and sexual deception. Food-deceptive species may mimic rewarding species or rely on perceptual exploitation, while sexually deceptive orchids mimic the scent of female pollinators and entice males to attempt to copulate with the flower that also provides some visual resemblance to female pollinators (Dafni 1984, Schiestl 2005).

Jersáková et al. (2006) reviewed mechanisms of pollination by deception in orchids. They considered that the most commonly adopted deception by orchids was generalized food deception, but brood-site imitation (see Section 8.4.5) and sexual deception were also common. These authors suggest that generalized food deception based on perceptual exploitation occurs in six times as many genera as specialized Batesian mimicry. They also mention rendez-vous attraction, where rewardless orchids mimic the appearance of flowers that female bees feed on, with the plant thus benefiting from visits of male bees searching for females. This mechanism warrants further investigation since (as Dressler 1982 suggested) it may be an important step in the evolution of sexual deception. Further work would benefit from exploring one or more of the model systems listed by Jersáková et al. to explore how much pollination occurs by males (to which rendez-vous attraction applies) and how much by females (to which it does not). In another putative form of deception, Brodmann et al. (2008) report that flowers of the orchids *Epipactis helleborine* and *E. purpurata* release green-leaf volatiles that attract social wasp species that would commonly prey on the caterpillars whose damage triggers very similar volatile bouquets. These plants offer a nectar reward to pollinators, and so the extent to which wasps are deceived by plants or come to associate the volatiles with this reward remains to be explored. In general, it is clear that potential examples of pollination by deception are taxonomically and mechanistically diverse. The review of Cozzolino and Scopece (2008) discusses an important difference between food and sexual deception. Sexually deceived pollinators are less likely to carry pollen between different flower species than food-deceived pollinators. Consequently, post-zygotic isolation barriers should be more important in food-deceptive flowers.

While the high proportion of generalized food-deceptive orchids illustrates the importance of perceptual exploitation, we know very little about which mechanisms are exploited. The best evidence that non-rewarding flowers exploit sensory biases of pollinators concerns scent mimicry in the sexually deceptive orchid *Ophrys exaltata*. While the orchid emits volatiles that are identical to that of the sexual pheromone of the pollinator, their relative proportions differ (Vereecken and Schiestl 2008). Importantly, imperfect mimicry does not reduce the attractiveness of the flowers in the way one might expect. On the contrary, male bees prefer flower bouquets to the sexual pheromone of the females (Vereecken and Schiestl 2008). This is likely to be explained by males' sensory biases to prefer unfamiliar pheromones, a tendency that may help to avoid inbreeding. They further argue that orchids are pollinator-limited but female bees are not, and so this preference in the male bees has a stronger selective influence on orchid odour than on female odour. Further exploration is warranted, linking variation between orchids and variation between female bees with whether male bees are responsive to this variation in ways that can be linked to improving their fitness. However, one important implication is that sensory biases may select for signal divergence and imperfect mimicry (Vereecken and Schiestl 2008), whereas Batesian mimicry (in the absence of strong penalties) selects for signal convergence.

Generalized food deception and Batesian mimicry are sometimes not clearly distinguished in the literature, perhaps because both strategies involve negative frequency-dependent selection (Ferdy et al. 1998; Anderson and Johnson 2006). The commoner a rewardless flowering species is, the higher the likelihood that pollinators learn to avoid that particular species. As counter-adaptations to reduce the effects of negative frequency dependent selection, rewardless plants often have pronounced longevity of their flowers as well as heightened variability in chemical or visual flower traits (Ayasse et al. 2000). In line with theoretical predictions, rare morphs in flower polymorphisms of unrewarding flowers are often selectively favoured and many rewardless species show high inter- and intra-individual variability in floral traits (Ayasse et al. 2000; Gigord et al. 2001; Vereecken et al. 2007).

Antagonistic co-evolution between rewardless flowers and pollinators seems likely. However, the interactions between orchids and their pollinators are often highly asymmetric and most evolution apparently occurs on the side of the plant (Nilsson 1992). Selection is unlikely to act strongly on pollinators; as mistaking an unrewarding flower for a rewarding one is a waste of time and energy, but does not inflict strong fitness costs compared with, say, a predator mistaking a harmless snake for a poisonous one. The typically low frequency of most unrewarding plants further reduces selection for learned avoidance. Finally, pollinators need to be tolerant towards variable nectar or pollen returns, since they are unable to distinguish from a distance a flower that has recently depleted by another pollinator from one that has not (Renner 2005). A normally rewarding flower may be temporarily unrewarding, and it is this unpredictability of nutritional returns that is exploited by rewardless flowers.

The concepts of Batesian mimicry and perceptual exploitation can be distinguished. A core tenet of Batesian mimicry is that the signals of the mimic should adverge to that of the model. Moreover, there are conditions that pertain only to Batesian mimicry: a strong spatio-temporal overlap between mimic and model, both species relying on the same pollinators, and pollinators that mistake one species for the other (Roy and Widmer 1999; Johnson et al. 2003b). If fitness costs associated with identifying each particular individual as a mimic or a model are low (as they are likely to be for pollinators), Batesian mimicry requires a tight spatial and temporal association between mimic and model (Joron 2008). In contrast to Batesian mimicry, generalized food deception does not require a specific model and does not require that pollinators confuse one species with another. Rather than mimicking a specific model, generalized food deception exploits pollinators' innate sensory biases that may either arise from constraints on the structure of the sensory recognition system, or that may be adaptive by allowing animals to forage more effectively on the most rewarding food sources (Raine and Chittka 2007).

8.4.2 Empirical evidence for floral mimicry

It is important to consider potential magnet effects when evaluating whether pollination by deception occurs. There is evidence that pollinators may be preferentially drawn to larger patches of flowers, and so a rewardless plant may benefit from close proximity to rewarding plants even if there is no deception (Johnson et al. 2003b) – this is the magnet effect. In an experimental study by Internicola et al. (2007), rewardless plants obtained more visits when

placed with similar-looking rewarding flowers than when with dissimilar-looking flowers. Similarly Gumbert and Kunze (2001) found that bees foraging on a rewarding species were more likely to visit an unrewarding species as well if the two were similar in appearance. Similar results were found by Gigord et al. (2002) and Johnson et al. (2003b). Given that floral communities vary spatially and temporally, many pollinators will not have strong innate preferences for particular species (although they are likely to have innate preferences that help them identify flowers from non-flowers). In general, rewardless flowers will benefit from visits by naive generalists (Gigord et al. 2002) regardless of their appearance relative to co-occurring rewarding flowers. However, some resemblance to local rewarding species will enhance rates of visits by experienced pollinators. There may be little selective pressure in this case for precise mimicry, providing mimics are sufficiently rare that their costs to the pollinator (in terms of wasted time) are insufficient to drive a need for the pollinator to learn to differentiate between rewarding and unrewarding species.

In an influential review, Roy and Widmer (1999) argued that no case of floral mimicry had been fully verified. Their key issue was that previous studies had failed to demonstrate that individuals of at least one species benefited because of sensory confusion with the other species causing a change in the behaviour of pollinators. They say, 'to test whether resemblance is adaptive in a Batesian system, one needs to establish that the mimic receives more visits and has higher fitness when the rewarding species is present than when it is absent.' However, this might simply be a magnet effect, so we feel that it is also essential to demonstrate that the fitness benefit is higher when the model is present that when the model is replaced by a plant that offers similar rewards but has quite a different appearance (see Johnson 2000).

In Müllerian floral mimicry a rewarding species gains enhanced pollination through its similarity to another rewarding species. Roy and Widmer (1999) also suggest that 'to test whether resemblance is adaptive in a Müllerian system, one should test whether individuals of the rarer species (which is thus termed the mimic) have higher fitness when they are most similar to the more common species (the model)'. One of the most convincing demonstrations of Müllerian mimicry in plants is the South American perennial herb *Turnera sidoides pinnatifida* in the study of Benitez-Vieyra et al. (2007). There is geographical variation in the colouration of this species, which matches that of geographical variation of co-occurring malvaceous species. This co-variation is difficult to explain as convergent evolution. In further support of mimicry, the flower colours of the putative model and mimic are not common in the local floral communities, nor are such colours seen in other subspecies of *T. sidoides* or more generally in the *Turneraceae*. The authors explain the apparent advergence of the herb towards looking like a mallow by the relative rarity of the herb in the environment. In choice tests, pollinators did not differentiate between the two plant types. The authors further demonstrated that the herb receives more visits by pollinators when growing with the mallow than when growing alone. We would welcome a further manipulative (transplant) study to explore how the rate of pollinator visitation compares in herbs that do and do not provide a close visual match to the locally abundant mallow.

Anderson et al. (2005) report similar between-population variation in the appearance of a model (*Zaluzianskya microsiphon*) and a Batesian mimic (*Disa nivea*). They report that the specialist long-proboscid fly pollinator did not appear to distinguish between flowers of the two species. They argue that the flower size and colour of these species are quite different from other species that utilize the same pollinator, and so the similarity is unlikely to be explained by convergence of both species to sensory biases of the pollinator. This is a promising case of Batesian mimicry and measurement of fitness benefits to putative mimic individuals would be worthwhile.

Peter and Johnson (2008) suggest that the non-rewarding South African orchid *Eulophia zeyheriana* benefits from mimicry of the rewarding plant *Wahlenbergia cuspidate*. The orchid seems to be pollinated only by solitary bees of a single genus, and bees are commonly found carrying pollen from both plant species, suggesting that the insects do not discriminate. Peter and Johnson present evidence that the colouration of the two species is very

similar to bees. This pale blue colour is entirely different from that of the orchid's close relatives. Manipulative experiments demonstrated that the orchid received fewer visits from pollinators when its colour was manipulated to look less like the 'model', relative to controls manipulated in ways that did not change colour but controlled for other possible cues. The orchid seems only to be found in close association with the 'model', and translocation experiments demonstrated that proximity to *W. cuspidate* increased visitation rate (Figure 8.3). This could, however, be a magnet effect, and we would welcome exploration of whether proximity to the 'model' provides increased visitation rates compared to proximity to a similar density of similarly rewarding but different-looking flowers that are also pollinated by the same insects, this control group might most easily be created by manipulation of the appearance of the 'model' species.

Another interesting study is the suggested mimicry of *Scabiosa columbaria* by both a rewarding orchid (*Brownleea galpinii*) and a rewardless orchid (*Disa cephalotes*) (Johnson et al. 2003b). Both mimics are reported to be rarer than the model, and in choice experiments pollinators did not seem to discriminate between the three species, but were less willing to land on orchids modified to reduce visual similarity to the model, or land on closely related orchids that differed in appearance from the model. The authors argue against convergence of all three species on a generalized appearance driven by inherent sensory biases in the pollinator by suggesting that across its range the model experiences a wide variety of different pollinator types but shows no variation in appearance associated with variation in the importance of different pollinators. The two mimic species are also suggested to only be found near the model.

We thus have a small number of recent studies that provide increasingly compelling evidence of both Batesian and Müllerian floral mimicries.

8.4.3 Pseudoflowers

Many rust fungi have outcrossing mating systems and require insect visitation for sexual reproduction. They attract insects with sugary rewards; however, in some cases mimicry of flowers has been

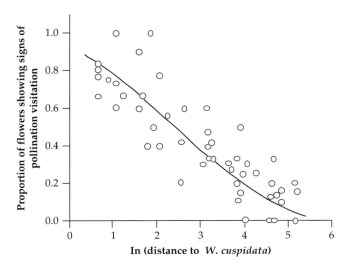

Figure 8.3 The fraction of *E. zeyherianana* flowers showing signs of visitation by an insect pollinator as a function of distance from a patch of *W. cuspidata*. Redrawn from Peter and Johnson (2008).

suggested. Roy (1993) considers the 'pseudoflowers' of the rust *Puccinia monoica* on its hosts: species of the genus *Arabis*. Infected leaves create structures that look like a flower (see Plate 14). However, they do not look or smell like flowers of *Arabis* species. Rather Roy suggests that they visually mimic buttercups (*Ranunculus inamoenus*) that grow in similar localities to *Arabis* species. Visitation rates to both pseudoflowers and buttercups are enhanced by increasing the local density of the other (Roy 1994). However, since both are rewarding, this could be a magnet effect (Johnson et al. 2003b), rather than any mimicry whereby pollinators are duped. Indeed, the rewarding nature of the pseudoflowers means there is no need for duping. Further, confusion by insects might be to the detriment of both insects and rust, since the pseudoflowers are associated with an order of magnitude more nectar than co-occurring flowers. The smell of the pseudoflowers is also quite different from that of the buttercups (Roy and Raguso 1997). However Roy (1993) demonstrates that both buttercup flowers and pseudoflowers achieve higher visitation rates when in a mixed group than either would achieve in a single-species group of the same size and density. She suggests that the reason for this may be that the different spe-

cies provide complementary attraction cues or visitor rewards. However, this result again does not strongly support the mimicry hypothesis for explaining the appearance of the fungi, and Roy concludes that 'the fungal pseudoflowers are not mimicking any particular species, but are more general flower mimics'. This is what we call perceptual exploitation because the element of confusion so central to mimicry is missing.

Floral mimicry has also been suggested for the effect of the mummyberry fungus (*Monilinia vaccinii-corymbosi*) on its host plant: blueberry (*Vaccinium* spp.) (Batra and Batra 1985). It is true that infected leaves attract insects that transport spores to host flowers. However, since the infected leaves provide a sugary food reward, the fact that the same insects that are attracted to them are also attracted to flowers is not sufficient evidence to suggest floral mimicry. Indeed, it is quite likely that pollinators can keep infected leaves and flowers apart based upon visual cues associated with morphology and location on the plant.

The anther smut fungus *Microbotryum violaceum* is an obligate parasite of many species of *Caryophyllaceae*. Flowers of infected plants are sterile and bear only fungal teliospores. These are transported by insects from infected to uninfected flowers. Although highlighted as a form of mimicry by Ngugi and Scherm (2006), infected flowers still produce nectar and there may be little need to dupe insects into visiting.

There is little evidence that any fungi profits by duping insects into responding to it as if it were a flower. In fact, it is easier to make a case for some plants mimicking fungi. Kaiser (2006) reviews studies on fungus gnat flowers that have an appearance and smell similar to mushrooms and are pollinated by female insects that normally deposit their eggs in fungal fruiting bodies. However, it is not clear whether a specific model is mimicked or whether only a generalized resemblance is required to dupe the insects. In the next section we explore such brood-site mimicry by plants further.

Before leaving fungi we mention one recent study that might involve (admittedly non-floral) mimicry. Sclerotia of an athelioid fungus (*Basidiomycoat, Agaricomycotina*) of the genus *Fibularhizoctonia* form 'termite balls' that are tended by Reticulitermes termites as if they were their own eggs. Since the termites normally destroy fungi that they find in their nest, this fungus enjoys competitor-free space. The fungus does not germinate in egg piles but only after they have become shrunken and deformed with age, which triggers the termites to discard them in a corner of the nest. There is likely to be dead plant material on such refuse piles that the fungus can exploit, gaining the energy to produce new termite balls, that are collected by the termites and taken to their egg pile. This exploitation of the termites' parental care involves chemical mimicry (Matsuura et al. 2009) and likely size, shape, and textural mimicry also.

8.4.4   Mimicry of carrion and dung

The dead horse arum (*Helicodiceros muscivorus*) appears to recruit pollinators by duping carrion-visiting blowflies into mistaking it for a dead animal. The pollinators are mainly *Calliphoridae* blowfly females that typically lay their eggs in carrion (Stensmyr et al. 2002), and such insects have been demonstrated to react similarly to the smell of rotting flesh and volatiles emitted by arum florets (Stensmyr et al. 2002). The plant also produces heat

at the same time as its distinctive odour, and such heat increases attraction of potential pollinators (Angioy et al. 2004), likely by mimicking heat generated by microbial decay of a dead animal.

Several species of the moss family Splachnaceae appear able to attract female flies that lay their eggs in or on carrion or dung (reviewed by Marino et al. 2009). Many of these mosses grow on faeces or carrion, but by the time the moss has grown to produce dispersive stages the substrate will no longer be attractive to flies as an oviposition site. However, it does seem that the spores of these mosses are dispersed by flies that are drawn by odour and visual cues from the mosses and subsequently are drawn to suitable sites both for their oviposition and for the moss to take hold. This is an exception to the trend discussed in Chapter 3, that plants that transport their propagules on the outside of animals do not attract those animals towards them.

Lev-Yadun et al. (2009) suggests that production of carrion or dung odours may provide protection from vertebrate herbivores, if these herbivores associate these cues with the presence of carnivores and thus heightened predation risk. This argument is logically plausible, but requires follow-up work to explore its ecological importance. In this case, we would be interested in observation of natural vertebrate herbivores in the vicinity of plants that are considered to be dung or carrion mimicking. Do they show different behaviour when downwind of such plants? Can any difference in behaviour be related to fitness gain to the plant?

8.5   Non-floral mimicry in plants

Mimicry in other plant systems is not as well investigated as in flowers. It may be that plants use insect-like spots simulating insect aggregations on their flowers to increase their attractiveness to pollinators (Johnson and Midgley 1997). Lev-Yadun and Inbar (2002) suggested that various plant markings and structures provide relief from herbivores through mimicry of infestation by herbivores (aphids and caterpillars) or mimicry of predators such as ants. Several apparent cases involve plants whose colouration may simulate the presence of herbivores by sporting insect-like spots, egg-like spots, or variegated patterns on their leaves that may mimic damage caused by herbivores (e.g., Gilbert 1975, Benson et al. 1976, Lev-Yadun and Inbar 2002, Soltau et al. 2009, Lev-Yadun 2009; and see Plate 15). Herbivores may avoid already-infected plants because competition may be higher on such plants or these plants might already be better defended through induced defences. Photographs can be very suggestive (e.g., Lev-Yadun and Inbar 2002), but we need manipulative experiments to explore whether removal of these patterns leads to changes in herbivore behaviour that leads to fitness costs for the plants. Again, only when such benefits have been shown need we probe whether they arise (at least in part) because of duping of herbivores. For example, if the benefit arises from misidentification of putative anti-mimicking patterns as ants, then we would expect herbivores that are vulnerable to ant attack to avoid the plants, but not herbivores that are indifferent to ants. Further, we might expect that predators of ants might be similarly fooled and preferentially attracted to such plants. As yet, in none of the examples above has an anti-herbivore benefit of the identified plant characteristics been demonstrated, let alone that such benefits arise through mimicry of the animals involved. This also applies to the possible mimicry of fungi infestation in plants sporting white colouration (Lev-Yadun 2006).

That some colourful seeds may mimic fleshy fruits has long been suggested (Van der Pijl 1972, McKey 1975, Foster and Delay 1998). Unrewarding seeds have low visitation rates compared to fruits whose flesh offers a reward to potential seed dispersers (Peres and van Roosmalen 1996, Galetti 2002). However, this hypothesis of fruit mimicry has been challenged by Peres and van Roosmalen (1996), who suggested that such seeds benefit from mutualism with terrestrial frugivores, and not deception of aerial frugivores. They found that coloured seeds are consumed and dispersed intact by terrestrial birds that may use them to aid the mechanical breakdown of smaller, softer seeds in their digestive system. However, it is unclear whether such mutualism is widespread, as seeds of other putatively mimicking species that were ingested by large frugivores had lower germination probabilities than control seeds (Galetti 2002). But this result is not strongly supportive of the hypoth-

esis that these seeds are adapted to dupe frugivores into dispersing them. It is important to note that mutualism with terrestrial frugivores and deception of aerial frugivores are not necessarily mutually exclusive mechanisms: Foster and Delay (1998) observed putative mimetic seeds taken both by aerial and terrestrial frugivores. A critical point is that model species for mimetic seeds have rarely been identified, let alone the overlap between mimic and putative model addressed and fitness benefits quantified. Further, obtaining ingestion by deception seems difficult when fleshy fruits will provide quite different tactile cues to hard seeds while being handled prior to ingestion. Observational study of allegedly-duped frugivores interacting with these seeds would be very welcome. Thus, at present, the case for colourful seeds mimicking fleshy fruits remains unproven.

Burns (2005) presented another case of apparent fruit mimicry, where this time mimic and model both are congeneric species that co-occur and both have fleshy fruits. One of the species, *Rubus spectabilis*, produces red and orange fruits, while its congener, *Rubus parviflorus*, only produces red fruits. Birds prefer fruits from the putative model *R. parviflorus* over *R. spectabilis*. In presence of the red fruits of the model, birds remove the red morph of the mimic, *R. spectabilis*, more quickly than the orange morph, suggesting that they may confuse the red fruits of the mimic with those of the model. However, birds also prefer the red fruits of the mimic over orange fruits in the absence of the model, suggesting that other processes might explain their preference (Gervais et al. 1999). Thus, while fruit mimicry is a plausible concept, evidence supporting it is currently scant.

8.5.1 Mistletoe host resemblance

Probably the most frequently-cited potential case of non-floral mimicry is the resemblance of parasitic mistletoes to their hosts. There are 72 species of Australian mistletoe, of which around half are considered to show vegetative similarity to their host (Barlow and Wiens 1977). Such similarity has also been reported in mistletoes in other parts of the world, and in other parasitic plants (Bannister 1989, Barlow and Wiens 1977). Not all mistletoes are host specialists, and the appearance of one species may be strikingly similar to some hosts but not others.

There may be reasons for similarity that have nothing to do with mimicry as protection against herbivores. Parasite and host are necessarily in a similar physical situation, so they may be exposed to similar environmental pressures and have separately evolved to a similar appearance. There may be genetic or other constraints that prevent convergence to the same appearance in some cases. However, it does appear (through comparison with closely related species) that the mistletoe has converged to the appearance of the host, rather than both species converging on a similar appearance. It might be that as the mistletoe taps into the systemic system of its host, it experiences host hormones, and it may be that biochemical triggers of development normally used by the host influence the appearance of the parasite. This was called the 'Host Morphogen Hypothesis' by Atsatt (1983). Canyon and Hill (1997) suggested that if a single mistletoe species had its appearance manipulated by the different hosts, in each case to resemble that host, then this would be consistent with this thesis. Unfortunately, this prediction is also made by other theories. Further, failure to observe such flexibility in host matching cannot be taken as strong evidence against this theory, since the hormonal biochemicals of some hosts may influence the parasite more effectively than others.

An explanation based on plant–animal interaction is that resemblance is driven by mimicry of the host by the mistletoe to gain protection from herbivores. This hypothesis predicts that mistletoes that closely match their hosts are more attractive (post-detection) than their host to herbivores (Canyon and Hill 1997). Canyon and Hill also suggest that there is a logically distinct Camouflage Hypothesis that suggests that close similarity between mistletoe and host arises because, in order to reduce exposure to herbivores, the mistletoe matches its background, which happens to be the host plant. We do not see this as logically distinct from the mimicry hypothesis because it only functions if the host species is less attractive to the herbivore than the mistletoe. We are also uncertain about their prediction from this hypothesis that non-cryptic mistletoes should suffer more from her-

bivores than cryptic (host-matching) ones. If a mistletoe (because of chemical defences say) was unattractive to herbivores then its low attack rate would reduce selection pressure for background matching. We think a more powerful test of this hypothesis would be through experimental manipulation of host appearance. If herbivore impact on the normally-host-similar mistletoe increases when the appearance of the host is altered, then this would suggest that the visual similarity brings a benefit (through mimicry or – equivalently in this case – background matching). As an alternative to experimental manipulation, one could also explore whether herbivory was lower when a given mistletoe species is on a host to which it is similar compared to a nearby host to which it is not.

The only study to compare these different hypotheses explicitly is that of Canyon and Hill (1997). This study involved comparison between *Amyema biniflora*, which is very similar in appearance to its host *Eucalyptus tessallaris*, and *Dendrophthoe glabrescens*, which is similar to one host (*E. tessallaris* again) but not similar to another of its hosts (*E. playtphylla*). The work was carried out in suburban parkland. Canyon and Hill found no difference between rates of herbivory on *D. glabrescens* on seven trees on which it was similar to the host compared to four trees of other species. Our ability to draw conclusions from this study is greatly weakened by the small sample sizes. They also found that herbivores attacked the mistletoes more than the hosts both when the mistletoes where host-similar and when they were not (Figure 8.4). Finally, giving some support to the Host Morphogen hypothesis, *A. Biniflora* exhibited some degree of variation in leaf width when on different host species. Thus, there is no strong evidence for herbivores influencing parasitic mistletoes to be visually similar to their hosts, but further multi-hypotheses studies like Canyon and Hill (1997) would be worthwhile. Such studies might utilize a range of less human-influenced locations in order to attract representative herbivore populations, and should use a larger sample size to increase statistical power.

One difficulty in exploring selection pressure on mistletoe appearance is in identifying the herbivores involved. Although Barlow and Wiens (1977) argue that vertebrate herbivores use vision more

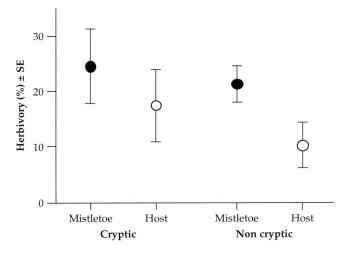

Figure 8.4 Rates of herbivory on *D. glabrescens* and on its host tree, both when on trees on which it was similar to the host (the cryptic situation) and trees of other species (non-cryptic). Illustrated are means and S.E. Redrawn from Canyon and Hill (1997).

often than insects when selecting plants, we do not find this a compelling case for disregarding insect herbivores as an important selective agent on mistletoe visual appearance. Bannister (1989) found that host-similar mistletoes had higher nitrogen concentrations (relative to their host) than non-similar ones, in a survey of 41 New Zealand species. The author argues that nitrogen content should be a good proxy for attractiveness to herbivores given that nitrogen is usually relatively scarce in plant material. The vertebrate herbivore population of New Zealand has changed dramatically in the last few centuries with the extinction of the Moas. However, we would still welcome work that explores the current herbivore pressure on the two types of mistletoe along the lines we outline above. Ehleringer et al. (1986) performed an essentially similar study to Bannister (1989), this time utilizing 48 Australian mistletoes, obtaining essentially similar results, with the conclusion that host-similar mistletoes had higher nitrogen concentrations than their hosts, whereas non-similar ones had lower concentrations than their hosts. Again, the vertebrate herbivore community in Australia has been much altered in the last few centuries, so we may be forced to accept that, especially for vertebrate herbivores, the selection pressures acting today on mis-

tletoe appearance may be quite different to those in the evolutionarily relevant past.

8.5.2 Mimicry of crops by weeds

It is well know that human agricultural practices can impose strong selection pressure on wild plants and animals on farmed land (for example, consider the rise of herbicide and pesticide resistance). Sometimes this selection can take the form of crop mimicry by weeds. A weed that looks like the crop is less likely to be weeded out by humans. Barrett (1983) details several convincing examples of such mimicry (e.g., the mimicry of maize *Zea mays* by a common weed in Central America, *Zea mexicana*). Although investigations have not been experimental, the case for mimicry is strong, with examples of mimetic weeds that look much more like their alleged model crop than they do like more closely related species. Barrett (1983) details examples where regional variation in crop appearance is mirrored by variation in the appearance of the mimic. Such systems warrant further study, to explore whether such variation is also apparent in control (non-mimetic) weed species in order to explore whether the variation in the putative mimic is driven by model crop appearance or by some confounding environmental factor. Study of the changing selection pressure on crop mimics in regions where human weeding is being replaced by chemical control of weeds would also be interesting.

Another phenomenon in crop mimicry by weeds is the mimicry of seeds (in size, shape, and weight) in order to make mechanical separation more difficult. Again, Barrett (1983) provides convincing examples. Although crop mimicry has not been intensively studied, it is conceptually straightforward, empirically well supported, and selection is likely to be strong and consistent; this is an unusual case in plant–animal interaction where the animal is *Homo sapiens* and so our visual impression of putative cases of mimicry is an appropriate one. Crop mimicry could be a compelling example of contemporary evolution on time scales that only comprise a few decades to several centuries.

8.5.3   Can plants avoid grazing by mimicking defended plants?

It has repeatedly been suggested (Launchbaugh and Provenza 1993 and references therein) that otherwise attractive plants might obtain protection from herbivores by mimicking more defended plants. However, evidence for such mimicry has not been forthcoming. We suspect that such mimicry will be uncommon or non-existent. The main reason for this is that (in general) herbivores can sample a defended plant at little cost or risk of injury. Mimicry becomes less effective if herbivores are able to sample individual plants and identify them as either models or mimics. Essential to such differentiation is that the herbivore has the ability to identify defended plants quickly. In complex natural environments it may be difficult to separate one plant from another visually or through olfaction, and so taste is likely to be a particularly important sense for herbivores to quickly categorize a plant as attractive or not. This would allow herbivores to spit out particular aversive material rather than swallow it, further reducing the costs and risks associated with sampling. Hence, we expect that herbivores will rely heavily on taste, since post-ingestion processes operate on too slow a timescale to allow effective categorization in individual plants or micro-habitats as attractive for feeding or not,

and also requires potentially harmful substances to be ingested. Use of sampling and taste discrimination should however make it difficult for a undefended plants to avoid herbivory by mimicry, since the compound-specialist receptors involved in taste likely make mimicry of toxin compounds by benign ones difficult.

It may be that Müllerian mimicry is possible: with defended plants that share the same potential herbivores also sharing the same defensive compounds, in order to ease the herbivore's task of successful identifying these plants as defended. However, there may be selection for novel compounds if novelty of taste aids aversive learning, and/or because a common defensive compound would provide strong selective pressure for the herbivore to evolve tolerance to that particular compound. Hence, we would not be surprised if there is little or no Müllerian (or Batesian) mimicry against herbivores.

8.6   Mimicry of plants by animals

The apparent mimicking of twigs, leaves, and flowers by animals appears to be a widespread phenomenon (see Skelhorn et al. 2010a). Although such mimicry is very compelling to human observers, until recently there had been no demonstration that any animal gained an anti-predatory advantage through being misidentified as a plant part. However, in a laboratory study, Skelhorn et al. (2010b) demonstrated an anti-predatory benefit to twig-mimicking in caterpillars that could not be explained by crypsis (where detection of the potential prey is hindered). These authors argue that this benefit is caused by masquerade (where the prey gains protection through misidentification as a twig). Demonstration of a similar benefit to masquerade in a field situation is still awaited. A lack of research hinders our interpretation of other examples of masquerade as plant parts: such as mantids that seem to have evolved to resemble plant parts; and fish, frogs, and insects that look like leaves (see examples in Cott 1940, Edmunds 1974).

Recently, Yanoviak and co-workers (2008) describe a striking case where fruits may be the model for parasite-induced changes in ants. Ants parasitized by a nematode developed 'conspicuous red gasters' that contained hundreds of yellowish nematode eggs. Comparing bird predation on infected ants with that of uninfected ants, in an experiment where both were experimentally restrained, the authors found that infected ants were removed more quickly. In a second experiment, they found that red and pink artificial fruits were more quickly removed than artificial fruits of other colours. Yanoviak et al. (2008) suggest that their paper presents the first case of fruit mimicry by an insect and one of the rare cases where model and mimic belong to different biological kingdoms.

Key assumptions of mimicry such as the spatial and temporal overlap between ants and the model fruit species, *Hyeronima alchorneoides*, have not been quantified. Likewise, it has not been established whether ants and fruits are consumed by the same animals. It is thus unknown whether animals are duped into misidentifying infested ants as fruits. The latter is unlikely because the colour similarity between infested ants and the model red fruits is not strong. When analysed according to a model of avian vision (Vorobyev and Osorio 1998), the colour contrast ($\Delta S = 10.6$) is large enough to indicate ease of discrimination for birds (see Siddiqi et al. 2004). Imperfect mimicry may flourish given that the penalties for misidentification are low, owing to the reduced defences of parasitized ants. However, mimicry is not consistent with data on avian discrimination abilities. Omnivorous and frugivorous birds distinguish between insects and fruits and adjust their behavioural responses to colour stimuli accordingly even if insects are immobile (Gamberale-Stille 2001; Gamberale-Stille and Tullberg 2001). Finally, discrimination between living insects and fruits is not only based on colour but also involves shape and movement. Thus, the contention that colour change in parasitized ants represents an example of fruit mimicry depends on a number of untested assumptions.

Rather than duping frugivores into misidentifying the ant's gaster as a fruit, the colour change of parasitized ants may simply increase their conspicuousness to potential predators of ants. This alternative hypothesis requires none of the untested assumptions above. It does not impose temporal or spatial restriction – like the co-occurrence between model and mimic – on the effectiveness of transport

of the parasites. Further, this hypothesis is not frequency-dependent on the relative number of models and mimics, whereas the effectiveness of fruit mimicry is frequency-dependent.

Seasonal variation in the availability of the model provides an easily manipulated setting to distinguish between mimicry and alternative hypotheses. If ant predation occurs when no red fruits are available, the fruit mimicry hypothesis would not be supported. More specifically, the mimicry hypothesis predicts that per-capita predation risk of infected ants would decline if their density increases, but would probably increase with increasing fruit density. Conversely, the conspicuousness hypothesis is consistent with predation risk of infected ants increasing with the number of infected ants (because predator learning about the resource would probably have a stronger effect than predator satiation), and being only very weakly affected by the density of red fruits.

8.7 Conclusions

On the basis of current research, it seems that crypsis and aposematism may be less common in plants than in animals. If this trend is indeed true, then the main reason for this is likely to be the requirements of photosynthesis constraining plants' morphology and strongly influencing colouration through the green pigmentation of chlorophyll. Animals' mobility gives greater control over the background against which they are viewed, making crypsis easier. Further, the sessile nature of plants may give viewers a greater ability to inspect them at close quarters so as to see through any disguise (again making crypsis less attractive). The ability of plants to tolerate herbivore damage more effectively than animal prey can tolerate the attentions of predators may also reduce evolutionary pressure for aposematism (Williamson 1982).

Both Vane-Wright (1976) and Roy (1994) suggest one reason why floral mimicry may be expected to be less common that in the animal world: in animals, the consequence of the signal is to warn away predators, whereas in flowers it is to invite potential pollinators. Hence, there may be greater potential for competition between floral mimics than antipredator animal mimics. Another reason for less use of mimicry in plants may be the much greater control that mobile animal mimics have over their positioning with respect to models (Williamson 1982). Also, the sessile nature of plants may give viewers a greater ability to inspect plants at close quarters, making mimicry (like crypsis) a less successful strategy. Further, it may be more practical for a herbivore to sample a small part of a mimic without dire consequences, Williamson (1982) argues that, as a generality, animal models in mimicry systems are more likely to be highly damaging to potential predators than model plants are to herbivores. A study investigating whether this conjecture is supported would be worthwhile. Further, sampling of a plant by a herbivore is likely to be less damaging to the plant than sampling of an animal by a predator. That is, just as we argued with crypsis, it may be that the greater ability of plants to tolerate attack by herbivores lessens selection pressure to reduce attack rate through mimicry.

Increasing consideration of animals' sensory and cognitive processes will undoubtedly benefit the study of plant mimicry. In particular, mimicry remains challenging to demonstrate because it requires a signal receiver to misidentify different stimuli, and that this misidentification has important fitness consequences. We emphasize that exploitation of animals' perceptual processes is an alternative, less restrictive strategy; and that experimental manipulation may be required to distinguish mimicry from perception exploitation. As evidenced by the progress in understanding floral mimicry in recent years, more careful consideration of the assumptions underlying mimicry, as well as those of inherent and learned biases, will be needed to understand the evolution of similarity of appearance involving plants and plant parts.

Rewardlessness in flowers does seem a challenge to explain. Even with all the adaptations for deception discussed in this chapter, rates of visitation by pollinators are low compared to closely related rewarding species (Schiestl 2005). However, rewardlessness is a commonly occurring trait, most notably in around a third (7–8,000 species) of orchids. Schiestl (2005) argues that the energetic saving of not producing a nectar reward is minimal and the real benefit of rewardlessness is enhanced gene flow. It is certainly true that pollinators are more likely to quit a patch and are more likely to fly longer distances to their next flower after visiting a rewardless flower (reviewed in Schiestl 2005). Hence the idea of rewardlessness promoting outcrossing is plausible and merits further investigation.

An interesting alternative defensive appearance strategy has been suggested by Lev-Yadun et al. (2004): within- and between-plant variation in colouration might be a means of undermining background matching by insect herbivores. We deal with within-leaf variation (variegation) in Section 7.7. Exploring whether other forms of within- and between-plant variation in colouration makes plants less attractive to herbivores and/or makes herbivores more apparent and vulnerable to their antagonists should be amenable to relatively straightforward experimentation where colouration is manipulated either by use of selective breeding, genetic manipulation or application of paints. If variation in appearance is used by plants against herbivores then it might be useful to ask whether animals could use a similar strategy against external parasites. Such a strategy requires the co-option of a third species to act as the plant's defence and attack the herbivores; this interesting general strategy is considered more fully in the next chapter.

CHAPTER 9

Chemical communication by plants about herbivores

9.1 Introduction

After the discussion of aposematic signalling in the last chapter, our aim in this chapter is to explore further situations where plants transfer information, either to other plants or to animals. We focus on information about herbivory. Herbivore pressure is often spatio-temporally variable, and plants respond to herbivores or heightened risk of herbivory with a variety of induced defences. Although many herbivores are very mobile (e.g., migratory locusts), herbivory is often spatio-temporally correlated, with attacks on neighbouring plants being indicative of increased risk of future herbivory to a focal plant. Hence, plants can benefit from being informed about herbivorous attacks on neighbours. Animals too might be interested in information about the location of herbivores, since these herbivores may be competitors or prey. A particularly tantalizing idea is that there may be commonality of interest between plants being attacked by herbivores and predators, parasites, and parasitoids of those herbivores; and thus plants might attempt to recruit these enemies as a means of defending themselves against herbivores.

The term 'indirect defence' was introduced by Dicke and Sabelis (1988) to describe situations where plants attract, nourish and/or house other organisms that in turn reduce herbivory on the plants. It has been known for centuries that some plants provide homes within their structure that encourage ants to take up residence on the plants. Because these ants are often very aggressive, they clear the plant of potentially harmful insects. Plants may further entice ants with the offer of nourishment through provision of extra-floral nectar or fruiting bodies. Such plant–

animal relationships will be explored later in this chapter, where we will focus on the communication involved in such interactions. Plants may also signal to other antagonists of herbivores; the antagonists benefiting from an increased rate of prey-finding and the plant benefiting from reduced herbivore pressure. Price et al. (1980) are credited with the first clear suggestion that understanding plant–herbivore interactions requires that we also consider antagonists of the herbivores. Early empirical evidence that antagonists used volatiles released by the host plant to locate their prey came from Dicke (1986). Although the intuitive attractiveness of this idea has seen its rapid adoption in textbooks and popular texts, there has been long-standing controversy over its ecological relevance. Since such communication is related to issues at the heart of this book, we will focus on them in this chapter. However, before we address the issue of plants communicating with animals we begin by considering plants communicating with one other through volatiles released in the air. Much of the challenge to understanding communication between plants and animals lies in the number of interacting plant and animal species that may exist in any one natural ecosystem. Thus this chapter will start with very simple situations and gradually increase in complexity. We start with the simplest situation of all, a single plant individual talking to itself.

9.2 Within-plant communication of attack risk

Plants possess a complex array of chemical and physical defences against the animals and microbes that exploit them as a food source. These defences are often induced by direct attack or cues of increased risk of attack. This use of induced (rather than constitutive) defences can be understood since defences are often costly, and attacks are unpredictable. Therefore plants can save on the costs of defences that would be wasted if no attacks occur, but risk increased damage before they can deploy their induced defences. Thus, induced defences are likely to be more attractive against microbes and invertebrate herbivores, rather than vertebrate herbivores that can do catastrophic damage to a plant before induced defences kick in.

From our viewpoint the interesting issue is that, when subject to a localized attack, the defensive response may occur not only in the part of the plant that is attacked, but also in other parts of the same plant. Such a response can be seen as adaptive, since a localized initial attack on one part of the plant is likely to be a reliable indicator of future attacks on other parts. This within-plant spreading of an induced defence may result from the systemic transport of defensive metabolites from the area under attack to other plant parts. However, this is not the whole story, and there can also be *de novo* creation of defensive metabolites in response to signals from the attacked areas (see Heil and Ton 2008 for a review). This chemical signal can be transmitted through the vascular system of the plant, or can involve the release of volatile chemicals into the air from affected parts of the plant and their detection by other parts (see Heil and Ton 2008 for examples of each).

Heil and Ton (2008) argue that one advantage of airborne signalling is that it allows more widespread and rapid transmission across the plant than signalling via the vascular system. It seems likely that an initially localized herbivorous attack will spread to spatially close parts of the plant. However there may often not be any direct vascular connection between adjacent leaves on a branch, and (especially in larger plants) leaves in close proximity can originate on quite separate branches. In these cases, transport through the air will often be more rapid than through the vascular system. Moreover, an airborne signal can more easily encode the proximity to the site of the current attack because diffusion through the air is gradual meaning that concentrations are naturally highest closest to their point of release but will reach all parts of the plant, at lower and lower concentrations with increasing distance from the point of origin. Such a gradual transmission is not necessarily found in vascular signals. Transport in the xylem will necessarily only follow the upward transpirational flow, whereas that in the phloem will be away from the photosynthetic surfaces. While utilization of both systems may allow all parts of the plant to be reached, the strength of the signal reaching different parts may not reflect variation across the plant in the risk of subsequent attack. In summary, airborne signals may be more effectively transmitted and may more readily carry accurate information on spatial variation across the plant in risk of attack. Heil and Karban (2010) predict that we might particularly expect volatile signals in anatomically complex plants and in arid habitats where water conservation considerations limit the speed of vascular communication.

Heil and Ton (2008) also argue that volatiles may allow faster signalling: diffusion of molecules and their carriage in air currents can allow signal transfer over timescales of minutes or hours, whereas the slow speed of fluid flow in the vascular system of the plant might cause information to spread on a timescale of days or weeks. Some volatiles are released immediately upon damage, hence the characteristic smell to humans of a freshly cut lawn. Heil and Ton (2008) suggest that this separation of timescales may be used to allow two-step regulation, where detection of the volatiles triggers not full-blown expression of the defence but priming for that defence (physiological change that allows faster and/or more strong expression of the defence when it is finally triggered; see Frost et al. 2008 for a review on priming). Mechanistically, priming could involve the building of precursors of the chemical defences (as in commonality of precursors of red leaf pigments and chemical defences, discussed in Section 7.4) used to defend the plant against herbivores. In the two-step regulation model, volatile cues would prime defences, but triggering of the full defences would be delayed until a signal is received systemically. To us, this model would be particularly sensible if volatile signals were somehow less reliable than vascularly delivered signals. This is possible if (1) the same volatiles used by the plant in defensive signalling are also emitted in other contexts (e.g., after mechanical damage caused by wind), or (2) the volatiles used by the plant can also occur in the air around it even when the plant itself is not under attack.

The ecological relevance of the first of these conjectures is difficult to currently evaluate, since the volatiles used in demonstrated cases of within-plant defensive signalling have rarely been identified. However, plant–animal signalling appears to involve herbivore-related volatile blends that are distinct from those produced in other circumstances (see Section 9.4). The second of the conjectures can occur if neighbouring plants are attacked and emit similar volatiles to the focal plant. Whether this is a valuable source of information about increased risk of attack or a source of decreased signal reliability will depend on how strongly attacks are correlated at the appropriate spatial scales. The spatial scales over which volatiles can be detected has yet to be fully evaluated, but it is likely to be small. In a recent review, Heil and Karban (2010) found only one study (see Section 9.3) that reported volatiles to be effective over distances of up to 60 cm, whereas in other studies the threshold was much lower and in the range of 10–15 cm. We thus suspect that volatiles from neighbours may often be a valuable source of information rather than a cause of false alarms (see Section 9.3). As Heil and Karban (2010) suggest, priming of responses may be a useful mechanism to prevent run-away processes of the spread of signals from plant to plant such that spatially-distant plants not in immediate danger do not trigger defences that they do not need.

Heil and Ton (2008) also suggest that *eavesdropping* by neighbours can be a potential cost to within-plant signalling by volatiles. That is, the detection of warning signals from a neighbour could sometimes allow a focal individual to prime or activate its defences to prepare for potential attack in such a way as to allow it to gain advantage over the neighbour, which will be felt at a later stage as increasing competition with the neighbour. However, it is normally the case that parts of the focal plant will be closer to the point of volatile emission on that plant than parts of neighbouring plants are, and so the risk of eavesdropping is substantially reduced by

the strong decline in signal detectability with increasing distance. Signal detectability will generally be determined by physical factors, although these may play a less important role within a range of 10–60 cm. Feasibly, the level of detected volatile that triggers a response from the plant may be under selective pressure. If attack on one plant gives little useful information on the future risk to neighbours then the triggering threshold should be high to avoid false alarms; whereas in cases where attack on one plant in a locality strongly suggests that other plants in that locality will be attacked in time, a lower threshold would be useful to provide as early warning as possible about attacks in the area (we discuss this further in Section 9.3).

The physiological costs of volatile production have not been evaluated in comparison to vascular signalling. It seems likely to us that greater quantities of signalling molecules may have to be produced for volatile communication, because many molecules will be carried by air currents away from the plant and be wasted. There are further ecological consequences of volatile production that may bring additional costs or benefits; these are associated with eavesdropping by animals. The within-plant signals may be detected by nearby herbivores that might be attracted or repelled by them (see Section 9.5) or antagonists of the currently-attacking herbivores (see Section 9.4).

Since herbivory involves the destruction of cells and consequent release of volatile substances into the air, the evolution of within-plant signalling by herbivore-related volatile emission can easily be understood; especially since photosynthetic surfaces require gas exchange and the detection of different molecules occurs in contexts such as detection of competition through shading (see Box 9.1). Since the same individual is both sender and receiver, there should be no conflict of interest in signal design and no difficulty in maintaining signal honesty. However, as discussed above, there are numerous costs and benefits to the use of such signals, and so we would not expect their use to be ubiquitous. Rather, their use may be facultative, and only used by plants in cases when the cost – benefit trade-off is advantageous. We would welcome more investigation of this, and of the informational content of such signals. Based on the responses of antagonists to herbivore-associated volatiles (see consideration of specificity of signals in Section 9.4), it seems possible that such signals can encode information about the identity of the herbivore. As discussed above, it seems likely that the intensity of the signal may code information on the proximity of the threat and the intensity of the herbivore attack, although these may be confounded. Although this idea has not been explored, it may be that temporal variation in the intensity of volatiles may encode some information of distance to the emitter. That is, volatiles from more distance sources might arrive more irregularly because they are more influenced by turbulent air flows than those from closer sites. In a related vein, the ratio of different volatiles may be informative of distance, because volatiles differ according to their specific weight in the ability to diffuse through the air. The charismatic nature of the idea of plants signalling to animals has likely caused relative and unjustified neglect of within-plant signalling. This neglect is noteworthy because within-plant signalling likely represents the precursor for the evolution of more complex signalling between plants and between plants and animals.

Let us now turn to between-plant signalling. Here there is likely to be greater conflict of interest between sender and receiver, since these will generally be genetically different individuals, and so the evolution and maintenance of such signalling is more challenging to understand.

Box 9.1 Detection of competition from other plants

Plants can detect competition by a number of different means: including the strength of light flux, the spectral composition of the incident light, volatile compounds, and root exudates. The best-studied trigger of shade-avoidance behaviours in plants is the R/FR ratio. High reflectance of FR (far-red) radiation combined with high absorption of R (red) light is a highly characteristic property of chlorophyll-containing tissues. Phytochromes can sense a reduction in the ratio of red light in comparison to far-red wavelengths that is characteristic of proximity to plant tissue. However, other aspects of the incident spectra (particularly the intensity of blue light) can be important, as well as the overall flux density and the intensities of FR and R radiation, even without change in their ratio (see Ballare 2009 and references therein). The situation is further complicated by evidence (e.g., Pierik et al. 2003, 2004) that plant volatiles may also be an important component of shade-avoidance behaviours.

A key challenge for a plant is to differentiate between competition between different parts of itself and competition arising from other plants. Novoplansky (2009) summarizes the various ways by which such differentiation might be achieved. For example, shading of leaves near the centre or lower in the plant are more likely to result from self-shading than shading of leaves near the apex of branches or higher in the plant; and there is evidence that plants respond accordingly in terms of triggering branch elongation only in response to shading of higher or peripheral leaves. Although shading at the beginning and end of the day are more likely to be triggered by other plants than shading when the sun is higher in the sky, there is not strong evidence for differential shade-avoidance behaviour in response to the time of day associated with shading. This may be because shading only at dusk or dawn is relatively less costly to the plant than shading when the sun is stronger and moves more slowly across the sky.

It may be that uncertainty about whether competition is occurring between parts of one plant or between different plants might provide an opportunity for deception. A plant might be able to gain advantage if its shading of a neighbour is misinterpreted by that neighbour as within-plant shading. However we know of no study that has explored whether any deception occurs, or whether avoidance of such deception has influenced the evolution of the triggers of shade-avoidance behaviours.

Shade avoidance behaviours can include changed direction of growth, stem and petiole elongation, changes in the shape of new leaves, changes in leaf orientation, reduced branching, and increased senescence of shaded leafs. Such shade-avoidance behaviours seem to be finely controlled, with Novoplansky (2009) listing several studies that show that such growth is curtailed if initial growth does not yield a reduction in shading. Novoplansky interprets this as plants seeking only to 'fight the battles that can be won', and so reducing investment in avoiding shading when initial investment yields no or little benefit. This author demonstrated that plants will grow preferentially towards the direction with lower light levels if the brighter light direction features an R/FR ratio indicative of shading. His interpretation is that such a response is optimal if current shading is likely to become more costly in future, as the shading individual grows. Similarly there is evidence of growth being sensitive to the species identity of neighbours and even their genetic closeness to the focal plant (see Novoplansky 2009), so competition detection in plants is clearly a complex, highly evolved trait. Competition detection can even occur in seeds, with timing of germination being demonstrated to be linked to local density of seeds or seedlings (Tielborger and Prasse 2009). However, understanding of the functional aspects of such traits will need to advance further before we can make progress in understanding evolutionary aspects, and the signal-reliability issues and signal-receiver conflicts that are the recurrent interests of this book.

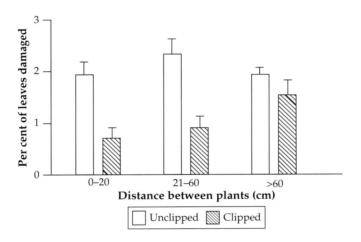

Figure 9.1 The (mean and SE) percentage of leaves damaged in sagebrush plants at different distances from either experimentally clipped or unclipped control conspecifics. Redrawn from Karban et al. (2006).

9.3   Between-plant communication of attack threat

That volatile production triggered by the damage done by herbivores on one plant can trigger increased anti-herbivore responses in a spatially-near plant has been known for some time and frequently demonstrated under laboratory conditions. However, field confirmation was slower in coming: Karban et al. (2004) demonstrated that wild tobacco (*Nicotiana attenuata*) growing in a field near to experimentally clipped sagebrush (*Artemisia tridentata*) became more resistant to herbivores than control plants near to unclipped sagebrush. Undamaged sagebrush individuals near to clipped conspecifics also experienced reduced herbivore damage relative to controls; these effects occurred only at relatively short interplant distances up to 60 cm (Karban et al. 2006; see Figure 9.1). Strangely, however, tobacco plants do not respond to experimentally clipped conspecifics (Karban et al. 2003). Based on previous laboratory studies, these results have been interpreted in terms of the emission and detection of volatiles. Plants' roots can be just as subject to attack as above-ground parts, and it may be that the mycorrhizal networks might provide an avenue for plant-to-plant communication about attack (Guerrieri et al. 2002; Dicke et al. 2003), although we await exploration of this.

The work of Karban et al. (2004) gives one example where the receiver of information about herbivores benefits. It is easy to see how this benefit could be widespread, since information about increased risk of attack that can be used to prime or induce defences in good time should often be advantageous. The use of within-plant signalling (as discussed in Section 9.2) would also make the initial evolution of such a response easy to understand. Consider, however, the consequences for the emitter. It may be that emission of the volatiles used by other plants is costly to the emitter. This cost could be the physiological cost of producing the volatiles but there could also be ecological costs. For example, it may be that if surrounding plants respond to emissions by triggering their own defences, these plants become less attractive to herbivores and so the emitter becomes relatively more attractive and thus subject to even greater herbivore attack. Alternatively, the enhanced warning given to the surrounding plants may not harm the focal individual directly but may increase the competitive ability of plants that compete with that focal plant. Where the focal plant does not benefit, the response of the surrounding plants must be seen as parasitic eavesdropping on cues (rather than signals) provided by the emitter. The limited distance over which volatiles are effective may be seen as adaptive to limit parasitic eavesdropping. This speculative idea awaits testing. We need to distinguish whether emissions are no more than a physical consequence of tissue damage by herbivores or are signals actively produced and aimed at other receivers (e.g., other parts of the same plant or antagonists of the herbivores).

On the other hand, the emitter may actually benefits from the response of the receiver. Firstly, it may be that if emitter and receiver are related then there may be kin-selected benefits to emitting a costly signal that allows related individuals to protect themselves. Additionally or alternatively, it may be that

induced reactions by the receivers (e.g., changes to their volatile emissions) help to repel further herbivores from the general area, thus reducing herbivore pressure on the focal emitter. Finally, as we will discuss later, a benefit may accrue because the receiver can help to lure antagonists of herbivores to the general area such that antagonist recruitment to the emitter increases. For example, the spatial spread of volatile production by different, neighbouring plants is likely to result in a similar spread in the local area from which antagonists of the herbivores can be recruited.

Thus, while we would welcome further demonstrations that receivers can respond to the emissions of nearby plants in ways that bring them a net benefit, we would particularly welcome evaluation of the fitness consequence to the emitter of the response of nearby individuals to the emissions. This could easily be done using the system of Karban et al. (2004) and making use of transgenically 'deaf' plants that cannot detect or respond to the relevant volatiles. If it is found that the emitter does gain a benefit from the responses of receivers, then the next challenge will be to explore signal honesty: that is, to explain why plants do not signal when they are not being attacked. It may be that there is no benefit to cheating in this way, since it is only attacked plants that can gain from the response of neighbours (for example in helping to attract antagonists), whereas cheats would invest in signalling for no gain. Note, however, that there may still be an incentive to cheat by exaggerating the extent of the threat. Further, our understanding is hampered because the physiological costs of signal production are unknown and may be relatively low.

9.3.1 Signalling by undamaged plants

Some undamaged plants have been demonstrated to emit volatiles similar to those emitted by nearby herbivore-infested plants (see references in Kobayashi and Yamamura 2007). Several theories have been put forward to explain this phenomenon.

Kobayashi and Yamamura (2007) used a simulation model to explore the conditions under which it might be evolutionarily stable for an undamaged plant to emit such volatiles, even if physiologically costly. Their fundamental hypothesis was that the signal might help nearby relatives that are being attacked by herbivores if it enhances the recruitment of antagonists of herbivores to those relatives. They show that evolutionary stability of such a signal is theoretically possible providing the cost of the signal is not too high and the attractiveness of the signal to the desired receivers is highly positively correlated with the local strength of the signal. Unsurprisingly, dispersal of plant offspring must be spatially restricted so as to produce some kin grouping, but this restriction fits well with the dispersal curves of most plants that have most seeds remaining in very close proximity to the mother plant (see Chapter 3). The risk of attack must also be sufficiently high that nearby kin are in sufficient danger to make investment in the signal worthwhile on average. More interestingly, the theory requires that the population be elastic, meaning that it is held by some top-down pressure below the maximum carrying capacity that would be set by between-plant competition. This pressure could be, but need not be, the herbivore. This restriction on the plant population density is necessary because, although the signal may benefit the focal individual through kin-selected benefits to related nearby individuals, those nearby individuals are also competitors for resources. Thus, improvement to their fitness may cause a reduction in the direct fitness of the focal individual. Hence, this type of signalling is more likely to flourish where such competition effects are reduced.

An alternative explanation for release of antagonist-attracting volatiles by undamaged plants may be the 'pre-attack extermination hypothesis' posited by Kobayashi and Yamamura (2003). This theory suggests that the focal plant can benefit if the herbivores currently infesting a neighbouring plant can be wiped out before they have a chance to spread to the focal individual. Using a mathematical model, the authors demonstrate the potential for evolutionary stability of such a signal provided (1) the cost of the signal is low compared to the cost of future infestation, (2) the attractiveness of the signal to antagonists of herbivores on an infected plant is positively correlated with the local density of the chemical signal, (3) plants with infested neighbours are more at risk from herbivores than those with uninfested neighbours, and (4) the lifespan of the plants is long

compared to that of herbivores. This theory makes the interesting prediction that both the uninfested signaller and the infested neighbour benefit from the signal. Thus, the originally infested plant might be selected to effectively signal its infestation to its neighbours, in order to trigger them into amplifying the SOS call to antagonists of herbivores.

Alternatively, or additionally, the 'pre-attack protection hypothesis' of Kobayashi et al. (2006) suggests that the focal plant may benefit from beginning to attract enemies of herbivores before these herbivores spread from already-attacked neighbouring plants. Again, the evolutionary stability of such signalling is demonstrated in a mathematical model. Stability of the signal requires that the signal (1) has some cost, but this cost is not too high; and (2) that the ability of the infected plant to attract antagonists of herbivores is not so great that future risk to currently uninfected plants is low. A key assumption of the model is that antagonists recruited before an attack occurs are retained long enough to be of later use to the plant. The authors argue that antagonists attracted from a distance will require significant time to search the plant for herbivores. They also suggest that retention over longer timescales might be achieved if the signal emission coincides with the induced provisioning of alternative food such as extra-floral nectar. Rose et al. (2006) demonstrated that volatiles associated to extra-floral nectar attracted parasitoid wasps. The cost to the plant of attracting some antagonists too early is unclear. If such antagonists subsequently do not respond to the plant, the local population of antagonists might sometimes be sufficiently mobile and wide-ranging that this will not greatly reduce later antagonist arrival rates when herbivores are present. Further, the costs of signalling before herbivores have arrived may be worthwhile if this delivers antagonists from the moment that herbivores arrive.

One factor that none of these models has considered is informational degradation of the signal and antagonists' responses to this. As soon as uninfested as well as infested plants signal to antagonists, informational degradation has the potential to make it more difficult for the antagonists to correctly identify infested plants. Such informational degradation will reduce the protection that antagonists can provide to plants, since they will find it more difficult to locate their prey. In the extreme case, this reduction in herbivore-finding may select for the antagonists to ignore the signal entirely. However, such degradation may not always be a problem: if infected and unaffected plants are in close proximity and complex air currents carry the volatiles. In such circumstances, even if only the infested plants signal, this may not allow antagonists to readily separate the infested from the uninfested without close examination. In this case, signalling by uninfested plants may increase the ability of antagonists to find the approximate vicinity of the infected plant without impairing the ability of the antagonists to identify the infested plants of interest to them. Kobayashi et al. (2006) note that uninfested plants released volatiles at a lower rate than infested plants. This may allow antagonists to differentiate between them. Further, perhaps this allows uninfested plants to respond to infested neighbours without responding to uninfested but signalling neighbours, and thus avoid the spread of the signal too far from the spatial proximity to the infested plant.

Other signalling explanations for such volatile emissions are also possible. For example, the volatile emissions triggered in non-infested plants may be within-plant communication aimed at priming or triggering anti-herbivore defences in response to enhanced likelihood of future attack. Secondly, they might be a signal to herbivores that the plant's defences have been primed or (deceptively, and less likely for that reason) that the plant is already infested.

It has yet to be demonstrated that currently uninfested plants gain any fitness benefits from volatile emissions. It may be that the apparent signalling of uninfested plants in response to volatiles emitted by infested neighbours may be a non-signalling by-product of priming or implementation of chemical defences, or even that is it nothing more that re-emission of compounds from nearby infested plants that are absorbed and then released by uninfested neighbours and are thus unrelated to anti-herbivore defence; see Himanen (2010).

We have a substantial body of modelling, but we very much need experiments to explore whether the phenomena that the models describe are ecologically relevant. Design of such experiments will be challenging, since the plant could subsequently become infested, and we need to separate the fitness benefits of volatile emission before and after infestation. If pre-infection emission is demonstrated to provide a fitness benefit, then we can explore whether this benefit derives from the response of herbivore antagonists to the emissions. Only if that is the case, need we face the challenge of exploring how the three mechanisms that have been the subject of mathematical modelling (and the others that we suggest above) contribute to such a benefit.

9.4 Signalling by infested plants to enemies of herbivores

Signalling can be triggered simply by a herbivore laying eggs on the plant, before any damage has occurred (Mumm and Hilker 2005). As well as in leaves, apparent signalling can also be seen in seeds (Steidle et al. 2005), bulbs (Aratchige et al. 2004), and roots (van Tol et al. 2001). The benefit to the plant of attracting predators of the herbivores or egg parasitoids seems self-evident and has been demonstrated in some cases (Van Loon et al. 2000, Fritzsche-Hoballah and Turlings 2001; see Figure 9.2). However the attraction of parasitoids of the plant-feeding stages is less obvious, since a parasitized herbivore may continue to feed. However Dicke and van Loon (2000) and van Loon et al. (2000) demonstrated a benefit to the plant in attraction of such a parasitoid: since parasitized herbivores showed a reduced uptake rate. We would not expect this effect to be general however, and sometimes parasitoids may have no benefit or even impose a cost to the host plant through their effect on infected herbivores (e.g., if these increased food uptake).

Early laboratory explorations of the attraction of antagonists raised questions as to whether attraction might be an artefact of unrealistic laboratory conditions. However, more recent field-based experiments have reduced these concerns. For example, Kessler and Baldwin (2001) quantified the volatiles emitted by plants of *Nicotiana attenuata* growing in natural populations and subject to herbivory by caterpillars, leaf bugs and flea beetles. They artificially added herbivore eggs to naturally growing plants and mimicked volatile release from some plants but not others. The volatile treatment

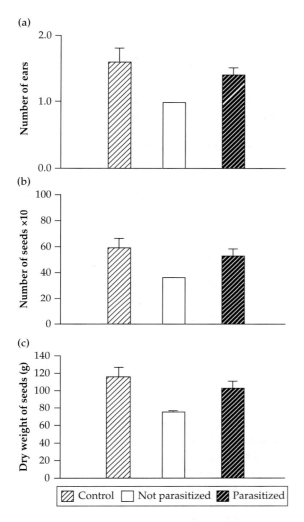

Figure 9.2 The (a) number of ears, (b) number of seeds, and (c) mean dry weight of seeds of maize seedlings manipulated in one of three ways: left as controls, infected with an unparasitized caterpillar, or infested with an already-parasitized caterpillar. Illustrated are means and S.E. Redrawn from Hoballah and Turlings (2001).

caused a two-fold reduction in the rate of oviposition by other herbivores and a five-fold increase in the rate of predation on the artificially added eggs. The authors estimate that naturally occurring volatile emissions would give plants the capacity to reduce their herbivore burden by 90 per cent in their study site. This is highly suggestive of a potential fitness benefit under natural conditions, but it falls short of making an unarguable case for such a ben-

efit. What would be required would be a follow-up experiment where recruitment of herbivores is allowed to occur naturally on plants, and then availability of natural herbivore antagonists is artificially decreased (or, less easily, increased) around some plants. Some measure of fitness (perhaps in terms of numbers of seed set) could then be compared between plants subject to different local antagonist availabilities. The indirect-defence hypothesis predicts a positive influence of antagonist availability on plant fitness. In such experiments, the attraction of using genetically-modified plants that are unable to produce putative signalling volatiles is that such volatiles can be artificially presented alongside some plants in each treatment group and not others. The indirect defence theory predicts that plants with added volatiles attract higher numbers of antagonists and achieve higher fitness than plants with similar initial herbivore burdens but without the volatiles. Such a design would allow us to identify whether volatiles, rather than, for example, the visual appearance of the herbivores, attracts antagonists.

Sabelis and De Jong (1988) used a mathematical model to argue that we should often expect plant populations to be polymorphic in their use of herbivore-related volatiles. They argue that polymorphism is likely to occur because a non-signaller can gain the benefits provided by nearby signallers and can save the physiological costs of signalling. Godfray (1995) argued that the signal requires a strategic cost in order to keep it honest. If the signal is very cheap, then plants may signal when they are very lightly infested or even if they are uninfested but there is a chance of infestation within the timescale of response to the signal. If this is the case, signalling will break down because all plants (regardless of their herbivore load) will signal and the signal will carry no useful information to the antagonists that will help them locate their herbivorous prey. Such a cost is easy to imagine, and may be the physiological cost of producing the volatiles and/or the ecological cost of attracting more herbivores to the focal plant and/or fewer herbivores to competitors.

We should consider whether the volatiles released when a plant is attacked by herbivores are best seen as a cue or a signal. That is, has the form of the emissions been shaped by their effect on herbivore antagonists? There is strong circumstantial evidence to suggest that they have, since there is evidence of facultative use of different emissions according to which will be most appropriate. For example, herbivore-damaged tobacco plants release different volatile blends by day and at night: the day-released volatiles attract parasitoids, whereas those released at night repel herbivores (De Moraes et al. 2001). Tobacco plants also suppress the production of nicotine and increase production of volatiles when attacked by a nicotine-insensitive specialized herbivore (Kahl et al. 2000).

The issue of specificity of herbivore-induced volatiles is a complex one. Some studies have demonstrated that different herbivores elicit different volatiles from the same plant species, and indeed simultaneous herbivory by different species can alter volatile emissions (Delphia et al. 2007). It has also been demonstrated that herbivore antagonists can be sensitive to such differences in order to preferentially exploit preferred prey (De Moraes et al. 1998 and reviewed by Takabayashi et al. 2006), but antagonists are not always sensitive to such differences (Turlings et al. 1995). It has also been demonstrated that parasitoids can gain information on the level of herbivore infestation purely from the volatiles produced (Tentelier and Fauvergue 2007).

Takabayashi et al. (1995) demonstrated that maize responded to attack by early instars of the armyworm (*Pseudaletia separate*) in a way that attracted the parasitic wasp (*Cotesia kariyai*), which attacks the armyworm; but such a response was not seen when attacked by later instars. This result fits with the observation by Tanaka et al. (1992), who showed that only parasitism in the early instars reduces the impact that this armyworm has on its host plant. Geervliet et al. (1998) reported that parasitic *Cotesia glomerata* individuals preferentially selected plants that were more heavily infested with their host *Pieris brassicae*, in an experiment where only volatile cues were available to the parasites. Hoballah and Turlings (2005) demonstrated that parasitoid wasps were able to distinguish the length of time for which leaves had been attacked by herbivores. There is also considerable evidence of differentiation between different host plants attacked by the same herbivore (e.g., Geervliet et al. 1996, De Moraes and Lewis 1999).

When measuring specificity, it is important to use behavioural assays of relevant antagonists rather than analytic instruments: because arthropod chemosensors may be more sensitive than our equipment (Dicke 1999), and because antagonists may not necessarily respond to differences that seem clear to our equipment. From the plant's viewpoint there may be little selection pressure to produce antagonist-specific signals, since there may be no cost to attracting other species alongside the target antagonists (Turlings et al. 1995); although there may be costs to attracting other herbivores. Selection pressures on the informational content of signals may also be complex (Takabayashi et al. 2006). If the plant is specific about the type of herbivore present, then this should make antagonists that particularly favour that herbivore more willing to respond to the signal. However, it may deter other antagonists from responding. This might cost the plant if these antagonists would still attack individuals of that herbivore type (even though they are not preferred prey) if the antagonists had invested in travelling to the plant while unsure of herbivore identity. The trade-off between being particularly appealing to the 'best' antagonists versus attracting a greater range of antagonists warrants further investigation.

It may not just be herbivores that can be combated by recruiting antagonists, competitors can be too. It has been demonstrated that some macro-algae attract herbivorous snails through the excretion of dissolved organic matter. These snails graze on micro-algae that are epiphytic to the macro-algae and compete with the host plants for light and nutrients, but the snails do not damage the macro-algae (Bronmark 1985, 1989).

It may be that even vertebrate antagonists can be used in indirect defences: Mantyla et al. (2008) demonstrated that birds removed more artificial prey from trees on which bags of herbivores were placed under leaves, compared to trees with empty bags. The bags were intended to deny birds visual cues of herbivory. However, trees with herbivores had lower net photosynthesis than control trees and this may have produced a change in leaf colour that birds cued upon. It is also possible that birds cued on the smell of the herbivores themselves rather

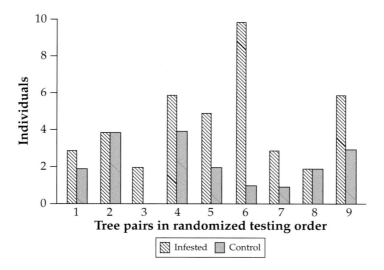

Figure 9.3 The first choice of willow warblers when given a choice between branches from control trees and branches from trees which had previously been experimentally exposed to herbivores. Birds were more strongly attracted to the manipulated trees than the controls. Redrawn from Mantyla et al. (2004).

than herbivore-related volatiles emitted by the plant. However, the authors did find a positive correlation between volatile emissions by the trees and birds' consumption of artificial prey. Previously Mantyla et al. (2004) had demonstrated a preference by insectivorous willow warblers (*Phylloscopus trochilus*) for branches from trees that had been attacked by sawfly larvae compared to unattacked controls in aviary-based choice trials (see Figure 9.3). Importantly, trees that were randomized to the attacked group had bags of sawfly larvae attached to some branches only, and unattacked branches from these trees were used in the subsequent bird-preference trials. Thus, the preference shown by the birds was not related to cues emanating directly from the larvae but rather from the whole tree's response to localized attack. These results are worthy of further research, especially as birds have the potential to be highly effective antagonists of herbivores (Van Bael et al. 2003) but also intra-guild predators of invertebrate antagonists, and so may have both positive and negative effects for the plant. Again, it would be interesting to separate between signals and cues in the triad between plants, herbivores and their predators. If herbivore attack leads to lower net photosynthesis and if this is associated with changes in leaf colour or lower ethylene

emission, then predators can respond to cues. Conversely, if predators respond to volatile emission produced in order to attract predators they are responding to signals.

9.4.1 The viewpoint of the plant

From the plant's viewpoint, the evolution of SOS signals is easy to understand. The mechanical damage produced by herbivory necessarily will induce volatile emissions. If these emissions act in a way that benefits the plant, they could be shaped by evolution to increase their effectiveness. We would not expect plants to always produce such signals, however, for several reasons. We can presume that production of volatiles for signalling purposes involves some physiological costs, and there may also be ecological costs to signalling, driven by eavesdropping by competitor plants and herbivores. The benefit may also be low if there are few appropriate antagonists around to attract, or if the antagonists are incapable of detecting the signal and responding appropriately. Further, as discussed by Sabelis and De Jong (1988), some plants may not signal if they can benefit from the signals of their neighbours and save on the costs of signalling.

Several factors may influence signal honesty. There is always the danger of plants signalling when they have not been infected. If there are no herbivores on plants that signal or on nearby plants, then we would expect antagonists to quickly learn to disregard the signal. However, since signalling is likely to exert some physiological and ecological costs, we would not expect plants to always signal when there is no immediate threat of herbivore attack; it is more plausible that as-yet-uninfested plants might signal when the risk of attack by herbivores is higher (perhaps because the focal plant has responded to infestation of neighbours); this has been discussed fully in Section 9.3.1. If this attracts antagonists to situations where the focal plant will often be infested soon and/or neighbours of the focal plant are already infested with suitable hosts, then the antagonist may still benefit from responding to the signal, and the signal should not be seen as dishonest.

Costs of signalling may be very different when comparing across species and when comparing across substances used to signal. Some volatiles are likely to be relatively cheap to produce from a physiological perspective, others will be more costly to build because they require further steps which may be limited by the availability of precursors. We could consequently predict (1) that costs will vary across species and across signalling substances and (2) that the specificity and the reliability of signalling will increase with increasing costs.

There has been a lack of empirical research on the influence of the timing of the initiation and termination of herbivore-related volatile emission – but see Puente et al. (2008a, b) for some mathematical modelling. Clearly, the sooner the plant starts signalling after damage has occurred, the better for it. Hence, we can understand why even oviposition by herbivores on a plant can trigger induced defences. If however, the herbivores are only vulnerable to parasitoids in their very early stages, signalling to such parasitoids may be worthless if *de novo* synthesis of the volatiles takes so much time that the herbivores have passed through the vulnerable stage before the signal can be effective in attracting parasitoids. If plants do not stop signalling when herbivores that are attractive to the antagonists have disappeared (e.g., because they have been eaten, or have

matured to an invulnerable stage) then the value of the signal to antagonists is reduced, and they will be more likely to ignore it. But since this cost will be paid by other plants as well as (or rather than) the focal signaller, we cannot expect this effect to drive the cessation of signalling. Rather, we have to expect signalling to be physiologically costly, such that it will be selected to stop when the plant has little or nothing further to gain from attracting further antagonists. All these assumptions and predictions still await empirical exploration. Shiojiri et al. (2009) experimentally investigated the timing of signalling by using plastic bags to prevent propagation of volatile compounds and experimentally manipulating bag removal. They found that volatiles were released for three days after experimental leaf clipping, or that immediately-released volatiles remained active over this period; but their experiment did not allow differentiation between these two possibilities.

There has been little consideration of whether plants are able to evaluate the likely effectiveness of SOS signals in their current circumstances, and then flexibly use herbivore-related volatiles only in advantageous situations (Dicke 2008). For example, such signalling may only be effective if suitable antagonists exist locally. The evaluation of local antagonist activity might be estimated by the plant, for example, from the uptake of nectar from extrafloral nectaries. However, the repeated demonstration of the triggering of herbivore-related volatiles in the complete absence of antagonists in the laboratory shows that such emissions are not always modulated by local antagonist availability.

9.4.2 The viewpoint of antagonists

Evolution of induced signalling by plants is also relatively easy to understand from the viewpoint of antagonists. Many arthropod antagonists will be pre-adapted for detecting such a signal due to the use of olfaction for tracking prey by the volatiles that the herbivores emit. Further, plants give off volatiles even in the absence of herbivore attack, and antagonists may be selected to respond to these if finding host plants is an effective way to find target herbivorous prey that have specific host plant requirements. However, we would not expect all antagonists to respond to all herbivore-related vola-tile cues all of the time. Antagonists are only likely to respond when the volatiles provide information that allows them to enhance their prey-finding. There may be situations where the cues do not give sufficiently specific information about the position or identity of the herbivores to justify the antagonist responding. Thus, plants may gain a fitness advan-tage if they can increase the informational quality of their signals, thus making it more attractive for appropriate antagonists to respond. We have seen that some signals do provide information on her-bivore type (even developmental stage, as well as species), and this may cause antagonists that are uninterested in such herbivores to ignore the signal. In situations where suitable herbivorous prey are locally commonplace and easy to find by other means, antagonists will have little reason to respond to signals. Again, we would expect the readiness of the antagonists to ignore a given signal will increase with the ease of finding prey by other means (including responding to the signals of other plants). Again, we see how vital it is to understand how sig-nalling by neighbours affects the distribution of antagonist arrivals across plants. If signals are use-ful for locating plants, and antagonists are rare, then this might lead to competition and arms race situa-tions between plants as they strive to attract antago-nists specifically to themselves.

9.4.3 The viewpoint of herbivores

It has generally been implicitly assumed that the herbivore is not an active agent in this interaction between plant and antagonists, but this is unlikely to be the case for several reasons. Firstly, if some plant types are more effective at attracting antago-nists than others, then this can be expected to influ-ence host plant selection by herbivores. Secondly, herbivores might be expected to be selected to dis-rupt the effectiveness of signalling to antagonists. They might be able to do this by disrupting the syn-thesis or emission of the volatiles concerned. This might most plausibly be done by triggering other responses in the plant that interfere with signal gen-eration, and this idea warrants research. Musser et al. (2002) demonstrated that the saliva of the cater-pillar *Helicoverpa zea* acted to reduce the amount of

toxic nicotine released by the tobacco plant *Nicotiana tabacum*; we see no reason why the production or emission of volatile warning signals could not also be similarly inhibited. If simultaneous attack by dif-ferent herbivore types can lead to less effective sig-nalling by the plant than if one herbivore type attacked alone, herbivores might gain some respite from indirect defences by selecting plants that are also being attacked by another types of herbivore. However, this must be set against the possibility that the other herbivore has already triggered induced defences in the plant that would also be effective against the later-arriving herbivore. Further, we can expect herbivores themselves to respond to herbiv-ore-induced volatiles, as we discuss below.

The evolution of plant–antagonist communica-tion is particularly complex because of the number of potential eavesdroppers. We have already dis-cussed other plants in this context (Section 9.3), but herbivorous arthropods are well known to use vola-tiles emitted by undamaged plants to aid them in host location (Visser 1986), so they seem pre-adapted to respond to herbivore-related emissions. The response of herbivores to herbivore-related vola-tiles can be either to be attracted or repelled by them (De Moraes et al. 2001; see also references in Carroll et al. 2006). Attraction may be the logical choice when suitable host plants are otherwise hard to find and/or if herbivores gain foraging or anti-antago-nist benefits from aggregating with others. This seems to occur in the case of the codling moth (*Cydia pomonella*) that was attracted to apples that were already infested with conspecifics on the basis of herbivore-related volatiles (Landolt et al. 2000; see Figure 9.4). This attraction occurred despite the authors' view that moth larvae would be disadvan-taged by sharing a fruit with others (compared to being alone on a fruit).

Repulsion may be logical if avoidance of conspe-cific or heterospecific competitors is advantageous: for example, to reduce the effects of already-induced chemical defences by the plant, indirect defence through the luring of antagonists of the herbivores, competition between herbivores, or even cannibal-ism by conspecifics (Carroll et al. 2006). Carroll et al. (2006) demonstrated that sixth-instar fall army-worms (*Spodoptera frugiperda*) preferred the vola-tiles emitted by conspecific-attacked plants over

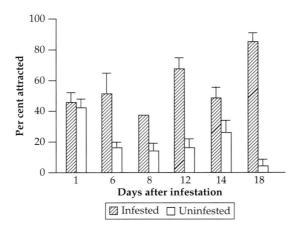

Figure 9.4 The mean and SE of the percentage of neonate codling moth larvae attracted to either uninfested or infested apples (in Y tube experiments) after different lengths of time since initial infestation. Redrawn from Landolt et al. (2000).

unattacked plants. They argue that caterpillars should do better on unattacked plants, but that the behaviour may be adaptive if the challenge of finding a host after being knocked from it is very great in natural situations, and finding a suboptimal plant is preferable to not finding any suitable host. They argue that this may be important to larval stages of many herbivores that have limited mobility compared to flying adults.

Using transgenic plants that were unable to release certain volatiles, Halitschke et al. (2008) demonstrated that volatiles that were associated with leafhopper damage attracted antagonists of herbivores to attacked *Nicotiana attenuata* plants under natural conditions. However, they further demonstrated that the same volatiles were also used in host-finding and feeding initiation by another flea beetle herbivore.

Responses of herbivores to herbivore-related volatiles can be expected to significantly impact on the evolution and maintenance of indirect defences. Responses can be an additional cost or benefit to the plant depending on whether herbivores are attracted or repelled. From the antagonist's point of view, the benefit of acting in response to herbivore-related volatiles will be influenced by whether the volatiles also increase or decrease the number of further herbivores attracted to the emitting plant. Furthermore, the behaviour of the antagonists will alter the costs

and benefits for the herbivores of avoiding or being attracted to the plant. Thus, evaluation of whether we would expect such indirect defences to evolve in any particular situation is very challenging because of the number of different agents involved and the complex web of influences that they have on each other (see Section 9.5). It may be that herbivore responses to herbivore-related volatiles will influence the form of signalling to antagonists as well as whether any signalling at all occurs. For example, herbivore responses might select for less specific volatile emissions, thus reducing information available to the herbivores.

The use of plant-related volatiles by herbivores to locate hosts should also affect our exploration of plant-to-plant signalling. If one plant is attacked and the emission of volatiles by that plant is linked to decreased subsequent herbivore damage on nearby plants, then the signalling explanation is that the volatiles acted as a signal to the neighbouring plants that caused them to invest more heavily in anti-herbivore defences. Another explanation is that the volatiles are a cue rather than a signal. In this scenario, volatile emissions are costly to the plant, but are an unavoidable consequence of attack. These volatiles act to attract herbivores to the emitting plant that would otherwise have exploited the neighbouring plants. These effects are not mutually exclusive, and demonstration of putative plant-to-plant signalling should also evaluate the importance of this additional mechanism.

Volatile chemical cues can also guide parasitic plants towards their hosts. In an ingenious study, Runyon et al. (2006) demonstrated that the parasitic plant *Cuscuta pentagona* used volatile cues emitted by its tomato plant host (*Lycopersicon esculentum*) to orientate towards its host. Such orientation took place toward volatiles extracted from the plant in the absence of other cues and was host-plant-specific (being towards tomato volatile cues versus wheat ones). Clearly the volatiles here are a cue rather than a signal. Although the parasite may be highly evolved to respond to them, the volatiles were clearly not selected for that purpose because the tomato plant pays a fitness cost through parasitism. Rather, the volatiles are produced as a by-product of physiological functions in the tomato plant.

9.5 The complex environment in which indirect defences operate

Evaluation of the effectiveness of indirect defences is all the more challenging because of the complex nature on the environment in which such signals will likely operate in nature. There will often be many signalling plants, diffusion through the air is contingent on habitat structure and wind, and the response of antagonists to detecting multiple signal sources is not well understood. There are likely to be several different types of antagonist attracted, and these antagonists may differ in their effectiveness. It may be possible for plants to evolve antagonist-specific signals that target only the most effective antagonist type, but such specialism may not be best for the plant when this antagonist type is comparatively rare locally; and the plant may do better by attracting greater numbers of generalists even if these are less effective antagonists. Further, there may be competition or even interference between antagonists on the plant, and this may even lead to intra-guild predation of some antagonists by others (Gnanvossou et al. 2003). Herbivore-related volatiles may also interfere with the attraction of potential pollinators. Firstly, it may mask cues that pollinators use, or pollinators may be selected to find herbivore-related volatiles aversive (because the antagonists attracted by them pose a threat to these pollinators as well as to herbivores), and lastly the antagonists may reduce pollination directly by consumption of pollinators (see Bruinsma et al. 2008 for further discussion). Finally, if plants respond systemically to herbivore attack, the quality of nectar may be lower in plants that have invested in greater defences. In this case, pollinators could use herbivore-related volatiles as an informative cue of the expected quality of nutritional rewards offered by the plant.

Plants can often be simultaneously infested by more than one herbivore type. This may have the effect of reducing the effectiveness of the signalling to antagonists, if the different herbivores require the attraction of different antagonists (Takabayashi et al. 2006). Further, the greater range of antagonists drawn by a diversity of herbivores may increase the risk of intra-guild predation between the antagonists. There may be consequences for the plant, even if herbivory by different species is not simultaneous. It may be that defences initiated by the first attackers interfere with (or enhance) the defences launched in response to the second type of attacker. Shiojiri et al. (2001, 2002) studied two herbivores of cabbage (*Brassica oleracea*): the diamondback moth (*Plutella xylostella*) and the cabbage white butterfly (*Pieris rapae*). They found that diamondback larvae suffered lower levels of parasitism by their specialist parasitoid when they infested a plant that was also infested by cabbage whites (compared to plants with a similar total number of larvae and similar levels of damage; see Figure 9.5a). They also found that diamondback adult females preferred to oviposit on plants with cabbage white larvae compared to ones with diamondback larvae or uninfested

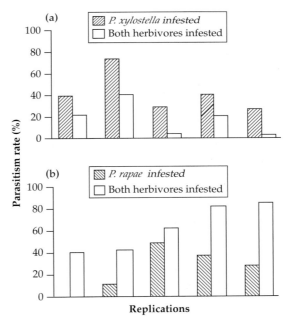

Figure 9.5 Rates of parasitism of (a) Diamondback *P. xylostella* when either alone on a plant or when the plant is infested with both *P. xylostella* and cabbage white *P. rapae*; (b) *P. rapae* either alone on a plant or on a plant infested with both *P. xylostella* and *P. rapae*. Diamondback larvae suffered lower levels of parasitism by their specialist parasitoid when they infested a plant that was also infested by cabbage whites (compared to plants with a similar total number of larvae and similar levels of damage). In contrast, cabbage white larvae were more likely to be parasitized by their specialist parasitoid wasp when in mixed groups on a plant than when in same-sized single-species groups. Redrawn from Shiojiri et al. 2002.

ones. In contrast, cabbage white larvae were more likely to be parasitized by their specialist parasitoid wasp when in mixed groups on a plant than when in same-sized single species groups (see Figure 9.5b). The effects of vulnerability to parasitism were linked by the authors to changes in the blend of volatiles emitted by the cabbage. De Boer et al. (2008) found that both lima beans and cucumber plants emitted a different blend of volatiles when attacked by spider mites and caterpillars simultaneously than when attacked by either alone. They also demonstrated that a predatory mite specializing on the spider mite preferred the volatiles induced by a mixed-attack compared to either of the two single-species emissions. Zhang et al. (2009) demonstrated that the indirect defences of the lima bean plant (*Phaseolus lunatus*) induced by spider mite (*Tetranychus urticae*) infestation were less effective at attracting predatory mites (*Phytoseiulus persimilis*), which prey on the spider mites, if the plant was also infested with whitefly (*Bemisia tabaci*). Thus, the complexity of herbivore guilds can be expected to have complex effects on the effectiveness of indirect defences by plants (see Zhang et al. 2009).

A further complication is that plants may have direct chemical defences (where the chemicals themselves affect the herbivore) as well as induced indirect ones (that impact on herbivores through the agency of a third species), and these defences may interact in complex ways. For example, plant toxins may make the plant less attractive to some herbivores, but other herbivores may be able to tolerate and sequester the toxins – turning them to their own defence against the antagonists that the herbivore-related volatiles may attract. Thaler (1999) reported on such a system, and concluded that although the direct defence did reduce the effectiveness of the indirect defence, it did not entirely remove it. However, further studies that explore the interaction between direct and indirect plant defences would be very welcome.

Another complication for the real-world application of SOS signals to antagonists of herbivores is that (except in agricultural settings) plants will often live in close proximity to plants of different species and this may influence the effectiveness of such signals. Thiery and Visser (1986) demonstrated that the Colorado potato beetle (*Leptinotarsa decemlineata*) was attracted to the odour of potato plants (*Solanum tuberosum*). However, if this odour was blended with that of wild tomatoes (*Lycopersicon hirsutum*) or cabbage (*Brassica oleracea*) then this attraction disappeared. There was no evidence that the other plants were repellent to the beetle, but rather that their odour masked that of the target host plant. Such a masking effect might also be plausible for responses to herbivore-related volatiles in a heterogeneous plant environment, and exploration of this would be worthwhile. In a greenhouse, Gols et al. (2005) demonstrated that parasitoids exposed to host-infested *B. oleracea* plants took longer to find host-infested plants in the presence of white mustard plants than in their absence. Khan et al. (1997) demonstrated several effects of planting odorous grass (*Melinis minutiflora*) in African maize fields. The grass constitutively emits a compound that maize emits only in response to caterpillar damage. Maize in mixed plantings experienced reduced herbivory both because the pest species was repelled and also because the pest's main parasitoid antagonist was attracted. Of course, the mix of plants in a given locality will also affect the population densities of different herbivores and their animal antagonists, and this will also influence selection pressures on indirect defences. How the dynamics of the ecological scale of communities influence the evolution of communication between plants, herbivores, and predators merits further attention.

9.6 Signalling in ant–plant mutualisms

Tropical plants from more than 100 genera provide accommodation for ants in exchange for the removal of herbivorous insects and their eggs from host plants, as well as epiphytes, pathogenic fungi, and even competing plants. This is a classic example of a mutualism, and both partners are generally more dependent on each other than in the indirect defences previously discussed. Plants invest in specialist structures to house (called domatia) and sometimes feed the ants, and the ants in turn may have an obligate association with the plants (not nesting anywhere else). But even here there can be conflicting interests between plant and insects. Frederickson (2009) reported observations of ants destroying the inflorescences of their host, and interpreted this in terms of the ants stopping the plants investing in reproduction, in order to encourage plant growth. She experimentally removed flowers from another tree species that ants did not naturally sterilize, and observed that the manipulated trees grew faster, produced more domatia, and housed larger, more fecund ant colonies, compared to control plants.

Bruna et al. (2008) explored the response of ant-mutualists to plant volatiles. The ant species *Pheidole minutula* is a host-specialist that shows increased activity in response to extracts of its own host (*Maieta guianensis*) but not to those of sympatric myrmecophytes and non-myrmecophytes. Another ant (*Azteca* sp.) also showed increased activity to extracts of its own host (*Tococa bullifera*) but also to those of other sympatric species (although to a lesser extent). This difference between species was linked to the use by *Azteca* sp. of plants nearby their primary host for the establishments of satellite nests. Edwards et al. (2006) demonstrated in the laboratory that host plants can emit volatiles that attract dispersing ant queens. These authors suggest that ant–plant symbioses originated from parasitic ants tending aphids on plants. Such worker ants have been commonly shown to use volatiles emitted by the plant wounds caused by the aphids to locate aphids. Over evolutionary time, the plants 'won the ants over' using alternate food sources (turning them from parasites into mutualists) and still used the volatiles to lure them, with the responding individuals switching from workers to

queens. This argument makes the interesting, and so far untested, prediction that the volatiles used to attract queens should be similar to aphid-induced volatiles.

Brouat et al. (2000) demonstrated that, even in the absence of herbivores, ants biased their patrolling of the host plant towards young leaves that were more vulnerable to herbivorous attack. The authors identified volatile cues as being an important aid to such biased patrolling, and explained the behaviour in terms of the long-term benefits to the ant colony of plant growth, rather than an increased food intake of ants on young leaves (Edwards et al. 2007). Agrawal (1998), Wood and Wood (2004), Romero and Izzo (2004). and Mayer et al. (2008) all demonstrate the ability of ants to move towards areas of their host plants where herbivore damage was occurring, and that this effect was driven at least in part by chemical cues emitted by the plant (see Figure 9.6). It is important to note in this case that the ants in ant–plant mutualism rarely use the herbivores as a source of food and so the biased distribution of the ants cannot simply be explained by ants maximizing their own food intake. Rather the evolution of this response has to be seen in terms of the longer-term benefit that the ant colony gains from increased host growth (and thus more ant homes and sometimes food; Edwards et al. 2006). It also appears that chemical cues are key to the ants' ability to differentiate between their host and other encroaching plants (Davidson and McKey 1993; Yu and Davidson 1997).

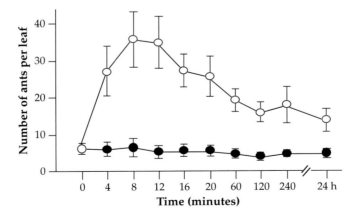

Figure 9.6 Time course of ant recruitment to control leaves (filled circles) or leaves that had been experimentally damaged (open circles). Illustrated are means and S.E. Redrawn from Agrawal (1998).

There has been little investigation of within-plant spatial distribution of antagonists of herbivores, and ant–plant mutualisms might be an ideal system to explore the consequences of this. Previous studies have demonstrated the use of plant-produced volatiles in biasing the distribution of ants across the plant (towards already attacked leaves or those particularly vulnerable to attack; see Edwards and Yu 2008). However, the fitness benefit to plant and ants of this behaviour has not been explored. If the volatiles can be synthesized and reproduced artificially, then the distribution of ants can be manipulated by experimenters and the fitness consequences for both ants and plant evaluated.

9.7   Deception of herbivore antagonists by plants

Brodmann et al. (2008) demonstrated the mimicry of herbivore-related volatiles by the flowers of unattacked orchids (*Epipactis helleborine* and *E. purpurata*). The volatiles in question were similar to those that are emitted by cabbage leaves attacked by caterpillars (*Pieris brassicae*) and have been demonstrated to be attractive to social wasp species that prey on this caterpillar (*Vespula germanica and V. vulgaris*). The orchids appear to be pollinated exclusively by social wasps. This work is very suggestive that the plant enhances pollination by use of the volatiles to lure the wasps. However, it may best be described as sensory exploitation rather than deception, since the plant does offer a nectar reward to pollinators. This interesting study deserves a follow-up study where the effect of removing or masking the volatiles for plant pollination is measured. It might also be interesting to explore whether the wasp needs previous exposure to caterpillars collected from volatile-emitting plants before being drawn to the orchid, and to evaluate how valuable the nectar is to wasps.

Brodmann et al. speculate that the two orchids grow in dark forest understory where there may be low numbers of alternate pollinators, thus making the recruitment of social wasps more appealing. The costs to the wasp are likely to be small (in terms of wasted time and energy), and may be offset by consumption of nectar. Hence, this relationship is not difficult to understand evolutionarily. If the orchids become common in the environment, however, then we would expect wasps to learn to ignore the volatiles if they do not generally lead to a valuable food reward. However, social wasp colonies can involve hundreds of individuals, and there may be considerable pollination potential for orchids during the learning of all of these wasps.

9.8   Conclusions

When the idea that plants could send out SOS calls to the antagonists of herbivores that were attacking them was first introduced, there was both widespread excitement at the idea and scepticism about the ecological relevance of this effect. In recent years there has been growing evidence from field-based experiments to suggest that such indirect defences may be relevant to some plants. However, as we have discussed, the potential effectiveness of such signalling will be variable between systems and we would not expect such indirect defences to be ubiquitous. Hence, we recommend caution: yes, such interactions do appear to have the potential to benefit plants in some situations, but we are still a long way from understanding how commonplace this defence is in the world around us.

For each system where indirect defences are suspected, we must demonstrate that under natural conditions all the steps of the following sequence are met: (i) infestation of herbivores changes the volatile emissions of the plant; (ii) this change in volatiles attracts increased numbers of antagonists of the herbivores to the plant; (iii) these antagonists affect herbivores in ways that reduce their impact on the plant; and (iv) these effects have a net fitness benefit to the plant. The final point is not assured by the first three, since there may be other costs to the production of the herbivore-induced volatiles (physiological and ecological costs; see van der Meijden and Klinkhamer 2000 for more discussion). There may also be situations where local availability of antagonists is simply too low to have a measurable impact on herbivore damage to the plant. Not all antagonists may be suitable, and indeed some parasitized herbivores are more damaging to host plants than unparasitized conspecifics. Studies must consider that the effectiveness of produced volatiles can be relatively low in natural situations when odours are masked by those produced by the large number of heterospecific plants. Moreover, naturally growing plants can be simultaneously attacked by several herbivores, and a single herbivore can be attacked by more than one type of antagonist. This complexity may be important, Rasmann and Turlings (2007) demonstrated that a parasitic wasp and an entomopathogenic nematode were both attracted to maize plants damaged by their respective herbivorous hosts, but when both herbivores attacked a plant simultaneously attraction of the antagonists was considerably reduced.

Transgenic plants planted in natural situations will be a powerful tool with which to explore indirect defences. We can make 'mute' plants that are unable to release the signalling volatiles. This allows a situation where the volatiles can be artificially added (at the appropriate concentration) in the vicinity of some of these plants, with the others representing controls. We similarly have the ability to produce 'deaf' plants that cannot detect the volatiles, and these can be used to explore to what extent the response of receiver plants affects herbivores, their antagonists, and the plants (see Baldwin et al. 2006 for further discussion). We also have growing knowledge of the triggers of herbivore-related volatile release, and this can be used to trigger release without requiring the actual presence of the herbivores (for example, by application of herbivore regurgitate or methyl jasmonate; e.g., Thaler 1999).

Exploration of the frequency of occurrence of plants signalling about herbivores requires acceptance of the complexity of natural ecosystems, which will provide a diversity of both emitters and receivers of these signals with varying selective pressures.

CHAPTER 10

Sensory aspects of carnivorous plants

10.1 Introduction

Carnivorous plants fascinate because they upset the natural order of things: we expect plants to be the victims of animals, not their predators. On having the Venus flytrap (*Dionaea muscipula*) reported to him, Linnaeus expressed disbelief at the idea that it was carnivorous, describing this suggestion as 'against the order of Nature' (Barthlott et al. 2007). Darwin was more ready to think the unthinkable and devoted an entire book to carnivorous plants: the first substantial study of the subject (Darwin 1875). However, in over a century since, research on carnivorous plants has not been particularly intense. Carnivory by plants is an often-overlooked area of plant–animal interaction: the recent monographs *Plant–animal Interactions: An Evolutionary Approach* (Herrera and Pellmyr 2002) and *Insect – Plant Biology* (Schoonhoven et al. 2005) make no mention of it. For this reason, we spend a little space introducing some aspects of the biology of carnivorous plants and setting them within the context of evolutionary ecology. From this foundation, we then explore sensory aspects of this plant–animal interaction.

10.1.1 What are carnivorous plants?

There are over 600 species of carnivorous plants, spread across six subclasses of angiosperms and distributed worldwide. Despite having evolved independently perhaps as often as ten times, carnivory is a very uncommon trait in plants (occurring in less than 0.2 per cent of angiosperm taxa). Approximately 25 per cent of carnivorous plants are pitcher plants that have leaves modified into pitcher-shaped pitfall traps. These pitchers fill

with rainwater and/or plant-produced fluid. Pitcher plants belong to three unrelated families (*Cephalotoceae*, *Nepenthaceae*, and *Sarraceniaceae*). The other commonly-used capture device is adhesion to sticky fluids. The structures holding the adhesive may be either static or move in response to physical pressure from the victim (generally involving closing around the victim). Less common techniques are closing traps (most famously the Venus flytrap) and suction traps.

Although the term *insectivorous plant* is more common, we consider *carnivory* a more accurate term; although insects are by far the most common prey type, prey can extend from unicellular organisms to (exceptionally) rodents (Barthlott et al. 2007). Careful definitions of plant carnivory were presented by Givnish et al. (1984) and Anderson and Midgley (2003). Essentially, a carnivorous plant has two key features. Firstly, it must have 'morphological, physiological or behavioural features whose primary effect is attraction, capture or digestion of prey' (Anderson and Midgley 2003). Secondly, they must be able to absorb nutrients either directly from the dead animal or via an obligate and host-specific digestive mutualism.

10.1.2 Relationships between carnivorous plants and non-prey animals

Some animals play a part in plant carnivory without themselves being prey. Anderson and Midgley (2003) discuss the family *Roridulacae* as an example of plants that rely on an obligate digestive mutualism with an animal. These plants catch insects on their leaves thanks to hairs bearing sticky droplets, but the plants have no means to absorb nutrients from these trapped animals directly. Each of the two *Roridula* species has an obligate species-specific association with a hemipteran (*Pameridae*). These insects are able to walk unhindered over the leaves and consume the captured prey. The hemiptera can move freely on the plant because they possess an epicuticular greasy secretion that prevents contact between the sticky plant secretion and the insect cuticle (Voigt and Gorb 2008). Prey species have thinner and more fragmentary epicuticular waxy layers. The hemiptera defecate on the plant leaves and the faecal nitrogen is absorbed by the plant through the leaf cuticle.

Such mutalistic relationships appear vulnerable to cheating by animals that consume the food but do not defecate on the plant. Anderson and Midgley (2003) also report on a spider (*Synaema marolthi*) that is associated exclusively with the plant *Roridula dentate*. The plant has the unique species of hemipteran mutualist (*Pameridea marlothi*). The spider preys on the hemipteran, and spends large amounts of time away from the leaf in a nest, where it defecates. The nitrogen from spider defecation is thus lost to the plant. The spider thus doubly disadvantages the plant: it reduces the number of hemipteran mutualists (see Figure 10.1), and it removes nutrients that would otherwise be available to the plant. Given that hemiptera are mobile, and their relationship with the *Roridula* plants is obligate, one might expect selection pressure for *Roridula* to attract members of their partner species, but such attrac-

tion has not been explored. Additionally or alternatively, traits that made the plant, or hemipterans on it, less detectable to spiders (for example, by masking their odour) might reduce rates of colonization by spiders, and thus provide a benefit to the plant; but again this has not been explored.

Pitchers generally are filled with either rain water, or plant-secreted fluid, or a combination of the two. Such temporary water bodies often support microfauna (often termed infauna or inquiline communities). The benefits of different infauna to the plant are likely to be complex: some may aid in prey decomposition in a way that benefits the plants, making nutrients easier to absorb. However, they will also lock up some nutrients in their own growth and reproduction. Some infauna may prey on other infauna in a way that may either benefit or harm the plant. It also seems possible that some of the plant's prey species may be drawn to the trap in order to prey on the infauna, but at a risk of themselves falling victim to the trap. A recent modelling study (Mouquet et al. 2008) suggested that the actual role of a member of the infauna is variable from being mutualist to comensualist or parasite, depending on the fine detail of the population structure of the infauna and the rate of prey capture. Mouquet et al. also argue that even when the plant is capable of digesting prey itself, some infauna can still be a net benefit to the plant through rapid and effective breakdown of prey.

A very unusual interaction occurs between the carnivorous plant *Nepenthes lowii* and tree shrews of the species *Tupaia montana* in the montane forest of Borneo (Clarke et al. 2009). Immature plants form pitchers that catch arthropod prey, similar to related pitcher-plant species. However, mature plants form pitchers that lack features (such as a waxy inner surface) that are used to trap invertebrates; since that plant now turns to tree shrews as a source of nitrogen. The 'lid' of the pitcher contains functional nectaries, and tree shrews can best access these by sitting inside the pitcher. It is the defecation of these mammals inside the pitchers, while they eat, that provides the plant with nutrients, rather than carnivory of invertebrates. Presumably immature plants are not physically strong enough to produce pitchers that can bear the weight of a tree shrew. The authors suggest that the

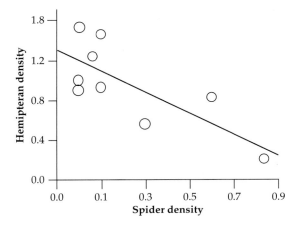

Figure 10.1 The densities of spider and hemipteran at 9 different study sites. Redrawn from Anderson and Midgley (2002).

relative scarcity of ants (and perhaps other insects) at the high altitude of this plant species has driven the development of this novel solution to nitrogen limitation by the plant.

10.1.3 Where are carnivorous plants found, and why are they uncommon?

The scarcity of carnivory by plants suggests that carnivory is not an efficient way for a plant to gather energy compared to photosynthesis. Carnivory is driven by a need to find an alternative source of nitrogen (and other nutrients), rather than a means of gathering energy; hence there is a strong link between carnivory and nutrient-poor water or soil. Although the consumption of organic nitrogen appears clearly beneficial, the rarity of carnivorous plants points to carnivory not being the most economic way for plants to secure nutrients (Benzing 1987). Carnivory is generally confined to habitats in which key nutrients are more limiting for plant growth than availability of light and water (Ellison et al. 2003). This conjecture is supported by the facultative expression of traps for carnivory that many species exhibit. Such plants do not produce traps in seasons when growth is not primarily limited by nutrient supply. As such, it is not surprising that carnivorous plants are associated with very specific habitats. Most carnivorous plants are restricted to wet, well-lit environments with very poor substrate (soil, rock, sand, or water) quality in terms of nutrient availability. Carnivory requires investment in some structure to capture prey; this investment must be traded off against investment in specialist photosynthetic structures and/or root structures (Givnish et al. 1984). Thus carnivorous plants often have a reduced root systems, and so a limited ability to gather water unless it is abundant in the substrate. High humidity may sometimes encourage carnivory by reducing evaporation, to the benefit of carnivorous plants that utilize aqueous fluids or sticky substances to trap their prey. Additionally, Moore (1991) speculated that kleptoparasitism by ants may be another selective pressure driving carnivorous plants to wetter habitats that are less congenial to ants. However, ants are also important prey for many carnivorous plants (e.g., Bauer et al. 2009).

Since construction of prey-gathering structures often comes at the expense of light-gathering structures, carnivory is generally associated with conditions offering plentiful light. Another restriction for carnivorous plants is that they must occupy a micro-habitat where prey is available (Alcalá and Dominguez 2003). Working in Spain, Zamora (1995) demonstrated a relationship between micro-climate and trapping success. This relationship was complex because micro-climate affected not only the abundance of prey but also the probability of prey being captured once it made contact with the plant. This occurred because the stickiness of fluids used to capture prey was affected by evaporation. There were fewer insects in warmer, sunnier locations, but trapping was more effective there because higher evaporation caused the leaf secretions to be more viscous. This illustrates complexity in the effect of micro-habitats on trap effectiveness, with this effect likely being specific to the plant and animal species involved in the interaction. Also, as Moore (1995) pointed out, one might expect quite different patterns of insect behaviour in cooler latitudes, with insect attraction towards sunnier spots.

10.1.4 Evolution of plant carnivory

The diversity of trapping mechanisms and their spread across families of extant plants points to several independent evolutionary developments of plant carnivory (Ellison and Gotelli 2009). Although carnivorous plants employ diverse trapping structures, all of these structures can be viewed as modifications of pre-existing characters in N-autotrophic plants (Benzing 1987). Some of these characters function as protection against abiotic factors in autotrophic plants, like anthocyanins as pigments and epicuticular waxes that lubricate traps and prevent insects' escape. Both anthocyanins and epicuticular waxes reduce the amount of light reaching the chloroplasts and are therefore often regarded as protection against high illumination (see Chapter 8). Alcalá et al. 2010 demonstrate that the leaf secretions of the carnivorous plant *Pinguicula moranensis* function both as a sticky trap for small arthropods but also as a deterrent to herbivory by insects, and so the second of these functions might have driven the initial evolution of sticky secretion. This is an extremely interesting idea because carnivory is then an exaptation of an originally defensive function against insect herbivory (see Section 1.8 for similar exaptations from defence to attraction in flowers). Other characters like nectar excretion and odour are used to attract pollinators in autotrophic plants. Bromeliads sport phytotelma for water and nutrient storage, which may easily be adapted to capture insect prey in the water body. Tracing the origins of adaptations for carnivory will further our understanding of the evolution of carnivory. In particular, we may be able to understand how pre-adaptations favoured the evolution of carnivory and which traits became adapted to function as traps.

For example, many non-carnivorous plants have the ability to absorb nutrients from dead insects that may be blown onto their leaves (Benzing 1987). Animals may be trapped on leaves and die after having been imprisoned by the surface tension of rainwater that has puddled in a depression on the leaf. One can see how there could be selection on the plants to enhance the frequency of such events until we end up with the pitfall traps of the pitcher plants. Similarly, one can imagine how small animals could become adhered to waxes secreted for reasons unrelated to prey capture, die, and provide a useful source of nitrogen. Again, incremental evolution towards current adhesive traps can be imagined without difficulty. Together these make up the two most common trapping mechanisms. There is no doubt that the evolution of bladder traps and snap traps are much less readily imagined, and this might be linked to their relative scarcity (see Section 10.3.2)

10.1.5 Carnivorous plants as a model of co-evolutionary processes

The predictable environmental effects on the interaction between carnivorous plants and the animals that interact with them in a mutualistic or antagonistic manner make an ideal system in which to study the co-evolutionary mosaic (Thompson 2005b). This framework predicts that the interactions among species are not necessarily the same across a gradient of geographic variation. The type of interaction between two organisms might even change from being mutualistic to antagonistic in

different parts of their range. Likewise, in some parts of the range, species might be well adapted to each other, whereas in others they may interact only loosely and without mutual adaptations. As yet, the co-evolutionary mosaic has been explored in the mutualistic and antagonistic interactions between plants, pollinators, and herbivores (Thompson and Fernandez 2006), as well as in the interactions between conifers and the animals that exploit their seeds (Benkman 1999). Exploring the co-evolutionary mosaic in the interaction between carnivorous plants, their prey, and possible mutualists is particularly exciting because variation in the external environment can have a very complex suite of effects on carnivorous plant fitness. For example, it is possible to imagine how micro-climate will not only affect the overall abundance of insects but also the size distribution of insects available. Similarly, micro-climate will affect trap stickiness which in turn will differentially affect the catchability of different sizes of prey. Since micro-climate will also affect water and light available for photosynthesis, there will be a complex of trade-offs in the production of traps in carnivorous plants.

An example of the variability of the interaction between carnivorous plants and animal species is given by Clarke and Kitching (1995). They show that the pitcher plant *Nepenthes bicalcarata* has a close association with ant species (particularly *Camponotus schmitzi*) that remove large prey from the pitchers and avoid accumulation of excess prey that may lead to putrefaction of the pitcher. They found that plants that naturally had ants were statistically less likely to suffer putrefaction of pitcher contents than plants without ants. Although ant presence was not manipulated, these authors demonstrated that experimentally introduced ants could thrive on plants on which they were initially absent. Thus they suggest that third variable effects are unlikely to affect their main experiment, with chance aspects of ant colonization being the major influence on whether or not a plant has ants. Putrefaction of the pitcher contents was demonstrated to kill the infauna that are needed by the plant to digest the prey. In the absence of large prey, however, the ants were observed to take larger items of the infauna, and thus the ant may switch from being a net benefit to the plant to a parasite, depending on how quickly the pitcher fills with prey. The ants gain a domatia provided by the plant, carbohydrate from the plant's nectaries, and protein from insects caught in the pitchers. The benefit to the plant of this association is not clear, but may be protection from herbivores rather than (or more than) any benefit of excess prey removal from pitchers. It may even be possible that a plant might catch prey not for its own benefit, but as a service to ant mutualists, although this remains no more than our speculation without empirical foundation.

10.1.6 Sensory aspects of plant carnivory

There are several different types of interaction that should be of interest to us in this book. The first and most obvious one is whether plants can lure their prey towards them in a way that increases capture rates. However, many carnivorous plants are pollinated by animals, and there is the potential for conflict between pollination and carnivory, in that the same individual animal may be able to function as both pollinator and prey, but the two tasks are mutually incompatible. This conflict is exacerbated in species that are pollen-limited such as *Pinguicula vallisneriifolia* (Zamora 1999). In this species, the interference between prey capture and pollinator visitation varies spatially because pollinator assemblage varies along an environmental gradient from sunny to shaded habitats. In sunny locations, pollinator abundance is relatively high and the conflict between carnivory and achieving pollination is less severe than in shaded locations with lower pollinator abundance. Further, as discussed earlier, there are animals that are immune to capture by the carnivorous plant but which are associated with the plant. This association may bring net costs or benefits to the plant or cause no change in plant fitness. For animals in the first two categories, there may be selection pressure on plants to exploit the sensory systems of animals with mobile self-directed life-history stages in order to either increase or decreases their likelihood of finding the plant. Such sensory interactions will be the focus of the remainder of this chapter.

10.2 Attraction of victims

10.2.1 Does attraction of prey to pitchers involve duping them?

A key paper in considering the attraction of prey to carnivorous plants is that of Joel (1988). Many previous influential studies (e.g., Wickler 1968, Wiens 1978, Pasteur 1982, Williamson 1982) had argued that pitcher plants function as Batesian mimics of flowers. That is, that insects were duped into landing on pitchers in the mistaken belief that they were flowers. Joel argues (in our view convincingly) against this theory.

Joel made their case based on a number of arguments:

1. There is no evidence that any insect visits a pitcher because it has been duped

In Batesian mimicry the predator has an aversion to some defended prey type (called the Model). Another species (called the Mimic) is undefended and would be attractive to predators, but the Mimic gains protection from these predators because of its resemblance to the Model. That is, the Mimic gains because the predator is unable to reliably differentiate between Models and Mimics. The previous works referred to above had suggested by analogy that pitchers benefit from some animals being unable to reliably differentiate harmful pitchers from rewarding flowers, and thus being sometimes actively drawn to pitchers in the mistaken belief that that the pitcher was a (potentially rewarding) flower. Joel is correct that no evidence existed in 1988, and no evidence exists now, that any insect is drawn to a pitcher because of misidentification of the pitcher as a flower. However, no evidence exists by default since no one has specifically tested this hypothesis. Such tests would not be difficult conceptually or practically in the laboratory. They would require comparison in vulnerability to pitchers across insects that were randomly allocated to treatment groups differing in exposure to flowers during a training period. In the field, it would be interesting to compare the distribution of insects caught in pitchers with the distribution of prey types available in the local habitat sampled by other means (nets or fly-papers, for example). If pitchers dupe flower-seeking insects, then one would expect the distribution of prey species in pitchers to be biased towards species that actively seek out flowers.

2. No specific models have been identified, only a generalized resemblance between the general class 'pitcher' and the general class 'flower'

Again, Joel was correct in 1988, and this statement remains correct now, but by default.

It is clear that for insects to be fooled into mistaking pitchers for flowers, the pitchers should be similar in appearance to locally available flowers and be active at the same time as these flowers. These predictions have not been tested. The appearance of pitchers seems highly evolved both morphologically and in terms of pigmentation. Given the great diversity of flowers and the specialization of many pollinators, if confusion between pitchers and flowers is to be effective, one might expect that pitchers have greater similarity (in some measures of morphology and pigmentation) to locally available flowers than to some control group of flowers. This has not been explored.

In the absence of such evidence, we consider a generalized similarity between flowers and pitchers to be (at best) weak evidence of pitchers profiting from mimicking flowers. If pitchers have characteristics in common with flowers then one interpretation of this is that the pitcher is mimicking a flower. However, it could be that these characteristics are an example of convergent evolution driven by exploitation of receiver biases (see Chapter 8). This argument is developed by Biesmeijer et al. (2005). These authors demonstrate that a stingless bee's close approach to a flower is influenced by a dark-centre preference, and that in approaching their nest entrance there is a preference for a dark centre with disrupted lateral ornamentation. Flowers often accentuate their dark centres by radial stripes. Biesmeijer et al. argue then that pitchers exploit bees' sensory bias for dark-centred objects because they often also present radiating stripes as visual stimuli. Hence, while the dark centre is an unavoidable consequence of pitcher morphology, radial stripes on the pitcher may increase capture success.

Again we await comparative analyses to test this suggestion.

It has been suggested by Joel (1988) and Wiens (1978) that the flowering period of non-carnivorous species and the peak insect catching time of carnivorous species in the same habitat are disjoint. It would be valuable to quantitatively test this suggestion. If there is evidence for such a temporal separation then this would argue against the suggestion of pitcher plants benefiting from duping prey, since prey could use time-of-season (apart from difference in size and shape) as a reliable cue to differentiate pitchers from flowers. Such an analysis would be particularly powerful if it could explore whether pitchers had produced any selection pressure on the timing of flowering of sympatric species. Such an effect would suggest that pitchers do impose significant costs on flowers by reducing pollination rates, just as Batesian mimics impose a predation cost on their models.

3. There seems to be no selection pressure on pitchers to be rare relative to flowers

In Batesian mimicry, the expectation is that the Mimic will be protected most effectively when it is rare relative to the Model. Each attack on a mimic brings a net reward, whereas each attack on a Model is costly to the predator. If the predator is unable to differentiate Models from Mimics, then attacking all individuals encountered becomes a more beneficial strategy for the predator as the density of Mimics increases relative to Models. Alternatively, if differentiation between Models and Mimics is possible but requires time-consuming investigation, then such investigation becomes more economic for the predator if Mimic density rises relative to Models. Lastly, if the aversion of the Model has to be learned, this learning will be hindered if the predator regularly encounters rewarding Mimics as well as punishing Models. For all these reasons, Batesian mimics are expected to be rare relative to their models. By analogy, if pitchers are to successfully dupe insects into misidentifying them as flowers then one would expect them to be rare relative to flowers. In suggested contravention of this, Joel (1988) cites reports of large and dense populations of carnivorous plants.

Most plant communities that contain carnivorous plants also contain a wide diversity of non-carnivorous plants (see Ellison et al. 2003 and references therein). However, this need not always be so: in wet and nutrient-poor ephemeral habitats in west Africa and elsewhere (termed inselbergs) short-lived carnivorous species are very common (Dorrstock et al. 1996), although carnivorous plants almost never dominate entire communities (Benzing 1987). We again lack a proper systematic exploration of whether pitcher plants in particular are found in areas where there are plentiful flowers that could act as models. However, it seems unlikely that pitchers have evolved to be flower mimics if potential prey can use cues from habitat to differentiate pitchers from flowers.

No study has tested whether variation in the relative density of pitchers and flowers has any effect on pitcher effectiveness. Indeed, there has been very little study of the effect of pitcher density on individual predation rate. Cresswell (1991) experimentally manipulated pitcher density downwards in *Sarracenia purpurea*, but found no adverse effect of density on prey capture rates. Gibson (1991a) did find an apparent competition effect where prey capture rate declines with pitcher density in a combination of natural variation and manipulation (see Figure 10.2). These experiments might prove useful templates for study of effects of the relative densities of pitchers and flowers on pitcher effectiveness.

4. Pitchers provide a nectar reward

To us, this is the most powerful of Joel's arguments. The suggestion is that there is no need to

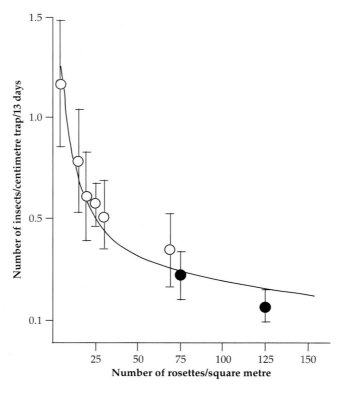

Figure 10.2 Insect capture rates (mean and 95 per cent confidence interval) as a function of the density of capture-structures (called rosettes) for the North American sundew *Drosera filiformis tracyi*. Filled circles are manipulated densities; open ones are naturally occurring. Illustrated are means and S.E. Redrawn from Gibson (1991a).

dupe insects into visiting pitchers, because for the insect the risk of being captured is more than compensated for by the potential food reward from nectar – especially because trapping efficiencies are usually fairly low. Many pitcher plants produce nectar (see Plate 16). Joel argued that nectar is basically a watery sugar solution, and that this should be cheap for plants living in wet, sunny environments to produce. However, given that traps are highly modified leaves which are less efficient in photosynthesizing (Hajek and Adamec 2009; Karagatzides and Ellison 2009), the relative costs of producing nectar are likely to be higher for carnivorous plants than for most autotrophic plants.

Newell and Nastase (1998) demonstrate that prey capture efficiency of the pitcher plant *Sarracenia purpurea* is very low. Only 1–2 per cent of those insects that visited the plant where captured. Since the plant provides nectar, this led the authors to speculate that the risk of loss of life might be worth the potential energy gain from the nectar for some visitors. The economics of reward versus risk have yet to be investigated to assess the evolutionary dynamics of the interaction between carnivorous plant and potential prey. From the plant's viewpoint this would require measurements of the fitness benefits of prey capture and costs of nectar production. From the prey's perspective we would need to estimate the benefit gained from nectar versus the mortality costs. Ant colonies might provide an ideal study system for this second aspect (ants were frequent visitors to the nectaries in the study of Newell and Nastase 1998). In line with the data of Newell and Nastase, Merbach et al. (2001) reported capture rates of ants on five species of *Nepenthes* pitcher plants, and estimated the percentage of ant individuals that feed from the nectaries and that were actually captured as ranging from 0.34 to 1.6 per cent. Dixon et al. (2005) found that only 1.3 per cent of wasps attracted to the pitcher plant *Darlingtonia californica* were captured.

Thus, the most powerful aspect of Joel's argument that pitchers are not flower mimics is that they offer nectar as an inducement to visitors, removing the requirement to dupe visitors. Many pitchers produce nectar, but we would welcome a systematic study of this to identify if there is anything different about the appearance (or more generally, the

ecology) of pitcher plants that do not produce nectar.

10.2.2 Other aspects of deception

Joel (1988) argued that insects that gain from visiting a pitcher should learn the association between pitcher appearance and reward, increasing their likelihood of visiting further pitchers. 'Those few insects that pay for the nectar with their lives cannot transfer their "knowledge" of the possible danger, simply because they die. Selection against visits will not develop because the proportion of "casualties" is limited.' Similarly, Juniper et al. (1989) argued that pitchers are flower-mimics but conclude: 'Since deceived insects are eliminated, insects experienced with the mimic will be rare and therefore avoidance will be negligible.' In both these cases, the authors seem not to have considered that aversion can be innate rather than learned. This is a well-demonstrated concept in a predator-prey context, where predators can have innate avoidance of potentially lethal prey, such as venomous snakes. In this case, selection has acted to develop unlearned aversion towards, or wariness of, snake-like objects. There is no reason why insects could not develop a similar aversion to pitchers, if these posed a risk of death without any reward to offset this risk. Indeed, there is experimental evidence that some flower-visiting bees have developed innate avoidance of flowers that are occupied by crab spider predators (Heiling et al. 2004). The crab spiders are clearly visible upon the flowers and various bee species are attracted to these contrasting floral displays from a distance. Native bees that share an evolutionary history with the crab spider innately avoid the spider by using olfactory cues, whereas introduced honey bees that does not possess a similar long history of sharing the same habitats with the crab spiders do not pay attention to olfactory cues.

At present there is no evidence that any animal has evolved counter-adaptations that have been selected in order to allow them to reduce their vulnerability to carnivorous plants. This may simply be a result of lack of study, but could also be a real phenomenon, again stemming from the particular habitat requirements of carnivorous plants. Although (as detailed by Joel 1988) carnivorous plants can be locally abundant, they are liable to be present in only a very small part of the spatial range of any insect population. As such, they are unlikely to be a significant source of mortality on the scale of populations for many insect species, and thus there would be negligible selection for traits that reduce this source of mortality. A corollary of this argument would be that if such avoidance adaptations are to be found, they will be in animals with very particular habitat requirements that strongly overlap those of the relevant carnivorous plant species.

Similar reasoning to the above causes us to predict that there is unlikely to be selection pressure for non-carnivorous plant species to avoid being misidentified as carnivorous species and thus avoided by potential pollinators. Firstly, we do not expect many (if any) animals to have developed adaptations to avoid carnivorous plants, making the need for such differentiation moot. Secondly, even if such avoidance did exist, then selection would only occur in plant species that were constrained to exist in very similar habitats to the carnivorous plants. Lastly, since most insects that visit carnivorous plants manage to escape, and some insects may actually be actively drawn to carnivorous plants by nectar rewards, there would be little selective pressure for non-carnivores to differentiate themselves from carnivores.

If Joel (1988) is correct, and carnivorous plants are not mimics, one might ask why carnivorous plants have not evolved to be deceptive, mimicking flowers or other attractive objects. For example, although the overwhelming majority of flowers do offer rewards to pollinators, some do not and exploit animals through deception (see Chapter 8.4). The answer may involve the general habitat requirements of carnivorous plants. Firstly, their association with wet and bright micro-habitats may make nectar very cheap to produce, although the costs have still to be quantified. If the cost of the nectar reward to animals is negligible for the plant then there is less need to deceive. However, orchids are also generally associated with wet and often bright environments, and mimicry is commonplace in the orchids as a means of exploiting pollinators. Although the visitation rates of unrewarding orchids are lower than those of rewarding ones, they benefit from not offering nutritional rewards through achieving

higher outcrossing rates; because animals more readily move away from the plant rather than visiting and pollinating the other flowers of that plant (Jersáková et al. 2006). Obviously, this selective pressure is absent in carnivorous plants. It may be that morphological and developmental constraints associated with trap structures being derived from leaves may hinder carnivorous plants from becoming very efficient mimics of flowers. Moreover, traps need to maintain a particular morphology and surface for efficiently trapping animal prey, which may impose further constraints on mimicry. For these reasons, we suspect that Joel's iconoclastic argument that pitchers are not deceptive, may well hold for carnivorous plants in general.

Joel (1988) suggested a case of mimicry *between* carnivorous plants: *Heliamphora heterodoxa* and *Brocchinia reducta*. Only the former produces nectar, whereas the other emits 'a sweet nectar-like odour'. Joel concludes that 'One can assume that this resemblance of *B. reducta* to *H. heterodoxa* in colour and in scent as well as in shape and size, is in fact a case of mimicry.' This is an intriguing idea that has not been followed up with empirical testing. Firstly, we would welcome confirmation that the presumed Mimic is closely associated spatially with the Model, and is confined within the spatial range of the model. If this is so, then exploration of whether the success of *B. reducta* in attracting insects can be related to (preferably experimentally manipulated) local density of *H. heterodoxa* would be worthwhile.

10.3 Signalling mechanisms

10.3.1 Visual signals

Since the prey capture structures of most carnivorous plants have evolved from leaves, it is no surprise that green is a very common colouration for these structures. However, pitchers and many sticky traps often sport red colouration presented in complex patterns; and some species are mainly red. Here we will summarize the evidence that such colouration acts as a prey attractant, before comparing the colouration of these structures to flowers.

Cresswell (1991) measured 87 pitchers of *Sarracenia purpurea* and explored the factors related to capture success. Whilst pitcher size had the strongest effect on success rate, when this was controlled for, the extent of red striping on the pitcher and production of nectar both positively affected capture rate. Although correlative, this study does suggest that pitcher colouration can affect prey attraction.

Ultraviolet reflective patterning appears quite common among carnivorous plants (Joel et al. 1985, Globner 1992). UV patterns are produced because mucilage and nectar reflect UV wavelengths well, while the green parts of the plants often only reflect small fractions of incident UV light (Joel et al. 1985). If nectar reliably reflects in the UV, carnivorous plants might visually advertise the presence of nutritional rewards in order to attract insects. Moran (1996) reported on the pitcher plant *Nepenthes rafflesiana* that produces two different types of pitcher (termed *upper* and *lower*). Both pitcher types have UV patterning associated with nectaries, and manipulative experiments demonstrated that these attract prey. The upper pitchers produce a fragrance that a further manipulative experiment demonstrated can enhance capture of flying prey (see also Di Giusto et al. 2008). Lower pitchers rest on the ground. Moran suggested that potential invertebrate capture rates are higher for these lower pitchers because of their positioning, and so the added attraction of fragrance is not required for them. Indeed he suggested that excessive prey capture should be avoided because of the production of toxins in putrefied prey that decay before they can be absorbed. This is supported by Clarke and Kitching (1993) who report anecdotally that 'putrification of the contents of termite-filled pitchers is common, killing most predatory larvae' (larvae that are needed to digest the insects so as to make the nutrients available to the plant).

Juniper et al. (1989) suggested that colour change during trap development should be a good indicator of the functioning of colouration in insect attraction. They suggested that the non-green pigmentation of many traps is not apparent until the trap is ready to capture prey. However quantitative information on this is sparse (Bauer et al. 2009). Juniper et al. discuss the observation of Schnell (1976) that the pigmentation of pitchers of *Sarracenia dionaea* is affected by illumination levels, but that the difference in pigmentation is not reflected in

trapping capability. Since illumination levels are likely to affect insect activity, such an observation is difficult to interpret. Juniper et al. also cited the observation of Edwards (1876) that the largest quantities of insect remains are found in pitchers that are deepest and richest in colour, although third factors or even reverse causation are not considered as explanations for this observation. We would suggest that quantitative evaluation of colour change during the lifetime of traps would be worthwhile, especially combined with manipulative experiments to explore the potential functioning of colouration. Also, analysing whether red traps are more likely to develop in plants receiving high irradiance compared to shaded plants of the same species, would be a fruitful approach to study whether anthocyanins in carnivorous plants develop as a protection against excess light (see Chapter 7). If so, this would be an example of how abiotic factors might channel the communication between plants and animals.

We (Schaefer and Ruxton 2008) experimentally explored the effect of colouration on prey attraction in the pitcher plant *Nepenthes ventricosa* whose pitchers show variation in colour from red to green. We experimentally painted some green and some red (using paints that gave low and similar odour emissions) and placed the plants out in a paired design. The red pitchers caught significantly more flying insects (see Figure 10.3), particularly those of the family *Symphyta* that possess a receptor sensitive to red light. Previously, contrast between pitcher elements was considered by many authors to be a primary factor in visual attractiveness to potential prey. Our results suggest that traps may not need to sport contrasting colour patterns to be attractive, it may be sufficient for them to differentiate themselves from the background (of other predominantly green plants). More manipulative studies of this nature would be welcome. Such experiments should take care to control for potential non-visual attractants; a recent study by Bennett and Ellison (2009), using the pitcher plant *Sarracenia purpurea*, argued that nectar availability is much more important than colour for attracting prey.

Although the traps of carnivorous plants can sport complex structures, and colourful patterns, they do not even approach flowers in their exuberance

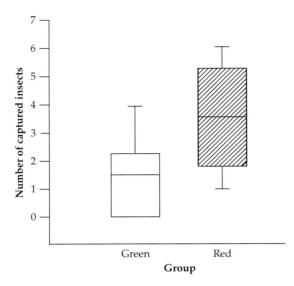

Figure 10.3 The median number (plus interquartiles and 10th and 90th percentiles) of insects caught by green and red painted pitchers. Illustrated are medians, interquartiles and 10ᵗʰ and 90ᵗʰ percentiles. Redrawn from Schaefer and Ruxton 2008.

of display. The most likely explanation for that difference is that flowers are developmentally more complex structures than traps even though both originate from leaves. Furthermore, traps are often long-lived and need to not just attract insects and other prey but also restrain and then digest them. These functional trade-offs may be part of the answer, but probably only a small part, since there are long-lasting flowers that are still stunning to behold (e.g., many orchids), and very elaborate and beautiful flowers that trap their pollinators for a short time before releasing them.

The fact that carnivorous plants are short on nutrients may mean that they do not have the resources to produce stunning displays. This argument is supported by studies showing that sexual reproduction is induced and enhanced by adequate prey supply (Juniper et al. 1989, Ellison and Gotelli 2001). To assess this argument, we would welcome a comparative study of the number, size, and colouration of the flowers of carnivorous plants to see whether these are more modest than appropriate comparator groups, as such a situation would add strength to the nutrient-shortage argument.

Another reason for the relative dullness of traps may be that much floral exuberance is aimed at differentiating species of flower from each other and attracting specific types of pollinator only, to aid successful cross-pollination between members of the same species (see Section 6.3). This selection pressure is absent in traps, which are less selective in their prey than many flowers are in their pollinators, and which do not generally need to differentiate themselves from other species. One exception might be if they do advertise their presence to plant-specific mutualists that may, for example, lay eggs in the pitchers.

Almost all work on potential use of attractants by carnivorous plants has focused on pitcher plants. For sticky traps, prey can sometimes escape if they can drag themselves to the edge of the sticky surface (Gibson 1991b). Thus one might expect selection pressure for plants to produce deceptive signals that induce prey to land in the centre of such structures, but we know of no study that has considered this idea. One possible way of doing so would be displaying dark centres as discussed earlier. However, dark centres do not seem to be a common feature of sticky traps. We would be interested in any study that explored differential landing rate of prey on different parts of sticky traps, and differential retention by the traps of those prey that land in different parts of the trap.

Many sticky traps have a highly reflective dew-like appearance that may be visible to, and attractive to, potential victims. Indeed such drops of sticky fluid might look like nectar to potential victims. Juniper et al. (1989) suggested 'Diptera, in particular, are attracted by glistening drops (Kugler 1956)', although Kugler studied flowers rather than carnivorous plants. We would be interested in an experiment that compared insect landing rates on unmanipulated leaves compared to others that have been sprayed to produce a surface of fluid droplets.

A feature of several pitcher plants is light-transmitting fenestrations (called Areoles) on the upper surfaces of the pitcher (see Plate 17). Juniper et al. (1989) suggested that these function to confuse insects within the pitcher as to the position of the aperture through which they could fly out, thus using sensory deception to aid retention of prey. The exhausted insect eventually falls into the liquid at the bottom of the trap. This seems plausible, but manipulative experiments would be welcome. Most simply we would be interested in comparing the rates of insects entering the pitcher and also subsequently escaping from it between controls with their Areoles intact, and a group that have had these painted over to become opaque.

10.3.2 Non-visual attractants

The volatiles released during prey decomposition have been suggested to attract additional prey (Kato 1993, Cresswell 1991 and reference therein), as has the decay of vegetable matter that may fall or be blown into traps (Gonzalez et al. 1991). There are, however, few experimental tests of these suggestions. The recent study by Bhattarai and Horner (2009) on the capture success of *Sarracenia alata* stands out for its careful design. In their study, plant morphology (that is, funnel area), which was correlated with plant height, affected the mass of captured prey. In contrast, colouration and the amount of nectar offered had no effect on prey capture success. Comparing the capture success of plants with those of similar-sized artificial traps, the authors showed that the capture success of plants was higher. This result is consistent with plants possessing attractants that increase prey capture. By using artificial traps that either contained the contents of natural pitchers or a control solution, the authors were able to show that volatiles emitted from the pitchers increased capture success. Yet, the capture success did not differ between a solution of pitcher contents, a solution of the extracts of pitchers, or a combination of the extract of pitchers and pitcher contents. This experiment suggests that insects react to odour in ways that increase capture rate. It remains to be seen whether the production of odour is an adaptation towards increasing prey capture or whether it is a necessary consequence of insects being digested by the plant and micro-fauna in the pitchers.

The adaptiveness of putrefaction odour is a long-standing question. Juniper et al. (1989) and Beaver (1983) argued that since pitchers only have a finite time window of effectiveness, by the time decomposition of prey has advanced to a stage where putrefaction odour is generated, the pitcher will have stopped trapping. If for no other reason, the toxins produced by putrefaction would have killed infauna in the traps and likely changed the chemical nature of the contained fluid in such a way as to make absorption of animal nutrients impossible for the plant. This argument appears convincing. Evaluation of timing of volatile release from decaying prey in comparison to the temporal distribution of prey captures would be instructive, as would exploration of whether later prey-captures in the lifetime of a trap are biased towards species that are attracted by decomposing animal or plant matter.

It is clear that there is a general trend for traps to be most effective when they are young (e.g., Karlsson et al. 1994). In part this may simply be because production of traps is timed to take advantage of any seasonal peak in prey availability. Further, traps may suffer physical damage over time that may reduce their trapping ability; this physical damage may be caused in part by previous prey captures (Karlsson 1987). However, it may also be that attraction declines over time, most obviously through reduced production of volatiles by the plant. We do not know of any study that has attempted to disentangle these factors. This could be done by monitoring volatile emissions over time, or by disentangling age and physical damage by placing physical protection around traps for varying lengths of time (manipulating past prey capture rate) and exploring the effect of this on the attraction and retention powers of differentially manipulated traps. Note also that Bauer et al. (2009) demonstrated an increase in capture rate during the first two weeks (representing approximately a quarter of a trap's lifetime) for pitchers of *Nepenthes raffesiana*.

Jürgens et al. (2009) reported high emission rates from the pitchers of *Sarracenia flava*, *S. leucophylla* and *S. minor*. They further suggested that the compounds involved were those more typically associated with fruits and flowers, rather than leaves. In a flight-cage experiment with *Lucillia cuprina* blowflies, they found that insects appeared more attracted to individual plants that had higher rates of volatile emissions (see Figure 10.4). This situation parallels that of flowers where an increase in odours enhances pollination rate (see Figure 6.11). These authors suggest that the composition of volatiles is indicative of fruit or flower mimicry. As discussed in Chapter 8, rather than mimicking specific flowers or fruits, in may be more likely that the pitcher plants benefit from exploiting a more generalized sensory bias in insects. However, we would welcome consideration of whether specific fruits or flowers can be identified that might act as models for the specific blends of scents emitted by these traps; and whether the insects caught in these traps are biased towards those that are attracted to these specific flowers and fruits. Further, it must be remembered that these traps offer nectar, and the similarity of volatiles with flowers and fruits might be a case of convergence on scents that are effective in being detected by insects.

The study of Zamora (1995) found that pitchers captured similar qualities of prey by day and by night, which he argues is suggestive that visual attraction was not important in his study species, *Pinguicula vallisneriifolia*. However, we see this conclusion as premature without evidence that prey availability was unchanging over the 24-hour cycle. In fact, we find it surprising that night-time gathering was as effective as it was, this may be because the study was conducted in Spain in summer. It may be that daytime temperatures were sufficiently

high that insects were more active at night than during the day for thermoregulatory reasons. In such circumstances, visual signals may have compensated for relatively low prey availability by day. In any case, this study raises the neglected issue of daily variation in trapping that deserves further study.

Zamora (1995) also found that plants experimentally placed in groups captured fewer but bigger prey than those placed singly. This reduction in captures is likely due to local depletion by nearby plants (so-called shadow interference, as studied in sit-and-wait animal predators). The increase in prey size is more difficult to explain. It may be a sensory effect, with larger prey being drawn to a stronger signal (produced by the aggregation of nearby plants). Alternatively it may be that large prey can wrest themselves free from one plant's trap (Zamora 1990), but on blundering into another trap shortly afterwards are less able to perform the feat again (perhaps through simple physical exhaustion). This is akin to the 'ricochet effect' that allows an aggregation of spiders' webs to catch large prey that a single individual could not. This is the most intriguing aspect of Zamora's study, and we would welcome further experimental exploration of the effect

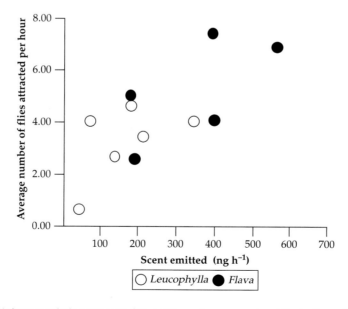

Figure 10.4 The relationship between volatile emission rate from a pitcher and average number of flies landing on the top of the pitcher. Redrawn from Jürgens et al. (2009).

of local carnivorous plant density on the types (as well as number) of prey caught. Another important aspect of this study is its demonstration of a strong effect of micro-habitat, which could be easily incorporated into models of a geographic co-evolution mosaic.

Juniper et al. (1989) present evidence from Bradshaw (1980) and Istock et al. (1983) that the mosquito *Wyeomyia smithii* that rears its larvae in pitchers is attracted by chemical rather than visual signals, but manipulative experiment on this would be very welcome. Williams (1976) discusses unpublished experiments on the landing rates of insects on boxes that contained the carnivorous plant *Drosera rotundifolia* compared to empty boxes. No difference was found. Although Williams interprets this as evidence of no use of chemical attractants by the plant, it may be that chemical attractants do not work in the absence of visual signals. Williams referred to his own field observation that insects are no more likely to land on this insectivorous plant than nearby non-insectivorous ones, but no data was provided. In trying to derive conclusions from distinct experiments on various plant and insect species, we have to remind ourselves that incongruences between studies may partly result from differences in the study design and partly from different responses of distinct plant and animal species. For example, we would not expect all potential prey to be similarly attracted by the same chemical compound.

There has been little investigation of chemical signals by carnivorous plants. It may be because of their very specialized habitat requirements that long-range chemical signals are not required, and potential prey that are looking for plants to extract nectar from can find the approximate location of suitable plants using habitat cues. Di Guisto et al. (2010) found 54 volatiles in blends of the pitchers of *Nepenthes rafflesiana* including terpenoids and benzenoids, and that the scents of the pitcher rim were particularly attractive for ants and flies. Further studies on the use of chemical attractants by carnivorous plants would be welcome.

Meyers and Strickler (1979) demonstrated potential tactile physical attractants in a carnivorous plant. *Utricula vulgaris* is a bladderwort that captures prey with a trapdoor mechanism. This species has structures termed antennae and bristles around their trapdoor, the experimental removal of which decreased prey capture rates. The plant traps micro-crustacean prey, animals that graze epiphytes off strands of filamentous algae. The plant structures are hypothesized to mimic these strands. Movement of the micro-crustaceans along the structures leads them directly to the trapdoor and capture.

Evidence of prey specialism, or even selectivity, is sparse in carnivorous plants (Ellison and Gotelli 2009). Merbach et al. (2002) reported on the apparent near-specialism on termites by the pitcher plant *Nepenthes albomarginata*. It takes it name from a fringe of edible white hairs on the rim of the pitcher. It seems that these are readily harvested by termites but at the cost that 'hundreds or even thousands' of termites fall into the pitcher, where they are trapped. The authors could not detect any long-range attractant emanating from the plant, suggesting that 'all contacts seem to happen by chance, with termites often missing pitchers less than 1 cm away from them'. There may be little need for a long-range attractant, such as the olfactory cues that are commonplace among *Nepenthes* species, because of the effective way the termites recruit nest mates to a food source. The pitcher only needs to be discovered by one termite for it to end up becoming literally filled with termites.

Several studies have reported different prey species distributions caught by sympatric carnivorous species (e.g., Thum 1986, 1988; Karlsson et al. 1987, Moran 1999, Gonzalez et al. 1991). However, the factors responsible for this difference are unclear. Nor is there a demonstration of potential for competition between species, or evidence of selective pressure to reduce this competition (Ellison and Gotelli 2009). Although sensory stimulation (either visual or olfactory) is one means by which differential attraction could occur, so is variation in micro-habitat, or physical aspects of the trap (Thum 1986). Moran (1999) does link different distributions of trapped species to known aspects of the visual and olfactory perceptions of different insects, but insufficient data were available to allow the author to achieve a convincing match; and independent testing of the insects' predicted preferences has not been carried out. At present, there is no strong evidence of selection for foraging niche separation between sympatric carnivorous species, and no evidence at all that such separation is maintained by different sensory signalling by the different species. However, this idea remains plausible and we would welcome any study that compared the appearance of, and volatile production by, a carnivorous plant between parts of its range that differ in potential for interspecific competition.

For the two species with snap-traps – the Venus flytrap *Dionaea muscipula* and the waterwheel plant *Aldrovanda vesiculosa* – there has been little study of prey attraction, although it has been noted that the Venus flytrap does produce secretions from the lip of the trap (Gibson and Waller 2009). For these species, a key sensory challenge is to detect and react to prey in the traps. The Venus flytrap features 'trigger hairs'. If a single hair is touched twice or two separate hairs are touched within 20 seconds of each other, then the trap snaps closed in under 0.3 of a second (Volkov et al. 2008). This 'two-touch' procedure is considered to reduce risk of false triggering by rain or windblown debris. According to Gibson and Waller, the trap initially closes only sufficiently to interlock the spines on the edges of the leaves, thus allowing small prey to escape, while retaining large prey. Further contacts of the trigger hairs are required to shut the trap fully and commence digestion, otherwise the trap reopens. Whilst this method of selecting only large prey is plausible, there is currently a lack of empirical investigation of such size selectivity (Ellison and Gotelli 2009). Molecular data suggests that the two snap-trapping species evolved from the same ancestor that utilized sticky 'flypaper' traps. Gibbon and Waller argue that snap-traps evolved to allow capture of larger prey than are possible with adhesive traps; we would welcome further data comparing the two trapping techniques in prey-size distribution. They further speculate that this complex set of adaptations has not become more common because large prey are also relatively rare; and the larger prey is, the more likely it is to be able to escape by burrowing through the walls of traps. More speculatively yet, if snap-traps are size selective, then it might be interesting to explore if trap closure timing is plastic. It might be that small prey are given less time to escape if the plant's previous experience suggests that only small prey are available and/or prey of any size are in short supply. Although the Venus flytrap is perhaps the most widely known carnivorous plant, much has yet to be discovered about its evolution and ecology.

10.4   Pollination

The overwhelming majority of carnivorous plants are pollinated by insects, and this introduces the potential for conflict between the use of insects as food and as pollinators. In many carnivorous plants there is spatial or temporal segregation between flowers and traps that should reduce such conflict. A stark example is provided by the widespread aquatic genus *Utricularia* that sets traps underwater but flowers above the water's surface. Further, many carnivorous plants are long-lived perennials, some being highly self-compatible, which eases the pressure to successfully pollinate in any given season. Indeed, it seems possible that if pollinators are abundant, then there may be synergy between flowers and traps with traps having a negligible impact of the success of pollination but flowers significantly increasing the success of traps by attracting insects to their vicinity. In support of this, Zamora (1999) demonstrated that experimental removal of flowers from *Pinguicula vallisneriifolia* greatly reduced prey capture rates (Figure 10.5). This despite the fact that the sticky leaf surfaces used to capture insects in this plant do not appear designed to share any flower-like characteristics. It should also be remembered that large pollinators (such as bees and butterflies) may be too large to be troubled by traps.

Whilst carnivorous plants often feature flowers with long peduncles that provide effective height separation from the traps, Anderson and Midgley (2001) challenged the assumption that this has evolved to reduce interference between pollination and predation. They suggested that carnivorous plants are generally low-growing (perhaps because their generally reduced root structure would not provide adequate anchorage for tall structures), and thus the long peduncles simply act to make the flowers more visible to flying pollinators, or to improve seed dispersal. In support of this, Anderson (2010) briefly reports on experiments that experimentally manipulated flower height in a species

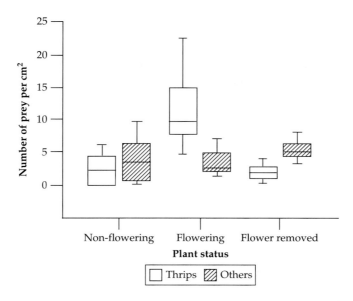

Figure 10.5 The median number of insects caught (plus 10th, 25th, 75th, and 90th percentiles) per 10 cm^2 of leaf (partitioned into thrips and other non-pollinating insects) for non-flowering *Pinguicula vallisneriifolia*, flowering individuals, and flowering individuals from which the flowers had been experimentally removed. Redrawn from Zamora (1999).

that naturally has flowers on the end of long racemes, and found reduced visits by pollinators when racemes were shortened. Comparative analysis to explore whether carnivory is associated with long peduncles once plant morphology has been taken into account should allow illumination of this issue. Similarly, we would welcome quantitative exploration of the suggested temporal separation of traps and flowers against appropriate null models.

Murza et al. (2006) recorded very little apparent conflict between pollination and predation in the carnivorous plant, *Drosera anglica*. The overlap of insect species involved in the two processes comprised only one species making up only 3 per cent of flower visitors and 0.5 per cent of prey. In this plant species, sticky leaf traps and flowers are available simultaneously, but there is spatial separation generated by long flower stalks. The results may be explained in part by a decrease in the effectiveness of the trapping mechanism with leaf age, so that for much of the time when traps and flowers appear to be available simultaneously the performance of the traps is considerably reduced. Plachno et al. (2007) reported on the pitcher plant *Heliamphora folliculate*.

This provides giant nectaries as a prey attractant, but the authors argued that this does not compromise pollination, because pollination in this species is completed exclusively by bumblebees that are attracted not by nectar but by pollen.

10.5 Sensory aspects of interactions with animals other than prey or pollinators

Inquilines that live in pitcher plants (at least in their juvenile stages) are known to respond to the age of the pitcher (Fish and Hall 1978, Heard 1994), although the cues used are not clear (Karlsson et al. 1994). Whether the plant benefits from communicating or miscommunicating to such species about trap age depends on whether these species have a positive, negative, or neutral effect on nutrient gathering by the plant. It may also be that such species do not always occur in sufficient numbers to influence the plant (see Bradshaw and Creelman 1984 for evidence of this). The cues used by mobile adult stages of inquiline species (e.g., mites, midges, and mosquitoes) in pitcher selection are not understood, so the potential for signalling or crypsis by the plant in this context has not been explored. We previously argued that potential prey of carnivorous plants are unlikely to have evolved to detect and avoid the traps of such plants. However, this is not so for inquiline species with mobile stages. Many of these have an obligate association with the plant, and thus such species should be highly selected to be effective at detecting and recognizing suitable plants, and where the species acts as a mutualist, the plant should be selected to aid such detection and recognition. Indeed, it may be that the colouration, patterning and volatiles produced by carnivorous plants are selected at least in part (but potentially predominantly) by the need to attract mutualists rather than prey. This is an area worthy of further research.

Ants are known to steal prey from carnivorous plants. Zamora (1990) argued that ants can take as much as 5 per cent of the prey captured, and have a significant negative impact on the plant. In order to consume the captured prey, the leaf of Zamora's study species (*Pinguicula nevadense*) folds over the prey in a matter of hours. This has the further benefit of restricting ants' ability to steal the prey after this time. Although Zamora states that 'the curling movement of the leaves makes the detection and extraction of prey by ants more difficult', he does not provide data or explanation as to how these two processes were disentangled. Again, the relationship between ants and carnivorous plants is likely to be complex and variable between species. For example, Bhattarai and Horner (2009) recorded that ants made up 42 per cent of all insects captured by *Sarracenia alata*, accounting for 94 per cent of prey mass. Thus, even if ants steal prey from carnivorous plants, their contact with the plant may benefit the plant through increased capture of ants.

Several species of spider commonly spin webs in the mouths of pitchers (Cresswell 1991). It is not known whether spiders are selective between pitchers in a way that enhances spiders' prey capture. However, this seems plausible since the ability of pitchers to attract prey is related to pitcher age in at least some species. Thus, it would be interesting to explore in the field whether spiders are preferentially associated with pitchers that attract higher rates of potential spider victims. Such an associa-tion needs not to be driven by the spiders identifying the characteristics of good sites, rather it could emerge from random site selection combined with site abandonment triggered by low capture rate. Separation of these alternatives, and (if appropriate) identification of the characteristics used by spiders in pitcher selection could be explored in the laboratory. Further, it has been speculated that such spiders have evolved body colouration that allows crypsis by background matching on the often brightly coloured rims of pitchers (Barthlott et al. 2007). This suggestion too has not been fully evaluated. Testing could involve measurement of the spectral matching between spiders and appropriate plant parts, and comparison of this with appropriate control spiders (perhaps species from the same ecosystem that do not exploit pitchers as web-building sites).

Some pitcher species (e.g., *Nepenthes bicalcarata*, *N. gracilis*, *N. rappesiana*) have a close association with small ant species of the genus *Campanotus*. The ants feed on plant-provided nectar and appear to be sufficiently nimble to avoid the trap. In the case of *N. bicalcarata* the ants are provided with a hollow space in the pitcher stalk in which to live. The benefit to the plant from this association is not clear, while Clarke and Kitchings (1993) suggested that ants may remove surplus prey from the traps (avoiding putrefaction and pathogen growth), Merbach et al. (2001) suggests more conventional protection through control of small herbivores, like aphids. Both mechanisms could be in operation, and warrant further investigation. More closely related to our interests, there has been no research done on how ants find plants to colonize.

10.6 Conclusions

It is clear from the limited study that has been undertaken that visual appearance of traps does affect their effectiveness. It would now seem appropriate to explore how effectiveness is modified and how it relates to plant fitness. For plants that offer nectar, the key function of appearance may be to advertise this, and to guide insects effectively to nectar sources. There may be a role for appearance in aiding the learning and generalization of previous rewarding experiences with nectar-providers.

Alternatively or additionally, the key function of appearance may be the simple detection (rather than recognition) of the rewarding trap structures in a way that causes them to stand out from their visual background. Selection pressure on the appearance of non-rewarding traps may be entirely the opposite, with selection for reduction of potential victims' ability to detect and/or recognize them. To summarize, previous research has demonstrated that appearance influences capture success; we must now move the field on to more mechanistic questions of how this influence is exerted. The sensory ecology of carnivorous plants' interactions with inquiline species have been inappropriately neglected in comparison to interactions with their prey. It may be that the colouration, patterning and volatiles produced by carnivorous plants are selected at least in part (but potentially predominantly) by the need to attract mutualists rather than prey. Fundamental questions about the sensory ecology of carnivorous plants remain unanswered.

CHAPTER 11

Final thoughts

11.1 Introduction

In this chapter we highlight some broad issues in plant–animal communication that emerge from integration of ideas across different chapters of this book. The communication between plants and animals is embedded into the interactions among multiple consumers (mutualistic and antagonistic animals, microbes) and multiple providers (different plants). This situation requires that we envision communication and its evolution as more complex than the traditional binary sender–receiver games that were so instructive in understanding signalling dynamics in intraspecific signalling. Understanding how the complexity of multiple species interacting with each other can affect the evolution of communication will undoubtedly be important for our general understanding of communication. This is because even within-species communication in courtship and dominance, the signalling systems that have been central to the development of communication theory in biology, is influenced by other species: like parasites and predators. We therefore think that plant–animal communication can be instructive to advance the concept of communication theory by integrating ecology and multiple players more fully into it. Central to any advance will be to examine the costs and benefits associated with communication. We therefore start (Section 11.2) by explaining the fundamental aspects associated with plant–animal communication and explain where they differ from those found in communication among animals. In the next section (11.3), we discuss adaptations of, and constraints on, communication in multiple-species interactions. In Section 11.4, we outline the implications of interactions among multiple species for the evolution of communication and suggest different scenarios thereof. In Section 11.5 we identify research avenues that address current gaps in our knowledge that we perceive as being particularly important for understanding how communication functions and how it evolves. Finally, in Section 11.6, we discuss the need for closer integration of the physiology of sensory systems with communication theory.

11.2 Principles of communication

It is tempting to conclude that communication between plants and animals is less specific than epigamic or dominance signalling in animals, because it involves a higher number of species (both mutualists and antagonists). However, communication between plants and animals spans the entire gradient from very specific systems where plants depend on a single pollinator (such as figs being pollinated by specific fig wasps) to very generalized systems that involve up to a hundred or more species of frugivores or flower visitors. Comparing communication in specialized and generalized interactions thus enables us to analyse how important specificity is in shaping communication. It is easy to envision that specificity is important because the selective pressures of one focal species are likely to be diluted in the interactions among multiple species. However, the dilution effect does not necessarily increase linearly with the number of species. It obviously depends on the species' impact on plant fitness and on the difference in sensory and cognitive abilities amongst animals. If these are small, and animals attend to the same traits, the dilution effect is likely to be small as well. Analysing whether dilution effects depend primarily on differences in sensory abilities or on the number of species interacting with a trait would be very much welcomed.

The costs and benefits involved in plant–animal communication are more asymmetric than they are in most communication systems among animals. In mate choice and dominance signalling among animals, the costs and benefits are not perfectly symmetric, but the currency is the same for senders and receivers. For example, in most species females invest more in the offspring and therefore experience higher costs of not choosing the optimal partner than do males. Although the benefits of communication during mate choice are not necessarily the same in females and males, they occur in the same dimension of offspring production. This is not the case for plant–animal communication. The fitness costs of failing to communicate to pollinators are tremendously high in an annual plant, whereas the fitness costs of failing to attend to a specific plant cue for the pollinators can be small (as it only misses a meal). This is the life–dinner principle of prey–predator relationships in the context of plant–animal interactions. The opposite is true for carnivorous plants, where the fitness costs are high for the animal and low for the plant. The asymmetry in the costs and benefits for plants and animals reduces the chances for a co-evolutionary process shaping communication among senders and receivers. We thus do not expect plant–animal communication to be on average strongly adapted to mutual selective pressure. Instead, asymmetry increases the chances of one-sided, advergent evolution of the organism that experiences higher costs or benefits. Based on an examination of the costs and benefits involved, we propose that advergent evolution is more likely to occur on the plant side in pollination, whereas it is more likely to occur in herbivorous insects in the context of oviposition. Integrating fitness-related measurements in the study of plant–animal communication will thus be pivotal for our understanding of the evolution of the various communication systems.

A further difference between plant–animal communication and between-animal communication is that receivers of plant communication are likely to be more tolerant towards variable returns than receivers in animal communication. Again, this is related to the relatively low fitness costs associated with making a mistake when attending to plant communication compared to, for example, attending to epigamic signals. Selection against cheating occurs among animals (Tibbetts and Dale 2004). Owing to the relatively low fitness costs of mistakes, selection against cheating should be less pronounced in plant–animal communication than it can be in animal communication. Pollinators, for example, need to be tolerant towards variable nutritional returns because they often cannot discriminate between flowers that have been recently depleted by other pollinators and those that have not. This tolerance may, for example, explain the repeated evolution of rewardlessness in flowers. There is also often variation in fruit nutritional contents both within and between individuals which can, at least, partly be attributed to variation in abiotic factors such as illumination and soil quality. However, it is not clear whether such environmentally induced variation is commonly more pronounced than that found, for example, among males advertising their qualities to females. If the association between communicative trait and nutritional return is to some extent variable, and animals are tolerant of that variability, it implies that cost-enforcing mechanisms like handicaps are unlikely to evolve in plant–animal communication. We therefore predict that the tolerance of animals interacting with plants results in a relatively low reliability of plant communication.

Another important aspect is that of verification. Verification can be an important aspect of communication systems because models showed that it can result in honest, cost-free signalling (Lachmann et al. 2001). The communication systems among animals differ in the degree to which verification is possible. Dominance signals can be immediately verified by challenging the opponent, whereas females cannot immediately verify male signals that advertise good genetic quality. Plant communication systems also differ in the extent to which animals are able to verify the association between plant communicative trait and plant quality. Frugivores that handle fruits so that they can taste their juices can immediately verify the association, whereas those that swallow the whole fruit need to rely on post-ingestive feedbacks (which are less precise and take longer). Likewise, herbivores that consume plant tissue are more likely to evaluate plant quality immediately and more precisely than females that oviposit on plants. Even if females can evaluate plant quality accurately before ovipositing, the quality of a plant may change over time and thus the current state may not be a good indicator of its future state at the time when offspring will feed on the plant. Although models predict that verification can be an important factor shaping the design of reliable signals, we lack empirical data on how verification may impact on communication not just in plants but more generally.

From the paragraphs in this section, it is evident that we still lack a conceptual framework of how factors like the specificity of communication, the asymmetry among partners in the costs involved, as well as the tolerance towards variable nutritional returns shape communication. What is possibly more worrying is that we are not even aware of how variable these factors are within and among communication systems. Explicitly considering this gap in our knowledge will be fundamental for turning communication theory into a more predictive tool.

11.3 Communication among multiple species

Given that communication has mostly been studied in intraspecific signalling systems, we need to examine the conceptual issues associated with communication among multiple species that can also be influenced by variation in abiotic conditions. The communication between plants and animals is perhaps more than many other communication systems (e.g., intraspecific social or epigamic communication in animals) susceptible to variation in the abiotic and biotic environment.

11.3.1 Conceptual framework

At its simplest, some physical aspects of a species' ecology present constraints on the forms of communication that are possible, or influence the relative effectiveness of different modalities. However, communication will exert selection pressure on cognitive and sensory abilities just as much as these abilities will shape the forms of communication used. This complex interaction of evolutionary influences and physical constraints likely explains, at least in part, the extraordinary diversity of types of communication that exists in the natural world. The open challenge that remains is to form a theory that can extract the general principles underlying this diversity and use these to form testable hypotheses for why specific modes of communication have evolved in the way that they have.

A very important corollary of this is that species and their communicative traits generally interact with other species as well as with the abiotic environment. Indeed, species are often embedded in a complex network of interactions with other species. Hence even communication systems that we traditionally think of as being within-species (e.g., courtship rituals) have the potential to influence individuals from other species (e.g., predators and parasites) and face selection from them. Chapters 4–6 on fruits and flowers emphasized that any communicative function must be understood by considering plant antagonists as well as mutualists. In short, we need to develop a theory of communication that acknowledges the complexity of ecologies and thus the multiple types of individuals that can influence and be influenced by a given communication.

The geographic mosaic (Thompson 2005a) provides an excellent conceptual tool to embrace the complexities of communication among multiple species (Section 1.8). This concept establishes that interactions among species are not necessarily the same across a gradient of geographic variation, which includes variation in both biotic and abiotic factors. In some parts of the range, species might be well adapted to each other (so-called co-evolutionary hotspots), whereas in others they may interact only loosely or without mutual adaptations (so-called coldspots; Thompson 2005b). Applying this concept to communication means that the relative importance of different selective factors acting upon the same trait is likely to vary among populations. This variation can be caused by differences in the composition and relative abundance of species interacting with traits as well as by differences in the abiotic factors acting upon traits. If the multiple species interacting with traits differ in their sensory and cognitive abilities, which they often do, they can exert temporally and spatially variable selection. The corollary of the geographic mosaic model is that the same trait can act as a signal in one population if communication is the dominant selective pressure and as cue in another population if communication is not the dominant selective pressure.

That the same trait can be a cue or a signal is not a trivial point because it complicates deriving conclusions from studies focussing on a single location. This is particularly worrying because, for example, only 7.5 per cent of studies on plant–pollinator interactions collected data on pollinator composition and pollinator-related traits in multiple populations (Herrera et al. 2006).

11.3.2 Communication and information can arise as by-products

An important aspect of interactions among multiple species is that olfactory, visual, and gustatory communication and information can arise as by-products. By-product communication is not easily compatible with our conversational use of communication, which comprises actions or structures whose existence arises only because of its influence on others. In its scientific use, this subset of communication is generally called signalling, and there is a clearer appreciation that the multiple and complex interactions of living organisms with each other and with the physical world all present a very complex set of sources of information to other individuals. There is a lot of inadvertent information around us. For example, animals can obtain information from observing others (Goodale et al. 2010) or from interacting with traits that have evolutionarily been shaped by the interactions of other species. In this case, information can be an emergent property originating from the interactions of many individuals from different species with the same trait. If the main selective pressure on a trait is unrelated to its communicative function, but that trait is still used in communication, then communication arises as a by-product. For example, both volatiles and pigments can carry incidental information (Steiger et al. 2011). The interesting aspect is then that by-product communication needs to be informative, otherwise a receiver would not respond to

that trait, and it would consequently not be conceived as a cue.

The required link between a cue and plant quality is not necessarily constant (as in an index). The implication is, however, that by-product communication relies on by-product information. This is exciting from a philosophical viewpoint because we have emphasized in Chapter 1 the need to separate information from communication. This is still practical in by-product communication because information as an inherent property of the cognition of the receiver is difficult to define. However, it is possible to measure a correlation between the phenotype of the sender (the plant) that an animal perceives, and the quality of the plant. We term this correlation phenotypic coupling. Note that phenotypic coupling differs from the term 'functional information' as defined by Hasson (1994), Maynard Smith and Harper (1995), and Carazo and Font (2010) because phenotypic coupling is relatively easy to measure, unlike functional information with its associated cognitive component. Ideally, phenotypic coupling is measured in animals' receptor space because the perceived phenotype may differ from the actual one owing to the sensory filters that animals possess. It is, however, the perceived phenotype that will matter in evolutionary terms, because this is the phenotype the receiver reacts to.

Interestingly, the inherent information found in cues can explain the evolution of signalling, that is, how a cue can be turned into a signal (see Section 11.4). On one hand, by-product communication can be entirely inert to selective pressures on the communicative function of a trait, while on the other hand cues can feasibly be moulded by selection upon communication. We thus need a framework of the selective pressures associated with such by-product communication, so we can begin to predict why in some circumstances by-product communication evolves into signalling and why it does not in others. Similarly, we need a framework to understand when by-product communication is reliable and whether – and if so under which circumstances – it may not be.

By-product communication entails that although plant–animal interactions are often mediated by the senses, the importance of sensory ecology for the evolution of plant–animal interactions is less clear than previously thought. Raguso (2009) argued consequently for a de-emphasis of sensory ecology in plant–animal interactions. Consistent with this view is that the communicative traits of plants are also influenced by abiotic factors and by organisms that do not communicate with that trait. There are two reasons for this phenomenon. First, the compounds that mediate interactions with animals are often derived from biosynthetic pathways that also regulate responses to abiotic factors. The sessile nature of plants requires that they respond to a plethora of stressors, which makes a shared biochemistry of stress responses particularly advantageous because it allows plants to respond more quickly to a second type of stressor after their biosynthetic pathway has been activated when responding to the first stressor. This mechanism can be termed biochemical pleiotropy. Second, the rewards offered by plants in their reproductive tissue consist mainly of metabolites derived from photosynthesis. The link between photosynthetic activity and the production of nutritional rewards offered to animals can turn plant traits that are linked to photosynthesis into informative cues even though plant did not evolve them to communicate. Hence, de-emphasizing sensory ecology does not make the study of plant communication intellectually less rewarding. On the contrary, it would be naïve to expect that the communication between plants and animals proceeds in isolation from the myriad of other processes that govern plant life. We therefore believe that the field of communication presents an exciting challenge to evaluate how distinct selective pressures are amalgamated in the evolution of plant communicative traits.

We will explain now why the evolution of communication is thus likely to be more dynamic and more complex than currently acknowledged in communication theory.

11.4 Scenarios for the evolution of communication

In the first and most basic scenario, communication is entirely a by-product of other selective pressures. Here, all communicative traits are cues. There are two mechanisms that may render cues informative, shared biochemistry of cues and nutrients (or secondary compounds) or ecological pleiotropy.

Given that shared biosynthesis can lead to two alternative scenarios, the link between plant quality and perceived phenotype is not constant across species. Shared biochemistry can result in a positive correlation between two substances because they can be synthesized more quickly from common precursors in a currently activated pathway. On the other hand, allocation trade-offs of the precursor forms might result in a situation where the production of one substance reduces the production of the other. Such trade-offs have been proposed – for example in the production of lignin and anthocyanins – but have rarely been studied in detail. Determining the genetic and biochemical basis of links between qualities and perceived phenotypes is important if we are to evaluate how reliable cues are and how common indices (which are based upon a necessary and constant link between quality and phenotype) are. Communication via indices may be common because in many communication systems the relationship between chemical properties and pigments or volatiles appears particularly robust. However, studies have rarely evaluated whether such relationships are found across different populations, let alone whether the possible indices are indeed unfakable. There is one further reason why consideration of biochemical or physiological mechanisms is important for understanding plant–animal communication. Plants often have systemic responses to herbivory involving raising their induced defences. Several studies have shown that these systemic responses can affect pollination and other ecological processes that are usually considered in isolation from herbivory.

Cues can be informative to one species because of their interactions with other species. Information can thus arise from ecological pleiotropy. For example, anthocyanins reduce fungal growth (Schaefer et al. 2008a) and can thus be correlated to fruit nutritional contents because fungi preferentially colonize nutritive fruits. It is feasible that dark fruit colour is predominantly selected by fungi, but that it is thereby informative to seed dispersers. We predict that ecological pleiotropy more commonly results in a phenotypic coupling of plant quality and perceived phenotype.

If cues are informative, they can represent starting points for signalling or intentional communication (the second scenario). In this scenario, signalling evolves as an exaptation from pre-existing cues. For example, using odours to attract pollinators may be an exaptation of an originally defensive function of floral odours (Pellmyr and Thien 1986). It goes without saying that consideration of selective pressures that originally shaped the trait before selection on communication started is important for understanding the evolution of communication in this scenario. However, a key aspect that has not been sufficiently considered in communication theory is that these selective pressures are unlikely to have ceased completely. It is more likely that they continue to influence the evolution of communication. Important research avenues follow from this scenario. Evaluating the relative contribution of selection upon communication is crucial to understanding the evolution of traits. It is conceivable that a trait is substantially influenced by selection upon communication even though its evolution is bounded by constraints associated with other functions. If selection upon communication shifts a trait value within such boundaries, we propose that this trait is a signal even though communication may not be the dominant selective pressure.

The second important research avenue that follows from multiple selective pressures is to determine how they canalize communication. Canalization can occur by deviating the evolution of traits from an optimal path towards increasing efficiency in communication. If the constraints on trait evolution are strong, the communicative function can be transferred to other traits such as accessory structures in flowers and fruits or other volatiles. Appreciation of the various selective pressures associated with multi-species interactions can thus contribute to understanding the functional mechanisms underlying the diversity of communicative traits.

Canalization can act on the design of communication and on its reliability. For example, pleiotropic effects in the melanocortin system lead to covariance between melanin-based colours and various behavioural traits in animals (Ducrest et al. 2008). Similar pleiotropic effects are known in plants where anthocyanin-recessive morphs are usually favoured by herbivores (Strauss and Whittall 2006). The covariance between quality-related traits and sensory traits induced by pleiotropy reduces the

potential for cheating. Hence, canalization likely represents an important and widespread mechanism enforcing the reliability of communication. Given the current debate on the relative importance of defining communication based upon information (Rendall et al. 2009, Scott-Phillips 2010, Carazo and Font 2010), we suggest that pleiotropy enables us to describe the association between perceived phenotype and reward that forms the basis for the functional information that animals can extract from communicative traits.

In the final scenario, signalling or true communication exists because selective pressure for communication is not strongly modified by other constraints. We suggest that if perceived phenotype and quality are produced by different pathways there is a higher probability of signalling (compared to production of the same pathway) because this phenotypic coupling requires cross-talk among pathways.

11.5 Integrating ecology into communication theory

We regard the need to evaluate how ecological factors (both biotic and abiotic) influence the design and reliability of communication as one of the most pressing conclusions from this book. We feel that a spatially explicit approach to communication is urgently needed. The geographic mosaic (Section 11.3) predicts that selection on communication is community-specific and spatially and temporally variable. However, there is still almost no theoretical framework for how senders diverge or converge on a community scale. Any response by a receiver to communicative traits will depend on the availability of alternative (food) resources. The availability of competing food resources includes a quantitative and qualitative aspect. The quantitative aspect is that a higher food supply enables consumers to become choosier (e.g., Schaefer and Schaefer 2006). The qualitative aspect can be analysed using market theory, which provides a powerful and underexploited tool for exploration of the signalling dynamics within communities. Market theory could easily be set against null models of no adaptive divergence or any divergence being driven by pleiotropic effects of other selective pressures.

An interesting research avenue would be to analyse whether the spatial (and temporal) distribution of plant rewards affects selection on the reliability of plant communication. In less mobile consumers, selection on reliability can be density-dependent and most pronounced in clumped resources, whereas density-dependent selection is likely to be lower in mobile foragers. Whether selection upon communication is thus predictably associated with an animals' mobility is a question that has not yet received attention.

Throughout the book we have emphasized that selection upon communication is variable among species and that this variability can be understood by consideration of a species' sensory ecology and the fluctuating selection inflicted by other species. The last paragraph, however, implies that other aspects of a species' ecology can be important in explaining differential selection of communicative traits. For example, differences in the cognitive abilities of consumers will also contribute to differential selection upon communicative traits. We need a framework that enables us to quantify such influences on communication and predict differences among species.

Network analyses provide a quantitative framework to analyse the interactions among multiple species (Bascompte and Jordano 2008). Network analyses revealed generally strong asymmetry between plants and animals, in that specialized animals interact with generalized plants and vice versa (Bascompte et al. 2003). If this pattern holds generally, it leads – in addition to the life–dinner principle (Section 11.2) – to one-sided evolution rather than mutual co-evolution because the selective pressures of specialists will often be diluted among the selective pressures of other animals. It seems tempting to conclude that communication is a functional mechanism contributing to explain network structure, and the limited evidence available so far suggests that this is indeed so (Olesen et al. 2007b, Junker et al. 2010).

11.6 Communication and sensory ecology

The gaps in our knowledge that we have summarized in the paragraphs above preclude immediate answer to the critical question of to what extent communication is really adapted to the sensory or perceptual world of receivers. It is obvious that answers will pertain to specific communication systems and may not easily be generalized. On one hand, plant communication systems are ideal to analyse the correlation between plant rewards (or defences) and traits such as reflectance spectra and volatile emission. On the other hand, we have shown that colour and scent are not inherent traits of the senders but rather traits that arise in the perceptual world of the receiver. Our current understanding of communication between plants and animals will thus depend critically upon progress in our current models of the sensory and perceptual spaces of different animals. The next frontier will be the cognitive dimension of animals interacting with plant traits, because it is their cognitive abilities that determine the information they obtain from plants' phenotype.

The basic principle of communication theory that selection should work towards greater efficiency applies to communication in all senses. We currently lack an encompassing framework to clearly predict how communication and its evolution differ among the senses. Odours are discrete traits, whereas reflectance is a continuous trait. Odour perception relies on tens to hundreds of receptor types, vision on less than a handful. It is feasible that the specificity of communication is greater in scent, but quantitative comparisons are lacking. The discrete nature of odours implies that olfactory conspicuousness can be very different from visual conspicuousness, if it relies on single compounds instead of blends.

It would be particularly welcome if studies started to develop an encompassing approach to communication by integrating the distinct communication systems of a plant (see Box 11.1). Ideally, analyses would determine the conspicuousness and the specificity of olfactory, visual, and tactile cues. The perceived intraspecific variability in plant communication could be compared in each sensory space within and among populations by comparing Euclidean distances among those sensory spaces.

Many of the most basic questions regarding plant–animal communication remain unanswered. What is selecting for the diversity of communica-

Box 11.1 Multimodal communication

Simultaneous communication in multiple modalities can be beneficial because it reduces the perceptual errors of receivers either owing to redundant information or owing to back-up signals that transmit differently in different environments. Also, multi-modal cues/signals can be associated with distinct qualities of the sender (Hebets and Papaj 2005, Bro-Jørgensen 2010).

Multi-modal stimulation is common in plant–animal communication. Flowers in particular send multiple signals or cues in distinct sensory modes such as vision and scent. For example, the wild tobacco hornworm moth, *Manduca sexta*, requires both a combination of visual and olfactory cues to stimulate feeding (Goyret et al. 2007). The combined use of distinct sensory modes is probably widespread in plant–animal communication, likely occurring in frugivory, herbivory, and carnivory as well.

The information obtained by the senses can be used sequentially or concomitantly. Insects that track odour plumes from afar and use additional visual cues for short-distance orientation are a good example for a predominant use of one sensory mode early in foraging and a concomitant use of two modes in the terminal foraging process. The relative importance of sensory modes is taxon-specific, but a few generalities of multi-modal communication probably apply to most animals. Here we lay out how they relate to future research avenues in plant–animal communication.

First, discrimination among differently rewarding targets is increased if animals can use cues in different sensory modes – as Kunze and Gumbert (2001) have shown for bumblebees. In this scenario, signals in each sensory mode are equivalent, indicating the same aspect (unlike scenario 4, see below). They are redundant stimuli increasing the accuracy of the response by the receiver. Feasibly, redundant stimulation could also serve as backup under fluctuating environmental conditions (scenario 5, see below). Increased discrimination of multi-modal cues/signals is probably expected not only for pollinators but other species as well. However, it remains as yet unclear whether and how better discrimination abilities of receivers affect plant trait evolution. For example, in light of the results of Kunze and Gumbert (2001), we might expect rewardless orchids to mimic rewarding species in both scent and colour, but the orchid *Orchis israelitica* mimics its model only in colour (Galizia et al. 2005).

In scenario 2, receivers' response to a trait in one sensory mode can be modified by their stimulation in another sensory mode. This is well known in humans, for example, where sweetness perception is affected by the colouration of the food (Bayarri et al. 2001). It seems as though such interactions between colour and taste perception are driven by the expectations of the receiver. For example, it has been hypothesized that green foods are perceived as less sweet by humans because green food items (including fruits) are commonly not very sweet. However, the consistent negative association between green colour and the perception of sweetness in humans is controversial (Clydesdale 1993). While interactions among the senses are probably widespread, they are mostly unknown in non-human animals. We consequently lack an understanding of whether such interactions among senses are evolutionary relevant. Working on how input in different sensory modes affects receivers' responses to stimuli will likely yield important insights into concepts such as mimicry, deception, and reliable signalling in plants.

In scenario 3, multi-modal stimulation of receivers increases their learning abilities and memory retention (Hebets and Papaj 2005). These effects are achieved by cross-modal integration of stimuli at several cognitive levels (reviewed in Bro-Jørgensen 2010). Memory retention of stimuli is particularly important in pollination because plants need to entice pollinators to visit conspecifics repeatedly. We therefore suggest that selection for recognizability via multi-modal stimulation, if it occurs, should be stronger in flower than in fruit or leaf communication.

In scenario 4, receivers may attend to multi-modal signals because they could indicate different aspects of reward quality. For example, receivers could feasibly select for multiple signals that indicate the status of the sender more reliably (Hebets and Papaj 2005). However, because mistakes when interacting with plant communicative

Box 11.1 Continued

traits do not usually inflict high fitness costs on animals, we do not expect that these select strongly and consistently across different communication systems for multiple signalling that increase reliability. Yet, animals could nevertheless attend to and select for cues in multiple sensory modes because biochemical or ecological pleiotropy may makes these cues reliable indicators of distinct plant qualities. For example, terpenoids could indicate plant defence status, whereas certain colours could indicate the quality and quantity of nutrients. Analysing whether animals react to multi-modal signals because these are more reliable or information rich is an under-studied topic.

In scenario 5, environmental fluctuations may select for multi-modal signals, because in unpredictable environments signalling in distinct modes may increase the likelihood of reaching the receiver and grabbing its attention (Bro-Jørgensen 2010). This hypothesis is particularly promising for plant–animal communication because sessile plants have no control over their wider environment (unlike animals they cannot seek out favourable conditions for communication). At present, it is difficult to evaluate whether this hypothesis applies to plant–animal communication, but it is clear that plant communicative traits can change according to the environment. For example, herbivory changes floral phenology, thereby leading to a shift in pollinators in the tobacco plant (Figure 6.14; Kessler et al. (2010)). The internal environment can also fluctuate (Bro-Jørgensen 2010). Again, this probably applies to plants where changes in rainfall patterns and illumination can greatly contribute to changes in the resources plants can devote to reproduction and defence.

tion? This question is particularly interesting in the many volatiles emitted by flowers and leaves. We are still far from understanding what is the proportion of volatiles that are adapted to communication relative to the proportion adapted to defence or those that are mere waste products. Other important questions will be whether specific compounds are more commonly the targets of selection than others or whether the ratio of different compounds within blends is the target of selection. Finally, is there diversifying selection on long-range vs. short-range functionality, in that long-ranging volatiles indicate location whereas short-ranging compounds are more intricately linked with quality aspects of the sender?

We have shown that communication is a neglected topic in many plant–animal interactions, for example the infauna of carnivorous plants, in fungi that also depend on pollinators, and in seed hoarders. Furthermore, many of the central concepts in animal communication such as aposematism, mimicry, camouflage, and sensory exploitation are still relatively unexplored among plants. For example, pollen mimicry may be widespread but is still not well known. There is much potential for conceptual

growth, particularly because plant traits are relatively invariant, the behavioural component of plants is relatively negligible during communication (at least when compared to communication among animals), and animals' interests in plant quality are easily predictable and measurable.

Finally, we highlight that plant communication systems can have impacts on the communication used among animals. First, the reactions of animals towards plant traits can be exploited by other signallers (a hypothesis already suggested by Allen 1879). For example, some fishes and primates are thought to have evolved red ornaments used in intraspecific signalling because mates exploited the pre-existing innate positive response towards red plant colours in the context of frugivory or herbivory (Smith et al. 2004, Fernandez and Morris 2007). Second, senders can exploit plant communication by mimicking plant traits. Examples of such mimicry include arthropods like mantids and crab spiders or fungi mimicking flowers or exploiting pollinators' innate biases to approach contrasting flower-like structures (see box 11.2). Also, communication systems of animals can be superimposed on plant traits. Possible

Box 11.2 Perceptual exploitation

Sensory exploitation has become an important model of signal evolution that has only recently been applied to plant–animal communication (Schaefer and Ruxton 2009). Sensory exploitation posits that receivers have pre-existing sensory biases for particular traits and that selection therefore favours any sender that evolves a trait matching these biases. Perceptual exploitation is a slightly more encompassing term than sensory exploitation because the biases for particular traits can also arise in the cognitive realm. Sensory biases are known to occur in pollinators and seed dispersers both for visual and olfactory traits in flowers and fruits (Schmidt et al. 2004, Naug and Arathi 2007, Raine and Chittka 2007, Vereecken and Schiestl 2008). In Chapter 8 we argue that – although still relatively unknown in the context of plant–animal communication – perceptual exploitation may be common and a more general model to explain the evolution of plant sensory traits than mimicry. The fact that very distinct animal have innate sensory biases, usually for contrasting and large displays, illustrates that perceptual exploitation may be common in visual communication.

Perceptual exploitation can also occur in animals that exploit existing communication systems between plants and animals. For example, crab spiders (*Thomisus spectabilis*) that lurk on flowers to capture pollinators exploit pollinators' innate preferences for contrasting floral displays. The exploitation of these preferences occurs because some crab spider species contrast strongly in the UV with their floral background, and more bees approached the flowers with contrasting spiders on them than control flowers (Heiling et al. 2003). Interestingly, the response of pollinators can co-evolve with the presence of crab spiders. *Austroplebia australis*, an Australian bee native to the range of *T. spectabilis*, approached flowers occupied by the spiders, but landed less often on them because they smelled the spider; whereas introduced honey bees that do not share a long evolutionary history with the spider landed on flowers occupied by spiders more often, because they did not use olfactory cues to avoid them (Heiling et al. 2004). This study shows that (i) perceptual exploitation can occur widely and (ii) multi-modal cues can be used to mitigate the risk of predation.

Finally, perceptual exploitation may contribute to explaining a core challenge in interspecific interactions, the cooperation between species that can not be explained by kin selection or tit-for-tat strategies (Edwards and Yu 2007). Perceptual exploitation may promote the evolution of interspecific cooperation by reduction of errors in asymmetric sanctions between cooperating partners, by enforcing cooperative behaviour rather than cheating, and by reducing the costs of association to a given partner (Edwards and Yu 2007). These are extremely interesting ideas that merit further studies.

examples are the chemical footprints left by pollinators on flowers, although current evidence suggests that these are cues rather than signals. However, here questions remain as to whether insects can perceive the time lag between visitations by other pollinators owing to a differential decay rate of certain volatile compounds in those footprints.

We hope that we have convinced that further development of current understanding of plant–animal communication is valuable, possible, and (above all) fascinating.

Glossary

Achromatic: Descriptor of visual stimuli indicating variation only in the total intensity of reflected light (cf. Chromatic).

Aposematism: Sensory stimulation (generally conspicuous visual appearance) that is usually associated with strong defences and functions to reduce the rate of attack by predators.

Biochemical pleiotropy: When a shared biosynthetic pathway leads to a covariance between substances (e.g., stimulus and underlying quality). Biochemical pleiotropy may or may not lead to indices (cf. Pleiotropy).

Chromatic: Descriptor of visual stimuli indicating variation in the spectral composition of the reflected light (cf. Achromatic).

Co-evolution: Reciprocal evolutionary change in each of two species to the selective pressures brought by the other species. Can also occur between more than two species, when it is termed diffuse co-evolution.

Colour: A visual trait that emerges in the perception of the animal, influenced by the spectral composition of reflected light from an object. Thus, in a strict sense, colour is not an inherent property of the observed object, although we have used it in this colloquial sense throughout the book.

Communication: Communication between one individual and another individual occurs if traits of the sender stimulate the sensory systems of the receiver in such a way as to cause a change in the behaviour of the receiver.

Crypsis: One mechanism of camouflage involving traits that reduce the rate at which the bearer is detected.

Cue: A feature of the world, animate or inanimate, that can be used as a guide to future action of an individual whose senses can detect the cue.

Deceit: Communication that benefits the sender but not the receiver.

Dichromacy: Requires two distinct photoreceptors and appropriate neural mechanisms.

Eavesdropping: Receivers react to communication in a way that benefits them and involves costs to the sender.

Ecological pleiotropy: The trait values are set by the interactions of multiple species with this trait. Ecological pleiotropy can render traits informative even though they are not selected to be informative (cf. Pleiotropy).

Exaptation: Traits that have a specific function now but had a distinct function originally.

Honest signalling: Signals are honest if on average they provide reliable information.

Index: An index occurs if trait values of a stimulus are necessarily and causally related to the quality being signalled, and which cannot therefore be faked.

Induced defences: Plant defences that are activated after an attack.

Mimicry: Resemblance between two species that is adaptive because (i) an animal confuses one species, the mimic, with the other species, the model, and (ii) this confusion confers fitness benefits to the mimic. If fitness benefits are limited to the mimic, this is termed Batesian mimicry, and if they also occur for the model, it is termed Müllerian mimicry.

Perception: Perception involves the processing of sensory stimuli by the brain of the receiver.

Perceptual exploitation: Perceptual exploitation occurs if receivers have pre-existing sensory or cognitive biases for particular traits and if selection therefore favours any sender that evolves a trait matching these biases.

Photo-inhibition: A persistent decrease in the efficiency of photosynthetic energy conversion in leaves. Occurs particularly when high light levels are coupled with environmental stress, such as cold temperatures or when shade-adapted plants are suddenly exposed to high light levels.

Photoprotection: Traits that protect against excess light. Photoprotection is often important in a combination of strong light and low temperatures (cf. Photo-inhibition).

Pleiotropy: A single gene is responsible for a number of distinct and seemingly unrelated phenotypic effects (cf. Ecological pleiotropy, Biochemical pleiotropy).

Post-ingestive feedbacks: Neuronal feedbacks between the viscera and the cognitive and sensory system of an animal.

Signal: Any act or structure that alters the behaviour of another organism, that evolved and/or is maintained through selection (at least in part) because of that effect, and which is effective because the receiver's response has also evolved (cf. Cue).

Tetrachromacy: Requires four distinct photoreceptors and appropriate neural mechanisms.

Trichromacy: Requires three distinct photoreceptors and appropriate neural mechanisms.

References

Abeles FB and Tanaka F 1990. Cellulase activtiy and ethylene in ripening strawberry and appel fruits. *Scientia Horticulturae*, **42**, 269–75.

Ackerman JD, Meléndez-Ackerman E and Salguero-Faria J 1997. Variation in pollinator abundance and selection on fragrance phenotypes in an ephiphctic orchid. *American Journal of Botany*, **84**, 1383–90.

Adam-Phillips L, Barry C and Giovannoni J 2004. Signal transduction systems regulating fruit ripening. *Trends in Plant Science*, **9**, 331–8.

Adler LS 2000. Alkaloid uptake increases fitness in a hemiparasitic plant via reduced herbivory and increased pollination. *American Naturalist*, **156**, 92–9.

Adler LS and Irwin RE 2005. Ecological costs and benefits of defenses in nectar. *Ecology*, **86**, 2968–78.

Aerts R 1996. Nutrient resorption from senescing leaves of perennials: Are there general patterns? *Journal of Ecology*, **84**, 597–608.

Agrawal AA 1998. Leaf damage and associated cues induce aggressive ant recruitment in a neotropical antplant. *Ecology*, **79**, 2100–12.

Aguirre LF, Herrel A, Van Damme R and Matthysen E 2003. The implications of food hardness for diet in bats. *Functional Ecology*, **17**, 201–12.

Alcalá RE and Dominguez CA 2003. Patterns of prey capture and prey availability among populations of the carnivorous plant *Pinguicula moranensis* (Lentibulariaceae) along an environmental gradient. *American Journal of Botany*, **90**, 1341–8.

Alcalá RE, Mariano NA, Osuna F and Abarca CA 2010. An experimental test of the defensive role of sticky traps in the carnivorous plant *Pinguicula moranensis* (Lentibulariaceae). *Oikos*, **119**, 891–5.

Allen G 1879. *The colour-sense: its origin and development.* Trübner & Co, London.

Allen JA and Knill R 1991. Do grazers leave mottled leaves in the shade. *Trends in Ecology & Evolution*, **6**, 109–10.

Allen RB and Lee WG 1992. Fruit selection by birds in relation to fruit abundance and appearance in the naturalized shrub *Berberis darwinii*. *New Zealand Journal of Botany*, **30**, 121–4.

Altshuler D 2001. Ultraviolet reflectance in fruits, ambient light composition and fruit removal in a tropical forest. *Evolutionary Ecology Research*, **3**, 767–78.

Alves-Costa CP and Lopes AV 2001. Using artificial fruits to evaluate fruit selection by birds in the field. *Biotropica*, **33**, 713–7.

Ames BN, Shigenaga MK and Hagen TM 1993. Oxidants, antioxidants, and the degenerative diseases of aging. *Proceedings National Academy of Sciences*, **90**, 7915–22.

Amico GC, Rodriguez-Cabal MA and Aizen MA 2011. Geographic variation in fruit colour is associated with contrasting seed disperser assemblages in a south-Andean mistletoe. *Ecography in press*.

Amsberry LK and Steffen JE 2008. Do contrastingly colored unripe fruits of the Neotropical tree *Ardisia nigropunctata* attract avian seed dispersers? *Biotropica*, **40**, 575–80.

Anderson B 2010. Did *Drosera* evolve long racemes to stop their pollinators from being eaten? Annals of Botany 106, 653,–657.

Anderson B and Johnson SD 2006. The effects of floral mimics and models on each others' fitness. *Proceedings of the Royal Society B*, **273**, 969–74.

Anderson B and Johnson SD 2008. The geographical mosaic of coevolution in a plant – pollinator mutualism. *Evolution*, **62**, 220–5.

Anderson B, Johnson SD and Carbutt C 2005. Exploitation of a specialized mutualism by a deceptive orchid. *American Journal of Botany*, **92**, 1342–9.

Anderson B and Midgley JJ 2001. Food or sex: pollinator–prey conflict in carnivorous plants. *Ecology Letters*, **4**, 511–13.

Anderson B and Midgley JJ 2002. It takes two to tango but three is a tangle: mutualists and cheaters on the carnivorous plant *Roridula*. *Oecologia*, **132**, 369–73.

Anderson B and Midgley JJ 2003. Digestive mutualism, an alternate pathway in plant carnivory. *Oikos*, **102**, 221–4.

Anderson JT, Rojas JS and Flecker AS 2009. High-quality seed dispersal by fruit-eating fishes in Amazonian floodplain habitats. *Oecologia*, **161**, 279–90.

Angioy AM, Stensmyr MC, et al. 2004. Function of the heater: the dead horse arum revisited. *Proceedings of the Royal Society B,* **271,** S13–15.

Aratchige NS, Lesna I and Sabelis MW 2004. Belowground plant parts emit herbivore-induced volatiles: olfactory responses of a predatory mite to tulip bulbs infested by rust mites. *Experimental and Applied Acarology,* **33,** 21–30.

Araujo-Lima C and Goulding M 1997. *So fruitful a fish: ecology, conservation and aquaculture of the Amazon's tambaqui.* Columbia University Press, New York.

Archetti M 2000. The origin of autumn colours by coevolution. *Journal of Theoretical Biology,* **205,** 625–30.

Archetti M 2007. Autumn colours and the nutrient retranslocation hypothesis: A theoretical assessment. *Journal of Theoretical Biology,* **244,** 714–21.

Archetti M 2009a. Classification of hypotheses on the evolution of autumn colours. *Oikos,* **118,** 328–33.

Archetti M 2009b. Decoupling vigour and quality in the autumn colours game: Weak individuals can signal, cheating can pay. *Journal of Theoretical Biology,* **256,** 479–84.

Archetti M and Brown SP 2004. The coevolution theory of autumn colours. *Proceedings of the Royal Society of London Series B,* **271,** 1219–23.

Archetti M, Döring TF, Hagen SB, Hughes NM, Leather SR, Lee DW, Lev-Yadun S, Manetas Y, Ougham HJ, Schaberg PG and Thomas H 2009. Unravelling the evolution of autumn colours: an interdisciplinary approach. *Trends in Ecology and Evolution,* **24,** 166–73.

Archetti M and Leather SR 2005. A test of the coevolution theory of autumn colours: colour preference of *Rhopalosiphum padi* on Prunus padus. *Oikos,* **110,** 339–43.

Armbruster WS 1997. Exaptations link evolution of plant–herbivore and plant–pollinator interactions: a phylogenetic inquiry. *Ecology,* **78,** 1661–72.

Armbruster WS, Antonsen L and Pélabon C 2005. Phenotypic selection on *Dalechampia* blossoms: honest signaling affects pollination success. *Ecology,* **86,** 3323–33.

Armbruster WS, Di Stillo VS, Tuxill JD, Flores TC and Velázquez Runk JL 1999. Covariance and decoupling of floral and vegetative traits in nine neotropical plants: a re-evaluation of Berg's correlation-pleiades concept. *American Journal of Botany,* **86,** 39–55.

Armbruster WS, Howard JJ et al. 1997. Do biochemical exaptations link evolution of plant defense and pollination systems? Historical hypotheses and experimental tests with *Dalechampia* vines. *American Naturalist,* **149,** 461–84.

Armbruster WS, Lee J and Baldwin BG 2009. Macroevolutionary patterns of defense and pollination in *Dalechampia* vines: adaptation, exaptation, and evolutionary novelty. *Proceedings National Academy of Sciences,* **106,** 18085–90.

Armbruster WS and Muchhala N 2009. Associations between floral specialization and species diversity: cause, effect, or correlation? *Evolutionary Ecology,* **23,** 159–79.

Arnold SJ 1992. Constraints on phenotypic evolution. *American Naturalist,* **140,** S85–107.

Ashan G and Pfanz H 2003. Non-foliar photosynthesis – a strategy of additional carbon acquisition. *Flora,* **198,** 81–97.

Ashman T-L and Majetic CJ 2006. Genetic constraints on floral evolution: a review and evaluation of patterns. *Heredity,* **96,** 343–52.

Atkinson MD and Atkinson E 2002. Sambucus nigra L. *Journal of Ecology,* **90,** 895–923.

Atsatt PR 1983. Mistletoe leaf shape: a host morphogen hypothesis. In: Calder M and Bernhardt P (eds) *Biology of Mistletoes.* Academic Press, Sydney

Augner M 1994. Should a plant always signal its defense against herbivores. *Oikos,* **70,** 322–32.

Augner M and Bernays EA 1998. Plant defence signals and Batesian mimicry. *Evolutionary Ecology,* **12,** 667–79.

Aukema JE and del Rio CM 2002. Where does a fruit-eating bird deposit mistletoe seeds? Seed deposition patterns and an experiment. *Ecology,* **83,** 3489–96.

Ayasse M, Schiestl FP, Paulus HF, Lofstedt C, Hansson B, Ibarra F and Francke W 2000. Evolution of reproductive strategies in the sexually deceptive orchid *Ophrys sphegodes:* How does flower-specific variation of odor signals influence reproductive success? *Evolution,* **54,** 1995–2006.

Babweteera F, Savill P and Brown N 2007. *Balanites wilsoniana:* Regeneration with and without elephants. *Biological Conservation,* **134,** 40–7.

Balazadeh S, Parlitz S, Mueller-Roeber B and Meyer RC 2008. Natural developmental variations in leaf and plant senescence in *Arabidopsis thaliana. Plant Biology,* **10,** 136–47.

Baldwin IT, Halitschke R, Paschold A, von Dahl CC and Preston CA 2006. Volatile signaling in plant–plant interactions: 'Talking trees' in the genomics era. *Science,* **311,** 812–5.

Ballare CL 2009. Illuminated behaviour: phytochrome as a key regulator of light foraging and plant anti-herbivore defence. *Plant Cell and Environment,* **32,** 713–25.

Ballare CL, Demkura PV, Abdala G and Baldwin IT 2009. Specific effects of solar UV-B radiation on plant defense mediated by the jasmonate signaling pathway. *Comparative Biochemistry and Physiology A, Molecular & Integrative Physiology,* **153A.**

Bannister P 1989. Nitrogen concentration and mimicry in some New Zealand mistletoes. *Oecologia*, **79**, 128–32.

Barkman TJ, Bendiksby M, Lim SH, Salleh KM, Nais J, Madulid D and Schumacher T 2008. Accelerated rates of floral evolution at the upper size limit for flowers. *Current Biology*, **18**, 1508–13.

Barlow BA and Wiens D 1977. Host – parasite resemblance in Australian mistletoes: case for cryptic mimicry. *Evolution*, **31**, 69–84.

Barrett SCH 1983. Crop mimicry in weeds. *Economic Botany*, **37**, 255–82.

Barthlott W, Porembski S, Seine R and Theisen I 2007. *The curious world of carnivorous plants*. Timber Press, Portland.

Bascompte J and Jordano P 2008. Plant–animal mutualistic networks: the architecture of biodiversity. *Annual Review of Ecology Evolution and Systematics*, **38**, 567–93.

Bascompte J, Jordano P, Melian CJ and Olesen JM 2003. The nested assembly of plant–animal mutualistic networks. *Proceedings of the National Academy of Sciences of the United States of America*, **100**, 9383–7.

Bassoli A 2004. 'Chemistry-Nature' still an open match for the discovery of new intensive sweeteners. *Agro Food Industry Hi-tech*, **15**, 27–9.

Batra LR and Batra SWT 1985. Floral mimicry induced by mummy-berry fungus exploits hosts pollinators as vectors. *Science*, **228**, 1011–3.

Bauer U, Willmes C and Federle W 2009. Effect of pitcher age on trapping efficiency and natural prey capture in carnivorous *Nepenthes rafflesiana* plants. *Annals of Botany*, **103**, 1219–26.

Bawa KS 1995. Pollination, seed dispersal and diversification of angiosperms. *Trends in Ecology and Evolution*, **10**, 311–12.

Bayarri S, Calvo C, Costell E and Durán L 2001. Influence of color on perception of sweetness and fruit flavor of fruit drinks. *Food Science Technology International*, **7**, 399–404.

Beaver RA 1983. The communities living in Nepenthes pitcher plants: fauna and food webs. In: Frank H and Lounibos LP (eds) *Phytotelmata: terrestrial plants as host of aquatic insect communities*. Plexus, Medford.

Becerra JX 2007. The impact of herbivore–plant coevolution on plant community structure. *Proceedings National Academy of Sciences*, **104**, 7483–8.

Benderoth M, Textor S, Windsor AJ, Mitchell-Olds T, Gershenzon J and Kroymann J 2006. Positive selection driving diversification in plant secondary metabolism. *Proceedings National Academy of Sciences*, **103**, 9118–23.

Bengtsson M, Jaastad G, Knudsen GK, Kobro S, Backman AC, Pettersson E and Witzgall P 2006. Plant volatiles mediate attraction to host and non-host plant in apple fruit moth, *Argyresthia conjugella*. *Entomologia Experimentalis et Applicata*, **118**, 77–85.

Benitez-Vieyra S, Hempel de Ibarra N, Wertlen AM and Cocucci AA 2007. How to look like a mallow: evidence of floral mimicry between Turneraceae and Malvaceae. *Proceedings Royal Society B*, **274**, 2239–48.

Benitez-Vieyra S, Ordano M, Fornoni J, Boege K, Domínguez CA (2010) Selection on signal-reward correlation: limits and opportunities to the evolution of deceit in *Turnera ulmifolia* L. Journal of Evolutionary Biology, **23**, 2760–2767.

Benkman CW 1999. The selection mosaic and diversifying coevolution between crossbills and lodgepole pine. *American Naturalist*, **153**, S75–91.

Bennett KF and Ellison AM 2009. Nectar, not colour, may lure insects to their death. *Biology Letters*, **5**, 469–72.

Benson W, Brown KJ and Gilbert L 1976. Coevolution of plants and herbivores: passion flower butterflies. *Evolution*, **29**, 659–80.

Benzing DH 1987. The origin and rarity of botanical carnivory. *Trends in Ecology and Evolution*, **2**, 364–9.

Berg RL 1960. Ecological significance of correlation pleiades. *Evolution*, **14**, 171–80.

Bergström G, Dobson HEM and Groth I 1995. Spatial fragrance patterns within the flowers of *Ranunculus acris* (Ranunculaceae). *Plant Systematics and Evolution*, **195**, 221–42.

Bernays EA and Simpson SJ 1982. Control of food intake. *Advances in Insect Physiology*, **16**, 59–118.

Bhattarai GP and Horner JD 2009. The importance of pitcher size in prey capture in the carnivorous plant, *Sarracenia alata* wood (Sarraceniaceae). *American Midland Naturalist*, **161**, 264–72.

Biesmeijer JC, Giurfa M, Koedam D, Potts SG, Joel DM and Dafni A 2005. Convergent evolution: floral guides, stingless bee nest entrances, and insectivorous pitchers. *Naturwissenschaften*, **92**, 444–50.

Bleecker AB and Kende H 2000. Ethylene: a gaseous signal molecule in plants. *Annual Review of Cell and Developmental Biology*, **16**, 1–18.

Bleiweiss R 2008. Phenotypic integration expressed by carotenoid-bearing plumages of tanager-finches (Thraupini, Emberizinae) across the avian visible spectrum. *Biological Journal of the Linnean Society*, **93**, 89–109.

Blendinger PG, Loiselle BA and Blake JG 2008. Crop size, plant aggregation, and microhabitat type affect fruit removal by birds from individual melastome plants in the Upper Amazon. *Oecologia*, **158**, 273–83.

Blüthgen N, Menzel F, Hovestadt T, Fiala B and Blüthgen N 2007. Specialization, constraints, and conflicting interests in mutualistic networks. *Current Biology*, **17**, 341–6.

Boege K and Marquis RJ 2005. Facing herbivory as you grow up: the ontogeny of resistance in plants. *Trends in Ecology and Evolution*, **20**, 441–8.

Borges RM, Gowda V and Zacharias M 2003. Butterfly pollination and high-contrast visual signals in a low-density distylous plant. *Oecologia*, **136**, 571–3.

Borowicz VA 1988. Do vertebrates reject decaying fruit? An experimental test with *Cornus amomum* fruits. *Oikos*, **53**, 74–8.

Botto-Mahan C, Pohl N and Medel R 2004. Nectar guide fluctuating asymmetry does not relate to female fitness in *Mimulus lutea*. *Plant Ecology*, **174**, 347–52.

Boughman, JW 2001. Divergent sexual selection enhances reproductive isolation in sticklebacks. *Nature* **411**, 944–948.

Boulay R, Coll-Toledano J, Manzaneda AJ and Cerda X 2007. Geographic variations in seed dispersal by ants: are plant and seed traits decisive? *Naturwissenschaften*, **94**, 242–6.

Boyd RS 2001. Ecological benefits of myrmecochory for the endangered chaparral shrub *Fremontodendron decumbens* (Sterculiaceae). *American Journal of Botany*, **88**, 234–41.

Bradbury JW and Vehrencamp SL 1998. *Principles of animal communication*. Sinauer Associates, Sunderland, MA.

Bradshaw HD and Schemske W 2003. Allele substitution at a flower colour locus produces a pollinator shift in monkeyflowers. *Nature*, **426**, 176–8.

Bradshaw WE 1980. Thermoperiodism and the thermal environment of the pitcher-plant mosquito *Wyeomyia smittii*. *Oecologia*, **46**, 13–7.

Bradshaw WE and Creelman RA 1984. Mutualism between the carnivorous purple pitcher plant its inhabitants. *American Midland Naturalist*, **112**, 294–304.

Brennan PA and Zufall F 2006. Pheromonal communication in vertebrates. *Nature*, **444**, 308–15.

Briscoe AD and Chittka L 2001. The evolution of color vision in insects. *Annual Review of Entomology*, **46**, 471–570.

Bro-Jørgensen J 2010. Dynamics of multiple signalling systems: animal communication in a world in flux. *Trends in Ecology and Evolution*, **25**, 292–300.

Brodmann J, Twele R, Francke W, Holzler G, Zhang QH and Ayasse M 2008. Orchids mimic green-leaf volatiles to attract prey-hunting wasps for pollination. *Current Biology*, **18**, 740–4.

Bronmark C 1985. Fresh-water snail diversity: effects of pond area, habitat heterogeneity and isolation. *Oecologia*, **67**, 127–31.

Bronmark C 1989. Interactions between epiphytes, macrophytes and fresh-water snails: a review. *Journal of Molluscan Studies*, **55**, 299–311.

Brouat C, McKey D, Bessiere JM, Pascal L and Hossaert-McKey M 2000. Leaf volatile compounds and the distribution of ant patrolling in an ant–plant protection mutualism: Preliminary results on Leonardoxa (Fabaceae: Caesalpinioideae) and Petalomyrmex (Formicidae: Formicinae). *Acta Oecologica-International Journal of Ecology*, **21**, 349–57.

Bruinsma M, Ijdema H, van Loon JJA and Dicke M 2008. Differential effects of jasmonic acid treatment of *Brassica nigra* on the attraction of pollinators, parasitoids, and butterflies. *Entomologia Experimentalis Et Applicata*, **128**, 109–16.

Bruna EM, Darrigo MR, Pacheco AMF and Vasconcelos HL 2008. Interspecific variation in the defensive responses of ant mutualists to plant volatiles. *Biological Journal of the Linnean Society*, **94**, 241–9.

Brunstrom JM and Mitchell GL 2007. Flavor – nutrient learning in restrained and unrestrained eaters. *Physiology and Behavior*, **90**, 133–41.

Buchholz R and Levey DJ 1990. The evolutionary triad of microbes, fruits, and seed dispersers: an experiment in fruit choice by cedar waxwings, *Bombycilla cedrorum*. *Oikos*, **59**, 200–204.

Bugnyar T and Kotrschal K 2002. Observational learning and the raiding of food caches in ravens, *Corvus corax*: is it 'tactical' deception? *Animal Behaviour*, **64**, 185–95.

Burger AE 2005. Dispersal and germination of seeds of *Pisonia grandis*, an Indo-Pacific tropical tree associated with insular seabird colonies. *Journal of Tropical Ecology*, **21**, 263–71.

Burkhardt D 1982. Birds, berries and UV. *Naturwissenschaften*, **69**, 153–7.

Burns KC 2005. Does mimicry occur between fleshy-fruits? *Evolutionary Ecology Research*, **7**, 1067–76.

Burns KC, Cazetta E, Galetti M, Valido A and Schaefer HM 2009. Geographic patterns in fruit colour diversity: do leaves constrain the colours of fleshy fruits? *Oecologia*, **159**, 337–43.

Burns KC and Dalen JL 2002. Foliage color contrasts and adaptive fruit color variation in a bird-dispersed plant community. *Oikos*, **96**, 463–9.

Byrne MM and Levey DJ 1993. Removal of seeds from frugivore defecations by ants in a Costa-Rican rain-forest. *Vegetatio*, **108**, 363–74.

Cahn MG and Harper JL 1976. Biology of leaf mark polymorphism in *Trifolium repens* L. 2. Evidence for selection of leaf marks by rumen fistulated sheep. *Heredity*, **37**, 327–33.

Caine NG and Mundy NI 2000. Demonstration of a foraging advantage for trichromatic marmosets (*Callithrix geoffroyi*) dependent on food colour. *Proceedings Royal Society B*, **267**, 439–44.

Campbell DR 2004. Natural selection in *Ipomopsis* hybrid zones: implications for ecological speciation. *New Phytologist*, **161**, 83–90.

Campbell DR and Aldridge G 2006. Floral biology and hybrid zones. In: Harder LD and Barrett SCH (eds) *Ecology and evolution of flowers*. Oxford University Press, Oxford.

Campitelli BE, Stehlik I and Stinchcombe JR 2008. Leaf variegation is associated with reduced herbivore damage in *Hydrophyllum virginianum. Botany-Botanique*, **86**, 306–13.

Canyon DV and Hill CJ 1997. Mistletoe host-resemblance: A study of herbivory, nitrogen and moisture in two Australian mistletoes and their host trees. *Australian Journal of Ecology*, **22**, 395–403.

Capon B and Brecht PE 1970. Variations in seed germination and morphology among populations of *Salvia columbariae* Benth. in southern California. *Ailso*, **7**, 207–16.

Carazo P and Font E 2010. Putting information back into biological communication. *Journal of Evolutionary Biology*, **23**, 661–9.

Carlo TA 2005. Interspecific neighbors change seed dispersal pattern of an avian-dispersed plant. *Ecology*, **86**, 2440–9.

Carlo TA and Aukema JE 2005. Female-directed dispersal and facilitation between a tropical mistletoe and a dioecious host. *Ecology*, **86**, 3245–51.

Carlo TA and Morales JM 2008. Inequalities in fruit-removal and seed dispersal: consequences of bird behaviour, neighbourhood density and landscape aggregation. *Journal of Ecology*, **96**, 609–18.

Carroll MJ, Schmelz EA, Meagher RL and Teal PEA 2006. Attraction of *Spodoptera frugiperda* larvae to volatiles from herbivore-damaged maize seedlings. *Journal of Chemical Ecology*, **32**, 1911–24.

Castro-Esau KL, Sanchez-Azofeifa GA, Rivard B, Wright SJ and Quesada M 2006. Variability in leaf optical properties of Mesoamerican trees and the potential for species classification. *American Journal of Botany*, **93**, 517–30.

Catoni C, Peters A and Schaefer HM 2008a. Life history trade-offs are influenced by the diversity, availability and interactions of dietary antioxidants. *Animal Behaviour*, **76**, 1107–19.

Catoni C, Schaefer HM and Peters A 2008b. Fruit for health: the effect of anthocyanins on humoral immune response and food selection in a frugivorous bird. *Functional Ecology*, **22**, 649–54.

Cazetta E, Schaefer HM and Galetti M 2008. Does attraction to frugivores or defense against pathogens shape fruit pulp composition? *Oecologia*, **155**, 277–86.

Cazetta E, Schaefer HM and Galetti M 2009. Why are fruits colorful? The relative importance of achromatic and chromatic contrasts for detection by birds. *Evolutionary Ecology*, **23**, 233–44.

Chalker-Scott L 1999. Environmental significance of anthocyanins in plant stress responses. *Photochemistry and Photobiology*, **70**, 1–9.

Chess SKR, Raguso RA and LeBuhn G 2008. Geographic divergence in floral morphology and scent in *Linanthus*

dichotomus (Polemoniaceae). *American Journal of Botany*, **95**, 1652–9.

Chittka L 1997. Bee color vision is optimal for coding flower color, but flower colors are not optimal for being coded – why? *Israel Journal of Plant Sciences*, **45**, 115–27.

Chittka L and Brockmann A 2005. Perception space – the final frontier. *PLOS Biology*, **3**, e137.

Chittka L and Döring TF 2007. Are autumn foliage colors red signals to aphids? *Plos Biology*, **5**, 1640–4.

Chittka L, Shmida A, Troje N and Menzel R 1994. Ultraviolet as a component of flower reflections, and the colour perception of hymenoptera. *Vision Research*, **34**, 1489–508.

Chittka L, Spaethe J, Schmidt A and Hickelsberger A 2001. Adaptation, constraint, and chance in the evolution of flower color and pollinator color vision. In: Chittka L and Thomson JD (eds) *Cognitive Ecology of Pollination*. Cambridge University Press, Cambridge.

Chittka L and Thomson JD (eds) (2001) Cognitive ecology of pollination. Cambridge University Press, Cambridge.

Chittka L and Waser NM 1997. Why red flowers are not invisible to bees. *Israel Journal of Plant Sciences*, **45**, 169–83.

Chittka L and Wells H 2004. Color vision in bees: mechanisms, ecology and evolution. In: Prete F (ed) *Complex worlds from simpler nervous systems*. MIT Press, Cambridge, MA.

Christensen R, Kleindorfer S and Robertson J 2006. Song is a reliable signal of bill morphology in Darwin's small tree finch *Camarhynchus parvulus*, and vocal performance predicts male pairing success. *Journal of Avian Biology*, **37**, 617–24.

Cipollini ML and Levey DJ 1991. Why some fruits are green when they are ripe: carbon balance in fleshy fruits. *Oecologia*, **88**, 371–7.

Cipollini ML and Levey DJ 1997a. Secondary metabolites of fleshy vertebrate-dispersed fruits: Adaptive hypotheses and implications for seed dispersal. *American Naturalist*, **150**, 346–72.

Cipollini ML and Levey DJ 1997b. Why are some fruits toxic? Glycoalkaloids in Solanum and fruit choice by vertebrates. *Ecology*, **78**, 782–98.

Cipollini ML and Levey DJ 1997c. Antifungal activity of Solanum fruit glycoalkaloids: implication for frugivory and seed dispersal. *Ecology*, **78**, 799–809.

Cipollini ML and Stiles EW 1992. Relative risks of microbial rot for fleshy fruits: significance with respect to dispersal and selection for secondary defense. *Advances in Ecological Research*, **23**, 35–91.

Clarke CM, Bauer U et al. 2009. Tree shrew lavatories: a novel nitrogen sequestration strategy in a tropical pitcher plant. *Biology Letters*, **5**, 632–5.

Clarke CM and Kitching RL 1993. The metazoan food webs from six Bornean Nepenthes species. *Ecological Entomology*, **18**, 7–16.

Clarke CM and Kitching RL 1995. Swimming ants and pitcher plants: a unique ant–plant interaction from Borneo. *Journal of Tropical Ecology*, **11**, 589–602.

Clydesdale FM 1993. Color as a factor in food choice. *Critical Reviews in Food Science and Nutrition*, **33**, 83–101.

Coberly LC and Rausher MD 2008. Pleiotropic effects of an allele producing white flowers in *Ipomea purpurea*. *Evolution*, **62**, 1076–85.

Cochrane EP 2003. The need to be eaten: *Balanites wilsoniana* with and without elephant seed-dispersal. *Journal of Tropical Ecology*, **19**, 579–89.

Coley PD, Bryant JP and Chapin FS 1985. Resource availability and plant antiherbivore defense. *Science*, **230**, 895–9.

Coleman SW, Patricelli GL and Borgia G 2004. Variable female preferences drive complex male displays. *Nature*, **428**, 742–5.

Connell JH 1971. On the role of natural enemies in preventing exclusion in some marine animals and in rain forest trees. In: den Boer PJ and Gradwell GR (eds) *Dynamics of populations*. Centre for Agricultual Publishing and Documentation, Wageningen.

Conner JK 2006. Ecological genetics of floral evolution. In: Harder LD and Barrett SCH (eds) *Ecology and evolution of flowers*. Oxford University Press, Oxford.

Cooper-Driver GA and Bhatttacharya M 1998. Role of phenols in plant evolution. *Phytochemistry*, **49**, 1165–74.

Cooper E, Whittall JB, Hodges SA and Nordborg M 2010. Genetic variation at nuclear loci fails to distinguish two morphologically distinct species of *Aquilegia*. *PLoS ONE*, **5**, 8655.

Cordeiro NJ and Howe HF 2003. Forest fragmentation severs mutualism between seed dispersers and an endemic African tree. *Proceedings of the National Academy of Sciences of the United States of America*, **100**, 14052–6.

Corlett RT 1996. Characteristics of vertebrate-dispersed fruits in Hong Kong. *Journal of Tropical Ecology*, **12**, 819–33.

Correa SB, Winemiller KO, Lopez-Fernandez H and Galetti M 2007. Evolutionary perspectives on seed consumption and dispersal by fishes. *Bioscience*, **57**, 748–56.

Costa-Arbulú C, Gianoli E, Gonzáles WL and Niemeyer HM 2001. Feeding by the aphid *Sipha flava* produces a reddish spot on leaves of *Sorghum halepense*: an induced defense? *Journal of Chemical Ecology*, **27**, 273–83.

Cott HB 1940. *Adaptive coloration in animals*. Methuen, London.

Cotton S, Fowler K and Pomiankowski A 2004. Do sexual ornaments demonstrate heightened condition-dependent expression as predicted by the handicap hypothesis? *Proceedings Royal Society B*, **271**, 771–83.

Cousens R, Dytham C and Law R 2008a. *Dispersal in plants: a population perspective*. Oxford University Press, Oxford.

Cousens R, Wiegand T and Taghizadeh M 2008b. Small-scale spatial structure within patterns of seed dispersal. *Oecologia*, **158**, 437–48.

Cozzolino S and Scopece G 2008. Specificity in pollination and consequences for postmating reproductive isolation in deceptive Mediterranean orchids. *Philosophical Transactions of the Royal Society B*, **363**, 3037–46.

Cozzolino S and Widmer A 2005. Orchid diversity: an evolutionary consequence of deception? *Trends in Ecology and Evolution*, **20**, 487–94.

Cresswell JE 1991. Capture rates and composition of insect prey of the pitcher plant *Sarracenia purpurea*. *American Midland Naturalist*, **125**, 1–9.

Crisosto CH, Crisosto GM and Mertheney P 2003. Consumer acceptance of 'Brooks' and 'Bings' cherries is mainly dependent on fruit SSC and visual skin color. *Postharvest Biology and Technology*, **28**, 159–67.

Cummings ME 2007. Sensory trade-offs predict signal divergence in surfperch. *Evolution* **61**, 530–545.

Cummings ME, Rosenthal GG and Ryan MJ 2003. A private ultraviolet channel in visual communication. *Proceedings Royal Society B*, **270**, 897–904.

Cunningham JP, Moore CJ, Zalucki MP and West SA 2004. Learning, odour preference and flower foraging in moths. *Journal of Experimental Biology*, **207**, 87–94.

Dafni A 1984. Mimicry and deception in pollination. *Annual Review of Ecology and Systematics*, **15**, 259–78.

Dafni A and Giurfa M 1998. Nectar guides and insect pattern recognition – a reconsideration. *Anais do encontro sobre Abelhas*, **3**, 55–66.

Dally JM, Clayton NS and Emery NJ 2006. The behaviour and evolution of cache protection and pilferage. *Animal Behaviour*, **72**, 13–23.

Dally JM, Emery NJ and Clayton NS 2004. Cache protection strategies by western scrub-jays (*Aphelocoma californica*): hiding food in the shade. *Proceedings Royal Society B*, **271**, S387–90.

Daly M, Jacobs LF, Wilson MI and Behrends PR 1992. Scatter hoarding by kangaroo rats (*Dopomys merriami*) and pilferage from their caches. *Behavioral Ecology*, **3**, 102–11.

Darst CR, Cummings ME and Cannatella DC 2006. A mechanism for diversity in warning signals: conspicuousness versus toxicity in poison frogs. *Proceedings National Academy of Sciences*, **103**, 5852–7.

Darwin C 1875. *Insectivorous plants*. John Murray, London.

Darwin C (1876) *The effects of cross- and self-fertilization in the vegetable kingdom.* Murray, London.

Davidson DW and McKey D 1993. Ant plant symbioses: stalking the Chuyachaqui. *Trends in Ecology and Evolution*, **8**, 326–32.

Davidson DW and Morton SR 1981. Myrmecochory in some plants (F. Chenopodiaceae) of the Australian arid zone. *Oecologia*, **50**, 357–66.

Davidson DW and Morton SR 1984. Dispersal adaptations of some Acacia species in the Australian arid zone. *Ecology*, **65**, 1038–51.

Davies KM 2009. Modifying anthocyanin production in flowers. In: Gould K, Davies K and Winefield C (eds) *Anthocyanins*. Springer, New York.

Davis CC, Endress PK and Baum DA 2008. The evolution of floral gigantism. *Current Opinion in Plant Biology*, **11**, 49–57.

Dawkins MS and Guilford T 1991. The corruption of honest signaling. *Animal Behaviour*, **41**, 865–73.

Dawkins R 1998. *Unweaving the rainbow*. Allen Lane, New York.

Dawkins R and Krebs JR 1978. Animal signals: information or manipulation. In: Krebs JR and Davies NB (eds) *Behavioural Ecology: an evolutionary approach*. Blackwell Scientific Publications, Oxford.

de Boer JG, Hordijk CA, Posthumus MA and Dicke M 2008. Prey and non-prey arthropods sharing a host plant: Effects on induced volatile emission and predator attraction. *Journal of Chemical Ecology*, **34**, 281–90.

De Moraes CM and Lewis WJ 1999. Analyses of two parasitoids with convergent foraging strategies. *Journal of Insect Behavior*, **12**, 571–83.

De Moraes CM, Lewis WJ, Pare PW, Alborn HT and Tumlinson JH 1998. Herbivore-infested plants selectively attract parasitoids. *Nature*, **393**, 570–3.

De Moraes CM, Mescher MC and Tumlinson JH 2001. Caterpillar-induced nocturnal plant volatiles repel conspecific females. *Nature*, **410**, 577–80.

Delph LF, Gehring JL, Arntz AM, Levri M and Frey FM 2008. Genetic correlations with floral display lead to sexual dimorphism in the cost of reproduction. *American Naturalist*, **166**, S31–41.

Delphia CM, Mescher MC and De Moraes CM 2007. Induction of plant volatiles by herbivores with different feeding habits and the effects of induced defenses on host-plant selection by thrips. *Journal of Chemical Ecology*, **33**, 997–1012.

Di Giusto B, Grosbois V, Fargeas E, Marshall DJ and Gaume L 2008. Contribution of pitcher fragrance and fluid viscosity to high prey diversity in a Nepenthes carnivorous plant from Borneo. *Journal of Biosciences*, **33**, 121–36.

Di Guisto B, Bessière JM, Guérolult, M. Lim LBL, Marshall, DJ, Hossaert-McKey M. and Gaume L 2010. Flower-scent mimicry masks a deadly trap in the carnivorous plant *Nepenthes rafflesiana*. journal of Ecology **98**, 845–56.

Dicke M 1986. Volatile spider-mite pheromone and host-plant kairomone involved in gregariousness in the spider-mite *Tetranychus urticae*. *Physiological Entomology*, **11**, 251–62.

Dicke M 1999. Are herbivore-induced plant volatiles reliable indicators of herbivore identity to foraging carnivorous arthropods? *Entomologia Experimentalis Et Applicata*, **91**, 131–42.

Dicke M 2008. Precise manipulation through a modeling study. *Journal of Chemical Ecology*, **34**, 943–4.

Dicke M, Agrawal AA and Bruin J 2003. Plants talk, but are they deaf? *Trends in Plant Science*, **8**, 403–405.

Dicke M and Sabelis MW 1988. How plants obtain predatory mites as bodyguards. *Netherlands Journal of Zoology*, **38**, 148–65.

Dicke M and van Loon JJA 2000. Multitrophic effects of herbivore-induced plant volatiles in an evolutionary context. *Entomologia Experimentalis Et Applicata*, **97**, 237–49.

Dierks A and Fischer K 2008. Feeding responses and food preferences in the tropical, fruit-feeding butterfly, *Bicyclus anynana*. *Journal of Insect Physiology*, **54**, 1363–70.

Dimitrova M, Stobbe N, Schaefer HM and Merilaita S 2009. Concealed by conspicuousness: distractive prey markings and backgrounds. *Proceedings of the Royal Society B*, **276**, 1905–10.

Dinerstein E and Wemmer CM 1988. Fruits rhinoceros eat – dispersal of Trewi nudiflora (Euphorbiaceae) in lowland Nepal. *Ecology*, **69**, 1768–74.

Dixon PM, Ellison AM and Gotelli NJ 2005. Improving the precision of estimates of the frequency of rare events. *Ecology*, **86**, 1114–23.

Dobat K 1985. *Blüten und Fledermäuse*. Waldemar Kramer, Frankfurt am Main.

Dobson HEM and Bergstrom G 2000. The ecology and evolution of pollen odors. *Plant Systematics and Evolution*, **222**, 63–87.

Dominy NJ 2004. Fruits, fingers, and fermentation: the sensory cues available to foraging primates. *Integrative and Comparative Biology*, **44**, 295–303.

Dominy NJ and Lucas PW 2002. Ecological importance of trichromatic vision to primates. Nature, **410**, 363–366.

Dominy NJ and Lucas PW 2004. Significance of color, calories, and climate to the visual ecology of Catarrhines. *American Journal of Primatology*, **62**, 189–207.

Dominy NJ, Lucas PW, Ramsden LW, Riba-Hernandez P, Stoner KE and Turner IM 2002. Why are young leaves red? *Oikos*, **98**, 163–76.

Donatti C, Galetti M, Pizo MA, Giumarães PR and Jordano P 2007. Living in the land of ghosts: fruit traits and the importance of large mammals as seed dispersers in the Pantanal, Brazil. In: Dennis AJ, Schupp EW, Green RJ and Westcott DW (eds) *Seed dispersal: theory and its implications in a changing world.* CAB International, Wallingford, UK.

Donoghue MJ and Doyle JA 1989. Phylogenetic studies of seed plants and angiosperms based on morphological characters. In: Fernholm B, Bremer K and Jornvall H (eds) *Hierarchy of Life – Molecules and Morphology in Phylogenetic Analysis.*

Donoghue MJ, Ree RH and Baum DA 1998. Phylogeny and the evolution of flower symmetry in Asteridae. *Trends in Plant Science,* **3**, 311–7.

Döring TF, Archetti M and Hardie J 2009. Autumn leaves seen through herbivore eyes. *Proceedings of the Royal Society B,* **276**, 121–7.

Döring TF and Chittka L 2007. Visual ecology of aphids – a critical review on the role of colours in host finding. *Arthropod – Plant Interactions,* **1**, 3–16.

Dormont L, Delle-Vedove, R., Bessière JM, Hoassaert-Mc Key M and Schatz B 2009. Rare white-flowered morphs increase the reproductive success of common purple morphs in a food-deceptive orchid. New Phytologist **185**, 300–310.

Dorrstock S, Porembski S and Barthlott W 1996. Ephemeral flush vegetation on inselbergs in the Ivory Coast (West Africa). *Candollea,* **51**, 407–19.

Dötterl S, Burkhardt D, Jürgens A and Monsandl A 2005. Sterioisomeric pattern of lilac aldehyde in *Silene latifolia,* a plant involved in a nursery pollination system. *Phytochemistry,* **68**, 499–504.

Dötterl S, Burkhardt D, Jürgens A and Mosandl A 2007. Stereoisomeric pattern of lilac aldehyde in *Silene latifolia,* a plant involved in a nursery pollination system. *Phytochemistry,* **68**, 499–504.

Dressler RL 1982. Biology of the orchid bees (Euglossini). *Annual Review of Ecology and Systematics,* **13**, 373–94.

Drew RAI 1988. Amino acid increase in fruit infested by fruit flies of the family Tephritidae. *Zoological Journal of the Linnean Society,* **93**, 107–12.

Ducrest AL, Keller L and Roulin A 2008. Pleiotropy in the melanocortin system, coloration and behavioural syndromes. *Trends in Ecology and Evolution,* **23**, 502–10.

Dufaÿ M, Hossaert-McKey M and Anstett MC 2003. When leaves act like flowers: how dwarf palms attract their pollinators. *Ecology Letters,* **6**, 28–34.

Dumont ER 1999. The effect of food hardness on feeding behaviour in frugivorous bats (Phyllostromidae): an experimental study. *Journal of Zoology,* **248**, 219–29.

Dumont ER and O'Neil R 2004. Food hardiness and feeding behavior in Old World fruit bats (Pteropodidae). *Journal of Mammalogy,* **85**, 8–14.

Dunn RR, Gove AD, Barraclough TG, Givnish TJ and Majer JD 2007. Convergent evolution of an ant–plant mutualism across plant families, continents, and time. *Evolutionary Ecology Research,* **9**, 1349–62.

Duthie C, Gibbs G and Burns KC 2006. Seed dispersal by weta. *Science,* **311**, 1575.

Dyer AG and Chittka L 2004. Bumblebees (*Bombus terrestris*) sacrifice foraging speed to solve difficult colour discrimination tasks. *Journal of Comparative Physiology A,* **190**, 759–63.

Dyer AG, Whitney HM, Arnold SEJ, Glover BJ and Chittka L 2006. Bees associate warmth with floral colour. *Nature,* **442**, 525.

Ebert D, Hayes C and Peakall R 2009. Chloroplast simple sequence repeat markers for evolutionary studies in the sexually deceptive orchid genus *Chiloglottis. Molecular Ecology Resources,* **9**, 784–9.

Eckart VM, Rushing NS, Hart GM and Hansen JD 2006. Frequency-dependent pollinator foraging in polymorphic *Clarkia xantiana* spp. *xantiana* populations: implications for flower colour evolution and pollinator interactions. *Oikos,* **112**, 412–21.

Edmunds M 1974. *Defence in animals: a survey of anti-predator defences.* Longman, Harlow.

Edwards DP, Arauco R, Hassall M, Sutherland WJ, Chamberlain K, Wadhams LJ and Yu DW 2007. Protection in an ant–plant mutualism: an adaptation or a sensory trap? *Animal Behaviour,* **74**, 377–85.

Edwards DP, Hassall M, Sutherland WJ and Yu DW 2006a. Assembling a mutualism: ant symbionts locate their host plants by detecting volatile chemicals. *Insectes Sociaux,* **53**, 172–6.

Edwards DP and Yu DW 2007. The roles of sensory traps in the origin, maintenance, and breakdown of mutualism. *Behavioral Ecology and Sociobiology,* **61**, 1321–7.

Edwards DP and Yu DW 2008. Tolerating castration by hiding flowers in plain sight. *Behavioral Ecology and Sociobiology,* **63**, 95–102.

Edwards H 1876. Darlingtonia californica. Text of a talk given to the California Academy of Sciences. *Proceedings of the California Academy of Sciences,* **6**, 161–6.

Edwards W, Dunlop M and Rodgerson L 2006b. The evolution of rewards: seed dispersal, seed size and elaiosome size. *Journal of Ecology,* **94**, 687–94.

Ehleringer JR, Ullmann I, Lange OL, Farquhar GD, Cowan IR, Schulze ED and Ziegler H 1986. Mistletoes: a hypothesis concerning morphological and chemical avoidance of herbivory. *Oecologia,* **70**, 234–7.

Ehrlen J and Eriksson O 1993. Toxicity in fleshy-fruits – a non adaptive trait? *Oikos,* **66**, 107–13.

Ehrlich PR and Raven PH 1964. Butterflies and plants: a study of coevolution. *Evolution,* **18**, 586–608.

Ellison AM and Gotelli NJ 2001. Evolutionary ecology of carnivorous plants. *Trends in Ecology and Evolution*, **16**, 623–9.

Ellison AM and Gotelli NJ 2009. Energetics and the evolution of carnivorous plants – Darwin's 'most wonderful plants in the world'. *Journal of Experimental Botany*, **60**, 19–42.

Ellison AM, Gotelli NJ, Brewer JS, Cochran-Stafira DL, Kneitel JM, Miller TE, Worley AC and Zamora R 2003. The evolutionary ecology of carnivorous plants. *Advances in Ecological Research*, **33**, 2–74.

Eltz T, Zimmermann Y, Pfeiffer C, Pech JR, Twele R, Francke W, Quezada-Euan JJG and Lunau K 2008. An olfactory shift is associated with male perfume differentiation and species divergence in orchid bees. *Current Biology*, **18**, 1844–8.

Endler JA 1980. Natural selection on color patterns in Poecilia reticulata. *Evolution*, **34**, 76–91.

Endler JA 1993a. The color of light in forests and its implications. *Ecological Monographs*, **63**, 1–27.

Endler JA 1993b. Some general comments on the evolution and design of animal communication systems. *Philosophical Transactions Royal Society London B*, **340**, 215–25.

Endler JA 2000. Evolutionary implications of the interaction between animal signals and the environment. In: Espmark Y, Amundsen T and Rosenqvist G (eds) *Animal Signals*. Tapir Academic Press, Trondheim.

Endler JA and Basolo AL 1998. Sensory ecology, receiver biases and sexual selection. *Trends in Ecology and Evolution*, **13**, 415–20.

Endler JA and Théry M 1996. Interacting effects of lek placement, display behaviour, ambient light and colour patterns in three neotropical forest-dwelling birds. *American Naturalist*, **148**, 421–52.

Eycott AE, Watkinson AR, Hemami MR and Dolman PM 2007. The dispersal of vascular plants in a forest mosaic by a guild of mammalian herbivores. *Oecologia*, **154**, 107–18.

Facelli JM 1993. Experimental evaluation of the foliar flag hypothesis using fruits of *Rhus glabra* (L). *Oecologia*, **93**, 70–2.

Fadzly N, Jack C, Schaefer HM and Burns KC 2009. Ontogenetic colour changes in an insular tree species: signalling to extinct browsing birds? *New Phytologist*, **184**, 495–501.

Faegri K and van der Pijl L 1966. *The principles of pollination ecology*. Pergamon, Oxford.

Fedriani JM and Boulay R 2006. Foraging by fearful frugivores: combined effect of fruit ripening and predation risk. *Functional Ecology*, **20**, 1070–9.

Feild TS, Lee DW and Holbrook NM 2001. Why leaves turn red in autumn. The role of anthocyanins in senescing leaves of red-osier dogwood. *Plant Physiology*, **127**, 566–74.

Fenster CB, Armbruster WS, Wilson P, Dudash MR and Thomson JD 2004. Pollination syndromes and floral specialization. *Annual Review of Ecology Evolution and Systematics*, **35**, 375–403.

Ferdy JB, Gouyon PH, Moret J and Godelle B 1998. Pollinator behavior and deceptive pollination: Learning process and floral evolution. *American Naturalist*, **152**, 696–705.

Fernandez AA and Morris MR 2007. Sexual selection and trichromatic color vision in primates: statistical support for the preexisting bias hypothesis. *American Naturalist*, **170**, 10–20.

Figuerola J and Green AJ 2002. How frequent is external transport of seeds and invertebrate eggs by waterbirds? A study in Donana, SW Spain. *Archiv für Hydrobiologie*, **155**, 557–65.

Fineblum WL and Rausher MD 1997. Do floral pigmentation and genes also influence resistance to enemies? The W locus in *Ipomoea pupurea*. *Ecology*, **78**, 1646–54.

Firestein S 2001. How the olfactory system makes sense of scents. *Nature*, **413**, 211–8.

Firn RD and Jones CG 2003. Natural products – a simple model to explain chemical diversity. *Natural Product Reports*, **20**, 382–91.

Fischbach RJ, Staudt M, Zimmer I, Rambal S and Schnitzler JP 2002. Seasonal pattern of monoterpene synthase activities in leaves of the evergreen tree *Quercus ilex*. *Physiologia Plantarum*, **114**, 354–60.

Fischer KE and Chapman CA 1993. Frugivores and fruit syndromes – differences in patterns at the genus and species level. *Oikos*, **66**, 472–82.

Fish D and Hall DW 1978. Succesion and stratification of aquatic insects inhabiting leaves of insectivorous pitcher plant, *Sarracenia purpurea*. *American Midland Naturalist*, **99**, 172–83.

Fleming TH 1992. How do fruit- and nectar-feeding birds and mammals track their food resources? In: Hunter MD, Ohgashi T and Price PW *Effects of resource distribution on animal–plant interactions*. Academic Press, New York.

Fleming TH, Geiselman C and Kress WJ 2009. The evolution of bat pollination: a phylogenetic perspective. *Annals of Botany*, **104**, 1017–43.

Fleming TH, Venable DL and Herrera LGM 1993. Opportunism vs. specialisation: the evolution of dipersal strategies in fleshy-fruited plants. In: Fleming TH and Estrada A *Frugivory and seed dispersal: ecological and evolutionary aspects*. Kluwer, Springer, Dordrecht.

Flörchinger M, Braun J, Böhning-Gaese K and Schaefer HM 2010. Fruit size, crop mass, and plant height explain differential fruit choice of primates and birds. *Oecologia*, **164**, 151–61.

Forest F, Chase MW, Persson C, Crane PR and Hawkins JA 2007. The role of biotic and abiotic factors in evolution of ant dispersal in the milkwort family (Polygalaceae). *Evolution*, **61**, 1675–94.

Forget PM 1992. Seed removal and seed fate in *Gustavia superba* (Lecythidaceae). *Biotropica*, **24**, 408–14.

Forsman A, Ahnesjö J, Caesar S and Karlsson M 2008. A model of ecological and evolutionary consequences of color polymorphism. *Ecology*, **89**, 34–40.

Foster MS and Delay LS 1998. Dispersal of mimetic seeds of three species of Ormosia (Leguminosae). *Journal of Tropical Ecology*, **14**, 389–411.

Fox JF 1982. Adaptation of gray squirrel behavior to autumn germination by white oak acorn. *Evolution*, **36**, 800–809.

Fraser PD, Truesdale MR, Bird CR, Schuch W and Bramley PM 1994. Carotenoid biosynthesis during tomato fruit development. *Plant Physiology*, **105**, 405–13.

Frederickson ME 2009. Conflict over reproduction in an ant–plant symbiosis: why *Allomerus octoarticulatus* ants sterilize *Cordia nodosa* trees. *American Naturalist*, **173**, 675–81.

Frey FM 2007. Phenotypic integration and the potential for independent color evolutioin in a polymorphic spring ephemeral. *American Journal of Botany*, **94**, 437–44.

Frost CJ, Mescher MC, Carlson JE and De Moraes CM 2008. Plant defense priming against herbivores: Getting ready for a different battle. *Plant Physiology*, **146**, 818–24.

Fuentes M 1994. Diets of fruit-eating birds: what are the causes of interspecific differences? *Oecologia*, **97**, 134–42.

Furuta K 1986. Host preference and population-dynamics in an autumnal population of the maple aphid, *Periphyllus californiensis shinji* (Homoptera, Aphididae). *Journal of Applied Entomology*, **102**, 93–100.

Furuta K 1990. Early budding of *Acer palmatum* caused by the shade; intra-specific heterogeneity of the host for the maple aphid. *Bulletin of Tokyo University Forests*, **82**, 137–45.

Galen C 1989. Measuring pollinator-mediated selection on morphometric floral traits: bumblebees and the alpine sky pilot, *Polmonium viscosum*. *Evolution*, **43**, 882–90.

Galen C 2000. High and dry: drought stress, sex-allocation trade-offs, and selection on flower size in the alpine wildflower *Polemonium viscosum* (Polemoniaceae). *American Naturalist*, **156**, 72–83.

Galen C and Cuba J 2001. Down the tube: pollinators, predators, and the evolution of flower shape in the alpine skypilot, *Polemonium viscosum*. *Evolution*, **55**, 1963–71.

Galetti M 2002. Seed dispersal of mimetic fruits: parasitsm, mutalism, aposematism or exaptation. In: Levey DJ, Silva WR and Galetti M (eds) *Seed dispersal and frugivory: ecology, evolution and conservation*. CAB international, Wallingford.

Galetti M, Alves-Costa CP and Cazetta E 2003. Effects of forest fragmentation, anthropogenic edges and fruit colour on the consumption of ornithocoric fruits. *Biological Conservation*, **111**, 269–73.

Galetti M, Donatti CI, Pizo MA and Giacomini HC 2008. Big fish are the best: seed dispersal of *Bactris glaucescens* by the pacu fish (*Piaractus mesopotamicus*) in the Pantanal, Brazil. *Biotropica*, **40**, 386–9.

Galizia CG, Kunze J, Gumbert A, Borg-Karlson A-K, Sachse S, Markl C and Menzel R 2005. Relationship of visual and olfactory signal parameters in a food-deceptive flower mimicry system. *Behavioral Ecology*, **16**, 159–68.

Galizia CG and Menzel R 2000. Odour perception in honeybees: coding information in glomerular patterns. *Current Opinion in Neurobiology*, **10**, 504–10.

Gamberale-Stille G 2001. Benefit by contrast: an experiment with live aposematic prey. *Behavioral Ecology*, **12**, 768–72.

Gamberale-Stille G and Tullberg BS 2001. Fruit or aposematic insect? Context-dependent colour preferences in domestic chicks. *Proceedings of the Royal Society B*, **268**, 2525–9.

Gammans N, Bullock JM, Gibbons H and Schonrogge K 2006. Reaction of mutualistic and granivorous ants to Ulex elaiosome chemicals. *Journal of Chemical Ecology*, **32**, 1935–47.

Gamon JA and Surfus JS 1999. Assessing leaf pigment content and activity with a reflectometer. *New Phytologist*, **143**, 105–17.

Garcia-Plazaola JI, Hernandez A and Becerril JM 2003. Antioxidant and pigment composition during autumnal leaf senescence in woody deciduous species differing in their ecological traits. *Plant Biology*, **5**, 557–66.

Garcia CM and Ramirez E 2005. Evidence that sensory traps can evolve into honest signals. *Nature*, **434**, 501–505.

Gautier-Hion A, Duplantier J-M, Quris R, Feer F, Sourd C, Decoux J-P, Dubost G, Emmons L, Erard C, Hecketsweiler P, Moungazi A, Roussilhon C and Thiollay J-M 1985. Fruit characters as a basis of fruit choice and seed dispersal in a tropical forest vertebrate community. *Oecologia*, **65**, 324–37.

Geervliet JBF, Vet LEM and Dicke M 1996. Innate responses of the parasitoids *Cotesia glomerata* and *C. rubecula*

(Hymenoptera: Braconidae) to volatiles from different plant–herbivore complexes. *Journal of Insect Behavior*, **9**, 525–38.

Geervliet JBF, Vreugdenhil AI, Dicke M and Vet LEM 1998. Learning to discriminate between infochemicals from different plant–host complexes by the parasitoids *Cotesia glomerata* and *C. rubecula*. *Entomologia Experimentalis Et Applicata*, **86**, 241–52.

Gegear RJ and Laverty TM 2005. Flower constancy in bumblebees: a test of the trait variability hypothesis. *Animal Behaviour*, **69**, 939–49.

Gervais JA, Noon BR and Willson MF 1999. Avian selection of the color-dimorphic fruits of salmonberry, *Rubus spectabilis*: a field experiment. *Oikos*, **84**, 77–86.

Getty T 2006. Sexually selected signals are not similar to sports handicaps. *Trends in Ecology and Evolution*, **21**, 83–8.

Gibson TC 1991a. Differential escape of insects from carnivorous plant traps. *American Midland Naturalist*, **125**, 55–62.

Gibson TC 1991b. Competition among threadleaf sundews for limited insect resources. *American Naturalist*, **138**, 785–9.

Gibson TC and Waller DM 2009. Evolving Darwin's 'most wonderful' plant: ecological steps to a snap-trap. *New Phytologist*, **183**, 575–87.

Gibson W 1993. Selective advantages to hemi-parasitic annuals, Genus Melampyrum, of a seed-dispersal mutualism involving ants. 1. Favorable nest sites. *Oikos*, **67**, 334–44.

Gigord LDB, Macnair MR and Smithson A 2001. Negative frequency-dependent selection maintains a dramatic flower color polymorphism in the rewardless orchid *Dactylorhiza sambucina* (L.) Soó. *Proceedings National Academy of Sciences*, **98**, 6253–5.

Gigord LDB, Macnair MR, Stritesky M and Smithson A 2002. The potential for floral mimicry in rewardless orchids: an experimental study. *Proceedings of the Royal Society B*, **269**, 1389–95.

Gilbert L 1975. Ecological consequences of a coevolved mutualism between butterflies and plants. In: Gilbert L and Raven P (eds) *Coevolution of animals and plants*. University of Texas Press, Austin.

Giles S and Lill A 1999. The effect of fruit abundance, conspicuousness and sugar concentration on fruit colour choice by captive silvereyes. *Ethology, Ecology & Evolution*, **11**, 229–42.

Giurfa M, Eichmann B and Menzel R 1996. Symmetry perception in an insect. *Nature*, **382**, 458–61.

Giurfa M, Vorobyev M, Brandt R, Posner B and Menzel R 1997. Discrimination of coloured stimuli by honeybees: alternative use of achromatic and chromatic signals. *Journal of Comparative Physiology A*, **180**, 235–43.

Givnish TJ 1990. Leaf mottling- relation to growth form and leaf phenology and possible role as camouflage. *Functional Ecology*, **4**, 463–74.

Givnish TJ, Burkhardt EL, Happel RE and Weintraub JD 1984. Carnivory in the bromeliad *Brocchinia reducta*, with a cost – benefit model for the general restriction of carnivorous plants to sunny, moist, nutrient-poor habitats. *American Naturalist*, **124**, 479–97.

Globner F 1992. Ultraviolet patterns in the traps and flowers of some carnivorous plants. *Botanische Jahrbücher*, **113**, 577–87.

Glover BJ and Whitney HM 2010. Structural colour and iridescence in plants: the poorly studied relations of pigment colour. *Annals of Botany*, **105**, 505–11.

Gnanvossou D, Hanna R and Dicke M 2003. Infochemical-mediated intraguild interactions among three predatory mites on cassava plants. *Oecologia*, **135**, 84–90.

Goddard MR 2008. Quantifying the complexities of *Saccharomyces cerevisiae's* ecosystem engineering via fermentation. *Ecology*, **89**, 2077–82.

Godfray HCJ 1995. Communication between the first and third trophic levels: and analysis using biological signaling theory. *Oikos*, **72**, 367–74.

Goldsmith TH 1990. Optimization, constraint, and history in the evolution of eyes. *Quart Rev Biol*, **65**, 281–322.

Gols R, Bukovinszky T, Hemerik L, Harvey JA, Van Lenteren JC and Vet LEM 2005. Reduced foraging efficiency of a parasitoid under habitat complexity: implications for population stability and species coexistence. *Journal of Animal Ecology*, **74**, 1059–68.

Gómez JM 2004. Bigger is not always better: conflicting selective pressures on seed size in *Quercus ilex*. *Evolution*, **58**, 71–80.

Gómez C and Espadaler X 1998. Myrmecochorous dispersal distances: a world survey. *Journal of Biogeography*, **25**, 573–80.

Gong YB and Huang S-Q 2009. Floral symmetry: pollinator-mediated stabilizing selection on flower size in bilateral species. *Proceedings Royal Society B*, **276**, 4013–20.

Gonzalez JM, Jaffe K and Michelangeli F 1991. Competition for prey between the carnivorous Bromeliaceae *Brocchinia reducta* and *Sarraceneacea heliamphora nutans*. *Biotropica*, **23**, 602–604.

Goodale E, Beauchamp G, Magrath RD, Nieh JC and Ruxton GD 2010. Interspecific information transfer influences animal community structure. *Trends in Ecology and Evolution*, **25**, 354–61.

Goodwin TW and Goad LJ 1970. Carotenoids and triterpenoids. In: Hulme AC (ed) *The biochemistry of fruits and their products*. Academic Press, New York

Gorb E, Voigt D, Eigenbrode SD and Gorb S 2008. Attachment force of the beetle *Cryptolaemus montrouzieri* (Coleoptera,

Coccinellidae) on leaflet surfaces of mutants of the pea *Pisum sativum* (Fabaceae) with regular and reduced wax coverage. *Arthropod – Plant Interactions*, **2**, 247–59.

Gorchov DL 1988. Does asynchronous fruit ripening avoid satiation of seed dispersers?: A field test. *Ecology*, **69**, 1545–51.

Gosper CR, Stansbury CD and Vivian-Smith G 2005. Seed dispersal of fleshy-fruited invasive plants by birds: contributing factors and management options. *Diversity and Distributions*, **11**, 549–58.

Gould KS, Davies K and Winefield C (eds) (2009) Anthocyanins. Springer, New York.

Gould KS 2004. Nature's Swiss army knife: the diverse protective roles of anthocyanins in leaves. *Journal of Biomedicine and Biotechnology*, **5**, 314–20.

Gould KS, Dudle DA and Neufeld HS 2010. Why some stems are red: cauline anthocyanins shield photosystem II against high light stress. *Journal of Experimental Botany*, **61**, 2707–17.

Gould KS and Lee DW 2002. *Anthocyanins in leaves*. Academic Press, London.

Gould KS, McKelvie J and Markham KR 2002. Do anthocyanins function as antioxidants in leaves? Imaging of H_2O_2 in red and green leaves after mechanical injury. *Plant Cell and Environment*, **25**, 1261–9.

Gould SJ and Lewontin RC 1979. The spandrels of San Marco and the Panglossian paradigm: a critique of the adaptionist programme. *Proceedings Royal Society B*, **205**, 581–98.

Goyret J, Markwell PM and Raguso RA 2007. The effect of decoupling olfactory and visual stimuli on the foraging behavior of *Manduca sexta*. *Journal of Experimental Biology*, **210**, 1398–405.

Goyret J, Markwell PM and Raguso RA 2008. Context- and scale-dependent effects of floral CO_2 on nectar foraging by *Manduca sexta*. *Proceedings National Academy of Sciences*, **105**, 4565–70.

Goyret J and Raguso RA 2006. The role of mechanosensory input in flower handling efficiency and learning by *Manduca sexta*. *Journal of Experimental Biology*, **209**, 1585–93.

Grafen A 1990. Biological signals as handicaps. *Journal of Theoretical Biology*, **144**, 517–46.

Grant KA and Grant V 1968. *Hummingbirds and their flowers*. Columbia University Press, New York.

Greig-Smith PW 1986. Bicolored fruit displays and frugivorous birds: the importance of fruit quality to dispersers and seed predators. *American Naturalist*, **127**, 246–51.

Gronquist M, Bezzerides A, Attygale A, Meinwald J, Eisner M and Eisner T 2001. Attractive and defensive functions of the ultraviolet pigments of a flower (*Hypericum calycinum*). *Proceedings National Academy of Sciences*, **98**, 13745–50.

Groom PK, Lamont BB and Duff HC 1994. Self-crypsis in *Hakea trifurcata* as an avian granivore deterrent. *Functional Ecology*, **8**, 110–17.

Guerenstein PG, Yepez EA, van Haren J, Williams DG and Hillebrand JG 2004. Floral CO_2 emission may indicate food abundance to nectar-feeding moths. *Naturwissenschaften*, **91**, 329–33.

Guerrieri E, Poppy GM, Powell W, Rao R and Pennacchio F 2002. Plant-to-plant communication mediating in-flight orientation of *Aphidius ervi*. *Journal of Chemical Ecology*, **28**, 1703–15.

Guimarães PR, Galetti M and Jordano P 2008. Seed dispersal anachronisms: rethinking the fruits extinct megafauna ate. *PLoS ONE*, **3**, e1745.

Gumbert A and Kunze J 2001. Colour similarity to rewarding model plants affects pollination in a food deceptive orchid, *Orchis boryi*. *Biological Journal of the Linnean Society*, **72**, 419–33.

Gumbert A, Kunze J and Chittka L 1999. Floral colour diversity in plant communities, bee colour space and a null model. *Proceedings Royal Society B*, **266**, 1711–6.

Hagen SB, Debeausse S, Yoccoz NG and Folstad I 2004. Autumn coloration as a signal of tree condition. *Proceedings of the Royal Society B, Biological Sciences*, **271**, S184–5.

Hagen SB, Folstad I and Jakobsen SW 2003. Autumn colouration and herbivore resistance in mountain birch (*Betula pubescens*). *Ecology Letters*, **6**, 807–11.

Hajek T and Adamec L 2009. Mineral nutrient economy in competing species of Sphagnum mosses. *Ecological Research*, **24**, 291–302.

Halitschke R, Stenberg JA, Kessler D, Kessler A and Baldwin IT 2008. Shared signals–'alarm calls' from plants increase apparency to herbivores and their enemies in nature. *Ecology Letters*, **11**, 24–34.

Hallem EA and Carlson JR 2006. Coding of odors by a receptor repertoire. *Cell*, **125**, 143–60.

Hallem EA, Dahanukar A and Carlson JR 2006. Insect odor and taste receptors. *Annual Review of Entomology*, **51**, 113–35.

Halpern M, Raats D and Lev-Yadun S 2007. Plant biological warfare: thorns inject pathogenic bacteria into herbivores. *Environmental Microbiology*, **9**, 584–92.

Hamilton WD and Brown SP 2001. Autumn tree colours as a handicap signal. *Proceedings of the Royal Society B, Biological Sciences*, **268**, 1489–93.

Hampe A 2003. Large-scale geographical trends in fruit traits of vertebrate-dispersed temperate plants. *Journal of Biogeography*, **30**, 487–96.

Han Q, Katahata S, Kakubari Y and Mukai Y 2004. Seasonal changes in the xanthophyll cycle and antioxidants in sun-exposed and shaded parts of the crown of *Cryptomeria japonica* in relation to rhodoxanthin accumulation during cold acclimation. *Tree Physiology*, **24**, 609–14.

Hanley ME, Lamont BB, Armbruster WS (2009) Pollination and plant defence traits co-vary in Western Australian *Hakeas*. New Phytologist, **182**, 251–260.

Hansen DM, Beer K and Müller CB 2006. Mauritian coloured nectar no longer a mystery: a visual signal for lizard pollinators. *Biology Letters*, **2**, 165–8.

Hansen DM, Olesen JM, Mione T, Johnson SD and Müller CB 2007. Coloured nectar: distribution, ecology, and evolution of an enigmatic floral trait. *Biological Reviews*, **82**, 83–111.

Hansen TF 2006. The evolution of genetic architecture. *Annual Review of Ecology Evolution and Systematics*, **27**, 123–57.

Harborne JB 1976. Function of flavonoids in plants. In: Goodwin TW (ed) *Chemistry and biochemistry of plant pigments*, 2nd edn. Academic Press, New York.

Harder LD and Barrett SCH (eds) 2006. Ecology and evolution of flowers. Oxford University Press, New York.

Hardesty BD, Hubbell SP and Bermingham E 2006. Genetic evidence of frequent long-distance recruitment in a vertebrate-dispersed tree. *Ecology Letters*, **9**, 516–25.

Harms KE, Wright SJ, Calderon O, Hernandez A and Herre EA 2000. Pervasive density-dependent recruitment enhances seedling diversity in a tropical forest. *Nature*, **404**, 493–5.

Hartmann T 1996. Diversity and variability of plant secondary metabolism: a mechanistic view. *Entomologia Experimentalis et Applicata*, **80**, 177–88.

Hasson O 1994. Cheating signals. *Journal of Theoretical Biology*, **167**, 223–38.

Håstad O, Victorsson J and Ödeen A 2005. Differences in color vision make passerines less conspicuous in the eyes of their predators. *Proceedings National Academy of Sciences*, **102**, 6391–4.

Hauser MD 1997. *The evolution of communication*. MIT Press, Cambridge, MA.

Heard SB 1994. Pitcher-plant midges and mosquitos: a processing chain commensalism. *Ecology*, **75**, 1647–60.

Heaton JW and Marangoni AG 1996. Chlorophyll degradation in processed foods and senescent plant tissues. *Trends in Food Science & Technology*, **7**, 8–15.

Hebant C and Lee DW 1984. Ultrastructural basis and developmental control of blue iridescence in *Selaginella* leaves. *American Journal of Botany*, **71**, 216–19.

Hebets EA and Papaj DR 2005. Complex signal function: developing a framework of testable hypotheses. *Behavioral Ecology and Sociobiology*, **57**, 197–214.

Heil M and Bueno JCS 2007. Plant volatiles carry both public and private messages. *Proceedings National Academy of Sciences*, **104**, 5467–72.

Heil M and Karban R 2010. Explaining evolution of plant communication by airborne signals. *Trends in Ecology and Evolution*, **25**, 137–44.

Heil M and Ton J 2008. Long-distance signalling in plant defence. *Trends in Plant Science*, **13**, 264–72.

Heiling AM, Cheng K and Herberstein ME 2004. Exploitation of floral signals by crab spiders (*Thomisus spectabilis*, Thomisidae). *Behavioral Ecology*, **15**, 321–6.

Heiling AM, Cheng K et al. 2006. Picking the right spot: crab spiders position themselves on flowers to maximize prey attraction. *Behaviour*, **143**, 957–68.

Heiling AM, Chittka L, Cheng K and Herberstein ME 2005. Colouration in crab spiders: substrate choice and prey attraction. *Journal of Experimental Biology*, **208**, 1785–92.

Heiling AM, Heberstein ME and Chittka L 2003. Crab-spiders manipulate flower signals. *Nature*, **421**, 334.

Heindl M and Winkler H 2003. Interacting effects of ambient light and plumage color patterns in displaying Wire-tailed Manakins (Aves, Pipridae). *Behavioral Ecology and Sociobiology*, **53**, 153–62.

Heithaus ER 1981. Seed predation by rodents on three ant-dispersed plants. *Ecology*, **62**, 136–45.

Herrera CM 1982. Defense of ripe fruit from pests: its significance in relation to plant–disperser interactions. *American Naturalist*, **120**, 218–41.

Herrera CM 1984. Avian interference of insect frugivory: an exploration into the plant–bird–fruit pest evolutionary triad. *Oikos*, **42**, 203–10.

Herrera CM 1985. Determinants of plant–animal coevolution: the case of mutualistic dispersal of seeds by vertebrates. *Oikos*, **44**, 132–41.

Herrera CM 1987. Vertebrate-dispersed plants of the Iberian peninsula: a study of fruit characteristics. *Ecological Monographs*, **57**, 305–31.

Herrera CM 1996. Floral traits and plant adaptation to insect pollinators: a devil's advocate approach. In: Lloyd DG and Barrett SCH (eds) *Floral biology: studies on floral evolution in animal-pollinated plants*. Chapman and Hall, New York.

Herrera CM 1998a. Long-term dynamics of mediterranean frugivorous birds and fleshy fruits: a 12-year study. *Ecological Monographs*, **68**, 511–38.

Herrera CM 1998b. Population-level estimates of interannual variability in seed production: what do they actually tell us? *Oikos*, **82**, 612–6.

Herrera CM 2002. Correlated evolution of fruit and leaf size in bird-dispersed plants: species-level variance in fruit traits explained a bit further. *Oikos*, **97**, 426–32.

Herrera CM and Bazaga P 2008. Population-genomic approach reveals adaptive floral divergence in discrete populations of a hawk moth-pollinated violet. *Molecular Ecology*, **17**, 5378–90.

Herrera CM, Castellanos MC and Medrano M 2006. Geographical context of floral evolution: towards an improved research programme in floral diversification. In: Harder LD and Barrett SCH (eds) *Ecology and evolution of flowers*. Oxford University Press, Oxford.

Herrera CM, Cerdá X, García MB, Guitian J, Medrano M, Rey PJ and Sánchez-Lafuente AM 2002. Floral integration, phenotypic covariance structure and pollinator variation in bumblebee-pollinated *Helleborus foetidus*. *Journal of Evolutionary Biology*, **15**, 108–21.

Herrera CM and Pellmyr O 2002. *Plant–animal interactions: an evolutionary approach*. Blackwell Publishing, Oxford.

Hessel K and Schmidt U 1994. Multimodal orientation in *Carollia perspicillata* (Phyllostomidae). *Folia Zoologica*, **43**, 339–46.

Heuschen B, Gumbert A and Lunau K 2005. A generalised mimicry system involving angiosperm flower colour, pollen and bumblebees' innate colour preferences. *Plant Systematics and Evolution*, **252**, 121–37.

Himanen SJ, Blande JD, Klemola T, Pulkkinen J, Heijari J and Holopainen JK 2010. Birch (*Betula spp.*) leaves adsorb and re-release volatiles specific to neighbouring plants – a mechanism for associational herbivore resistance? *New Phytologist*, **186**, 722–32.

Hoballah ME, Gübitz T, Stuurman J, Broger L, Barone M, Mandel T, Dell'Olivio A, Arnold M and Kuhlemeier C 2007. Single gene-mediated shift in pollinator attraction in *Petunia*. *Plant Cell*, **19**, 779–90.

Hoballah ME and Turlings TCJ 2005. The role of fresh versus old leaf damage in the attraction of parasitic wasps to herbivore-induced maize volatiles. *Journal of Chemical Ecology*, **31**, 2003–18.

Hoballah MEF and Turlings TCJ 2001. Experimental evidence that plants under caterpillar attack may benefit from attracting parasitoids. *Evolutionary Ecology Research*, **3**, 553–65.

Hoch WA, Singsaas EL and McCown BH 2003. Resorption protection. Anthocyanins facilitate nutrient recovery in autumn by shielding leaves from potentially damaging light levels. *Plant Physiology*, **133**, 1296–305.

Hoch WA, Zeldin EL and McCown BH 2001. Physiological significance of anthocyanins during autumnal leaf senescence. *Tree Physiology*, **21**, 1–8.

Hodges SA and Derieg NJ 2009. Adaptive radiations: from field to genomic studies. *Proceedings National Academy of Sciences*, **106**, 9947–54.

Hodgkison R, Ayasse M, Kalko EKV, Haeberlein C, Schulz S, Mustapha H, Zubaid A and Kunz TH 2007. Chemical ecology of fruit bat foraging behavior in relation to the fruit odours of two species of paleotropical bat-dispersed figs (*Ficus hispida* and *Ficus scortechinii*). *Journal of Chemical Ecology*, **33**, 2097–110.

Holbrook KM and Smith TB 2000. Seed dispersal and movement patterns in two species of Ceratogymna hornbills in a West African tropical lowland forest. *Oecologia*, **125**, 249–57.

Hollander JL, Wall SBV and Baguley JG 2010. Evolution of seed dispersal in North American Ephedra. *Evolutionary Ecology*, **24**, 333–45.

Holopainen JK 2008. Importance of olfactory and visual signals of autumn leaves in the coevolution of aphids and trees. *Bioessays*, **30**, 889–96.

Holopainen JK and Peltonen P 2002. Bright autumn colours of deciduous trees attract aphids: nutrient retranslocation hypothesis. *Oikos*, **99**, 184–8.

Holopainen, J.K., Heijari, J., Oksanen, E. & Alessio, G.A. 2010. Leaf volatile emissions of *Betula pendula* during autumn coloration and leaf fall. *Journal of Chemical Ecology* **36**, 1068–75.

Honkavaara J, Siitari H and Viitala J 2004. Fruit colour preferences of redwings (*Turdus iliacus*): experiments with hand-raised juveniles and wild-caught adults. *Ethology*, **110**, 445–57.

Horn MH 1997. Evidence for dispersal of fig seeds by the fruit-eating characid fish *Brycon guatemalensis Regan* in a Costa Rican tropical rain forest. *Oecologia*, **109**, 259–64.

Hossaert-McKey M, Gibernau M and Frey JE 1994. Chemosensory attraction of fig wasps to substances produced by receptive figs. *Entomologia Experimentalis et Applicata*, **70**, 185–91.

Hossaert-McKey M, Soler C and Schatz B 2010. Floral scents: their roles in nursery pollination mutualisms. *Chemoecology*, **20**, 75–88.

Howe HF and Estabrook GF 1977. Intraspecific competition for avian dispersers in tropical trees. *American Naturalist*, **111**, 817–32.

Howe HF and Smallwood J 1982. Ecology of seed dispersal. *Annual Review of Ecology and Systematics*, **13**, 201–28.

Hughes L 1991. The relocation of ant nest entrances: potential consequences for ant-dispersed seeds. *Australian Journal of Ecology*, **16**, 207–14.

Hughes L, Dunlop M, French K, Leishman MR, Rice B, Rodgerson L and Westoby M 1994. Predicting dispersal spectra: a minimal set of hypotheses based on plant attributes. *Journal of Ecology*, **82**, 933–50.

Hughes L and Westoby M 1992. Effect of diaspore characterstics on removal of seeds adapted for dispersal by ants. *Ecology*, **73**, 1300–12.

Hughes L, Westoby M and Johnson AD 1993. Nutrient costs of vertebrate-dispersed and ant-dispersed fruits. *Functional Ecology*, **7**, 54–62.

Hughes NM, Neufeld HS and Burkey KO 2005. Functional role of anthocyanins in high-light winter leaves of the evergreen herb *Galax urceolata*. *New Phytologist*, **168**, 575–87.

Hunt DM, Dulai KS, Cowing JA, Julliot C, Mollon JD, Bowmaker JK, Li W-H and Hewett-Emmert D 1998. Molecular evolution of trichromacy in primates. *Vision Research*, **38**, 3299–306.

Hurlbert A 1986. Formal connections between lightness algorithms. *Journal of the Optical Society of America*, **A3**, 1685–93.

Hurlbert AC 1998. Computational models of colour constancy. In: Walsh V and Kulikowski J (eds) *Perceptual constancy: why things look as they do*. Cambridge University Press, Cambridge, 283–322.

Hutchins HE and Lanner RM 1982. The central role of clark nutcracker in the dispersal and establishment of whitebark pine. *Oecologia*, **55**, 192–201.

Inbar M, Izhaki I, Koplovich A, Lupo I, Silanikove N, Glasser T, Gerchman Y, Perevolotsky A and Lev-Yadun S 2010. Why do many galls have conspicuous colors? A new hypothesis. *Arthropod – Plant Interactions*, **4**, 1–6.

Ings TC and Chittka L 2009. Predator crypsis enhances behaviourally mediated indirect effects on plants by altering bumblebee foraging preferences. *Proceedings Royal Society B*, **276**, 2031–6.

Internicola AI, Page PA, Bernasconi G and Gigord LDB 2007. Competition for pollinator visitation between deceptive and rewarding artificial inflorescences: an experimental test of the effects of floral colour similarity and spatial mingling. *Functional Ecology*, **21**, 864–72.

Irwin RE 2006. The consequences of direct versus indirect species interactions to selection on traits: pollination and nectar robbing in *Ipomopsis aggregata*. *American Naturalist*, **167**, 315–28.

Irwin RE, Strauss SY, Storz S, Emerson A and Guibert G 2003. The role of herbivores in the maintenance of a flower color polymorphism in wild radish. *Ecology*, **84**, 1733–43.

Istock CA, K. T and Zimmer H 1983. Habitat selection by the pitcher-plant mosquito *Wyeomyia smithii*: behavioural and genetic aspects. In: Frank JH and Lounibos LP (eds) *phytotelmata: terrestrial plants as host for aquatic insect communities*. Plexus, Marlton, NJ.

Iverson JB 1985. Lizards as seed dispersers. *Journal of Herpetology*, **19**, 292–3.

Izhaki I 1993. Influence of non-protein nitrogen on estimation of protein from total nitrogen in fleshy fruits. *Journal of Chemical Ecology*, **19**, 2605–15.

Izhaki I and Safriel UN 1989. Why are there so few exclusively frugivorous birds? Experiments on fruit digestibility. *Oikos*, **54**, 23–32.

Jacobs GH and Blakeslee B 1984. Individual variations in color vision among squirrel monkeys (*Saimiri sciureus*) of different geographical origins. *Journal of Comparative Psychology A*, **98**, 347–57.

Jansen PA, Bongers F and Hemerik L 2004. Seed mass and mast seeding enhance dispersal by a neotropical scatter-hoarding rodent. *Ecological Monographs*, **74**, 569–89.

Jansen PA, Bongers F and Prins HHT 2006. Tropical rodents change rapidly germinating seeds into long-term food supplies. *Oikos*, **113**, 449–58.

Janson CH 1983. Adaptation of fruit morphology to dispersal agents in a neotropial forest. *Science*, **219**, 187–9.

Janzen DH 1970. Herbivores and number of tree species in tropical forests. *American Naturalist*, **104**, 501–28.

Janzen DH 1977. Why fruits rot, seeds mold, and meat spoils. *American Naturalist*, **111**, 691–713.

Janzen DH 1980. When is it coevolution? *Evolution*, **34**, 611–12.

Janzen DH 1984. Dispersal of small seeds by big herbivores: foliage is the fruit. *American Naturalist*, **123**, 338–53.

Janzen DH and Martin PS 1982. Pleistocene seed dispersal. *Science*, **216**, 6.

Jerozolimski A, Ribeiro MBN and Martins M 2009. Are tortoises important seed dispersers in Amazonian forests? *Oecologia*, **161**, 517–28.

Jersáková J, Johnson SD and Kindlmann P 2006. Mechanisms and evolution of deceptive pollination in orchids. *Biological Reviews*, **81**, 219–35.

Jesson LK 2007. Ecological correlates of diversification in New Zealand angiosperm lineages. *New Zealand Journal of Botany*, **45**, 35–51.

Joel DM 1988. Mimicry and mutualism in carnivorous pitcher plants (Sarraceniaceae, Nepenthaceae, Cephalotaceae, Bromeliaceae). *Biological Journal of the Linnean Society*, **35**, 185–97.

Joel DM, Juniper BE and Dafni A 1985. Ultraviolet patterns in the traps of carnivorous plants. *New Phytologist*, **101**, 585–93.

Johnson M, Wall SBV and Borchert M 2003a. A comparative analysis of seed and cone characteristics and seed-dispersal strategies of three pines in the subsection Sabinianae. *Plant Ecology*, **168**, 69–84.

Johnson SD 2000. Batesian mimicry in the non-rewarding orchid *Disa pulchra*, and its consequences for pollinator behaviour. *Biological Journal of the Linnean Society*, **71**, 119–32.

Johnson SD 2006. Pollinator-driven speciation in plants. In: Harder LD and Barrett SCH (eds) *Ecology and evolution of flowers*. Oxford University Press, Oxford.

Johnson SD, Hargreaves AL and Brown M 2006. Dark, bitter-tasting nectar functions as a filter of flower visitors in a bird-pollinated plant. *Ecology*, **87**, 2709–16.

Johnson SD and Midgley JJ 1997. Fly pollination of *Gorteria diffusa* (Asteraceae), and a possible mimetic function for dark spots on the capitulum. *American Journal of Botany*, **84**, 429–36.

Johnson SD, Peter CI, Nilsson LA and Agren J 2003b. Pollination success in a deceptive orchid is enhanced by co-occurring rewarding magnet plants. *Ecology*, **84**, 2919–27.

Johnson SD and Steiner KE 2000. Generalization versus specialization in plant pollination systems. *Trends in Ecology and Evolution*, **15**, 140–3.

Jones CD, Osorio D and Baddeley RJ 2001. Colour categorization by domestic chicks. *Proceedings Royal Society B*, **268**, 2077–84.

Jordano P 1987. Avian fruit removal: effects of fruit variation, crop size, and insect damage. *Ecology*, **68**, 1711–23.

Jordano P 1988. Diet, fruit choice and variation in body condition of frugivorous warblers in Mediterranean scrubland. *Ardea*, **76**, 193–209.

Jordano P 1995*a*. Angiosperm fleshy fruits and seed dispersers: a comparative analysis of adaptation and constraints in plant–animal interactions. *American Naturalist*, **145**, 163–91.

Jordano P 1995*b*. Frugivore-mediated selection on fruit and seed size: birds and St. Lucie's Cherry, *Prunus mahaleb*. *Ecology*, **76**, 2627–39.

Jordano P, Garcia C, Godoy JA and Garcia-Castano JL 2007. Differential contribution of frugivores to complex seed dispersal patterns. *Proceedings of the National Academy of Sciences of the United States of America*, **104**, 3278–82.

Jorgensen EE 2001. Emission of volatile compounds by seeds under different environmental conditions. *American Midland Naturalist*, **145**, 419–22.

Jorgensen TH and Andersson S 2005. Evolution and maintenance of pollen-colour dimorphisms in *Nigella degenii*: habitat-correlated variation and morph-by-environment interactions. *New Phytologist*, **168**, 487–98.

Jorgensen TH, Richardson DS and Andersson S 2006. Comparative analyses of population structure in two subspecies of *Nigella degenii*: evidence for diversifying selection on pollen-color dimorphisms. *Evolution*, **60**, 518–28.

Joron M 2008. Batesian mimicry: Can a leopard change its spots – and get them back? *Current Biology*, **18**, R476–9.

Juniper BE, Robins RJ and Joel DM 1989. *The carnivorous plants*. Academic Press, London.

Junker RR and Blüthgen N 2008. Signals that attract mutualists but repel enemies: floral scents as defense against ants. *Evolutionary Ecology Research*, **10**, 295–308.

Junker RR, Höcherl N and Blüthgen N 2010. Responses to olfactory signals reflect network structure of flower – visitor interactions. *Journal of Animal Ecology*, **79**, 818–23.

Jürgens A, Dötterl S and Meve U 2006. The chemical nature of fetid floral odours in stapeliads (Apocynaceae – Asclepiadoideae – Ceropegieae). *New Phytologist*, **172**, 452–68.

Jürgens A, El-Sayed AM and Suckling DM 2009. Do carnivorous plants use volatiles for attracting prey insects? *Functional Ecology*, **23**, 875–87.

Kahl J, Siemens DH, Aerts RJ, Gabler R, Kuhnemann F, Preston CA and Baldwin IT 2000. Herbivore-induced ethylene suppresses a direct defense but not a putative indirect defense against an adapted herbivore. *Planta*, **210**, 336–42.

Kaiser R 2006. Flowers and fungi use scents to mimic each other. *Science*, **311**, 806–807.

Kaissling K-E and Priesner E 1970. Die Riechschwelle des Seidenspinners. *Naturwissenschaften*, **57**, 23–8.

Kalko EKV, Herre EA and Handley COJ 1996. Relation of fig fruit characteristics to fruit-eating bats in the New and Old World tropics. *Journal of Biogeography*, **23**, 565–76.

Karagatzides JD and Ellison AM 2009. Construction costs, payback times, and the leaf economics of carnivorous plants. *American Journal of Botany*, **96**, 1612–9.

Karageorgou P and Manetas Y 2006. The importance of being red when young: anthocyanins and the protection of young leaves of *Quercus coccifera* from insect herbivory and excess light. *Tree Physiology*, **26**, 613–21.

Karban R, Huntzinger M and McCall AC 2004. The specificity of eavesdropping on sagebrush by other plants. *Ecology*, **85**, 1846–52.

Karban R, Maron J, Felton GW, Ervin G and Eichenseer H 2003. Herbivore damage to sagebrush induces resistance in wild tobacco: evidence for eavesdropping between plants. *Oikos*, **100**, 325–32.

Karban R, Shiojiri K, Huntzinger M and McCall AC 2006. Damage-induced resistance in sagebrush: Volatiles are key to intra- and interplant communication. *Ecology*, **87**, 922–30.

Karlsson PS, Nordell KO, Eirefelt S and Svensson A 1987. Trapping efficiency of three carnivorous Pinguicula species. *Oecologia*, **73**, 518–21.

Karlsson PS, Thoren LM and Hanslin HM 1994. Prey capture by three *Pinguicula* species in a sub-arctic environment. *Oecologia*, **99**, 188–93.

Kato M 1993. Floral biology of *Nepenthes gracilis* (Nepenthaceae) in Sumatra. *American Journal of Botany*, **80**, 924–7.

Keasar T, Sadeh A, Gerchman Y and Shmida A 2009. The signaling function of an extra-floral display: what selects for signal development? *Oikos*, **118**, 1752–9.

Kelber A 2005. Alternative use of chromatic and achromatic cues in a hawkmoth. *Proceedings Royal Society B,* **272**, 2143–7.

Kelber A, Vorobyev M and Osorio D 2003. Animal colour vision – behavioural tests and physiological concepts. *Biological Reviews*, **78**, 81–118.

Kelling FJ, Ialenti F and den Otter CJ 2002. Background odour induces adaptation and sensitization of olfactory receptors in the antennae of houseflies. *Medical and Veterinary Entomology*, **16**, 161–9.

Kerner A 1895. *The natural history of plants: their forms, growth, reproduction and distribution.* Holt, London.

Kessler A and Baldwin IT 2001. Defensive function of herbivore-induced plant volatile emissions in nature. *Science*, **291**, 2141–4.

Kessler A and Halitschke R 2009. Testing the potential for conflicting seleciton on floral chemical traits by pollinators and herbivores: predictions and a case study. *Functional Ecology*, **23**, 901–12.

Kessler A, Halitschke R, Diezel C and Baldwin IT 2006. Priming of plant defense responses in nature by airborne signaling between *Artemisia tridentata* and *Nicotiana attenuata*. *Oecologia*, **148**, 280–92.

Kessler D, Diezel C and Baldwin IT 2010. Changing pollinators as a means of escaping herbivores. *Current Biology*, **20**, 237–42.

Khan ZR, AmpongNyarko K, Chiliswa P, Hassanali A, Kimani S, Lwande W, Overholt WA, Pickett JA, Smart LE, Wadhams LJ and Woodcock CM 1997. Intercropping increases parasitism of pests. *Nature*, **388**, 631–2.

Khan ZR, Pickett JA, van den Berg J, Wadhams LJ and Woodcock CM 2000. Exploiting chemical ecology and species diversity: stem borer and striga control for maize and sorghum in Africa. *Pest Management Science*, **56**, 957–62.

Klooster MR, Clark DL and Culley TM 2009. Cryptic bracts facilitate herbivore avoidance in the mycoheterotrophic plant *Monotropis odorata* (Ericaceae). *American Journal of Botany*, **96**, 2197–205.

Kneidel KA 1984. Influence of carcass taxon and size on species composition of carrion-breeding diptera. *American Midland Naturalist*, **111**, 57–63.

Knight RS and Siegfried WR 1983. Inter-relationships between type, size and color of fruits and dispersal in Southern African trees. *Oecologia*, **56**, 405–12.

Knudsen GK, Bengtsson M, Kobro S, Jaastad G, Hofsvang T and Witzgall P 2008. Discrepancy in laboratory and field attraction of apple fruit moth *Argyresthia conjugella* to host plant volatiles. *Physiological Entomology*, **33**, 1–6.

Knudsen JT, Eriksson R, Gershenzon J and Ståhl B 2006. Diversity and distribution of floral scent. *Botanical Review*, **72**, 1–120.

Kobayashi Y and Yamamura N 2003. Evolution of signal emission by non-infested plants growing near infested plants to avoid future risk. *Journal of Theoretical Biology*, **223**, 489–503.

Kobayashi Y and Yamamura N 2007. Evolution of signal emission by uninfested plants to help nearby infested relatives. *Evolutionary Ecology*, **21**, 281–94.

Kobayashi Y, Yamamura N and Sabelis MW 2006. Evolution of talking plants in a tritrophic context: Conditions for uninfested plants to attract predators prior to herbivore attack. *Journal of Theoretical Biology*, **243**, 361–74.

Koenig WD and Benedict LS 2002. Size, insect parasitism, and energetic value of acorns stored by acorn woodpeckers. *Condor*, **104**, 539–47.

Koenig WD, Knops JMH, L. DJ and Zuckerberg B 2009. Latitudinal decrease in acorn size in bur oak (*Quercus macrocarpa*) is due to environmental constraints, not avian dispersal. *Botany*, **87**, 349–56.

Kölreuter JG 1761. *Vorläufige Nachrichten von einigen das Geschlecht der Pflanzen betreffenden Versuchen und Beobachtungen*. Gleditschischen Handlung, Leipzig.

Komatsu H 1998. The physiological substrates of colour constancy. In: Walsh V and Kulikowski J (eds) *Perceptual Constancy: why things look as they do*. Cambridge University Press, Cambridge.

Kong J-M, Chia L-S, Goh N-K, Chia T-F and Brouillard R 2003. Analysis and biological activities of anthocyanins. *Phytochemistry*, **64**, 923–33.

Koricheva J 2002. Meta-analysis of sources of variation in fitness costs of plant antiherbivore defenses. *Ecology*, **83**, 176–90.

Korine C and Kalko EKV 2005. Fruit detection and discrimination by small fruit-eating bats (Phyllostomidae): echolocation call design and olfaction. *Behavioral Ecology and Sociobiology*, **59**, 12–23.

Korine C, Kalko EKV and Herre EA 2000. Fruit characteristics and factors affecting fruit removal in a Panamanian community od strangler figs. *Oecologia*, **123**, 560–8.

Kubitzki K and Ziburski A 1994. Seed dispersal in floodplain forests of Amazonia. *Biotropica*, **26**, 30–43.

Kugler H 1956. Über die optische Wirkung von Fliegenblumen auf Fliegen. *Berichte der deutschen botanischen Gesellschaft*, **69**, 387–98.

Kunze J and Gumbert A 2001. The combined effect of color and odor on flower choice behavior of bumble bees in flower mimicry systems. *Behavioral Ecology*, **12**, 447–56.

Kursar TA and Coley PD 1992. Delayed greening in tropical leaves – an antiherbivore defense. *Biotropica*, **24**, 256–62.

Kytridis VP, Karageorgou P, Levizou E and Manetas Y 2008. Intra-species variation in transient accumulation of leaf anthocyanins in *Cistus creticus* during winter: evidence that anthocyanins may compensate for an inherent photosynthetic and photoprotective inferiority of the red-leaf phenotype. *Journal of Plant Physiology*, **165**, 952–9.

Kytridis VP and Manetas Y 2006. Mesophyll versus epidermal anthocyanins as potential in vivo antioxidants: evidence linking the putative antioxidant role to the proximity of oxy-radical source. *Journal of Experimental Botany*, **57**, 2203–10.

Lachmann M, Szamado S and Bergstrom CT 2001. Cost and conflict in animal signals and human language. *Proceedings of the National Academy of Sciences of the United States of America*, **98**, 13189–94.

Lancaster JE, Lister CE, Reay PF and Triggs CM 1997. Influence of pigment composition on skin color in a wide range of fruit and vegetables. *Journal American Society for Horticulture Science*, **122**, 594–8.

Landolt PJ, Brumley JA, Smithhisler CL, Biddick LL and Hofstetter RW 2000. Apple fruit infested with codling moth are more attractive to neonate codling moth larvae and possess increased amounts of (E,E)-alpha-farnesene. *Journal of Chemical Ecology*, **26**, 1685–99.

Lankau RA and Strauss SY 2007. Mutual feedbacks maintain both genetic and species diversity in a plant community. *Science*, **317**, 1561–3.

Lanner RM 1988. Dependence of Great-Basin bristlecone pine on clark nutcracker for regeneration at high elevations. *Arctic and Alpine Research*, **20**, 358–62.

Launchbaugh KL and Provenza FD 1993. Can plants practice mimicry to avoid grazing by mammalian herbivores? *Oikos*, **66**, 501–504.

Leal IR and Oliveira PS 1998. Interactions between fungus-growing ants (Attini), fruits and seeds in cerrado vegetation in southeast Brazil. *Biotropica*, **30**, 170–8.

Lee D 2007. *Nature's Palette: the Science of Plant Color*. University of Chicago Press, Chicago.

Lee DW and Gould KS 2002. Anthocyanins in leaves and other vegetative organs: an introduction. In: Gould KS and Lee DW (eds) *Advances in Botanical Research*, 37, 2–17.

Lee DW, O'Keefe J, Holbrook NM and Feild TS 2003. Pigment dynamics and autumn leaf senescence in a New England deciduous forest, eastern USA. *Ecological Research*, **18**, 677–94.

Lee WG, Weatherall IL and Wilson JB 1994. Fruit conspicuousness in some New Zealand Coprosma (Rubiaceae) species. *Oikos*, **69**, 87–94.

Leimu R and Koricheva J 2006. A meta-analysis of genetic correlations between plant resistances to multiple enemies. *American Naturalist*, **168**, E15–37.

Leishman MR, Wright IJ, Moles AT and Westoby M 2000. The evolutionary ecology of seed size. In: Fenner M (ed) *Seeds: the ecology of regeneration in plant communities*. CABI International, Wallingford, UK.

Lepczyk CA, Murray KG, Winnett-Murray K, Bartell P, Geyer E and Work T 2000. Seasonal fruit preferences for lipids and sugars by American robins. *Auk*, **117**, 709–17.

Lev-Yadun S 2001. Aposematic (warning) coloration associated with thorns in higher plants. *Journal of Theoretical Biology*, **210**, 385–8.

Lev-Yadun S 2003a. Why do some thorny plants resemble green zebras? *Journal of Theoretical Biology*, **224**, 483–9.

Lev-Yadun S 2003b. Weapon (thorn) automimicry and mimicry of aposematic colorful thorns in plants. *Journal of Theoretical Biology*, **224**, 183–8.

Lev-Yadun S 2006. Defensive functions of white coloration in coastal and dune plants. *Israel Journal of Plant Sciences*, **54**, 317–25.

Lev-Yadun S 2009. Aposematic (warning) coloration in plants. In: Baluska F (ed) *Plant–environment interactions: from sensory plant biology to active plant behavior*, Springer, Berlin, 167–202.

Lev-Yadun S, Dafni A, Flaishman MA, Inbar M, Izhaki I, Katzir G and Ne'eman G 2004. Plant coloration undermines herbivorous insect camouflage. *Bioessays*, **26**, 1126–30.

Lev-Yadun S and Gould KS 2007. What do red and yellow autumn leaves signal? *Botanical Review*, **73**, 279–89.

Lev-Yadun S and Halpern M 2007. Ergot (*Claviceps purpurea*)–An aposematic fungus. *Symbiosis*, **43**, 105–108.

Lev-Yadun S and Inbar M 2002. Defensive ant, aphid and caterpillar mimicry in plants? *Biological Journal of the Linnean Society*, **77**, 393–8.

Lev-Yadun S and Mirsky N 2007. False satiation: the probable antiherbivory strategy of *Hoodia gordonii*. *Functional Plant Science and Biotechnology*, **1**, 56–7.

Lev-Yadun S and Ne'eman G 2004. When may green plants be aposematic? *Biological Journal of the Linnean Society*, **81**, 413–6.

Lev-Yadun S and Ne'eman G 2006. Color changes in old aposematic thorns, spines, and prickles. *Israel Journal of Plant Sciences*, **54**, 327–33.

Lev-Yadun S, Ne'eman G and Shanas U 2009. A sheep in wolf's clothing: do carrion and dung odours of flowers not only attract pollinators but also deter herbivores? *Bioessays*, **31**, 84–8.

Levey DJ 1987a. Seed size and fruit-handling techniques of avian frugivores. *American Naturalist*, **129**, 471–85.

Levey DJ 1987b. Sugar-tasting ability and fruit selection in tropical fruit-eating birds. *Auk*, **104**, 173–9.

Levey DJ, Bissell HA and O'Keefe SF 2000. Conversion of nitrogen to protein and amino acids in wild fruits. *Journal of Chemical Ecology*, **26**, 1749–63.

Levey DJ and Byrne MM 1993. Complex ant plant interactions: rain-forest ants as secondary dispersers and post-dispersal seed predators. *Ecology*, **74**, 1802–12.

Levey DJ, Moermond TC and Denslow JS 1984. Fruit choice in Neotropical birds: the effect of distance between fruits on preference patterns. *Ecology*, **65**, 844–50.

Levin RA, McDade LA and Raguso RA 2003. The systematic utility of floral and vegetative fragrance in two genera of Nyctaginaceae. *System Biology*, **52**, 334–51.

Leyva A, Jarillo JA, Salinas J and Martinezzapater JM 1995. Low-temperature induces the accumulation of phenylalanine ammonia lyase and chalcone synthase messenger RNAs of *Arabidopsis thaliana* in a light-dependent manner. *Plant Physiology*, **108**, 39–46.

Lindauer M 1975. *Verständigung im Bienenstaat*. Gustav Fischer, Stuttgart.

Livingstone M and Hubel D 1988. Segregation of form, colour, movement, and depth: anatomy, physiology, and perception. *Science*, **240**, 740–9.

Logan BA, Barker DH, DemmigAdams B and Adams WW 1996. Acclimation of leaf carotenoid composition and ascorbate levels to gradients in the light environment within an Australian rainforest. *Plant Cell and Environment*, **19**, 1083–90.

Lomáscolo SB and Schaefer HM 2010. Signal convergence in fruits: a result of selection by frugivores? *Journal of Evolutionary Biology*, **23**, 614–24.

Lomáscolo, S.B., Levey, D.J., Kimball, R.T., Bolker, B.M. & Alborn, H.T. 2010. Dispersers shape fruit diversity in *Ficus* (Moraceae). *Proc. Natl. Acad. Sci. U. S. A.* **107**: 14668–14672.

Lomáscolo SB, Speranza P and Kimball RT 2008. Correlated evolution of fig size and color supports the dispersal syndrome hypothesis. *Oecologia*, **156**, 783–96.

Lord JM and Marshall J 2001. Correlations between growth form, habitat, and fruit colour in the New Zealand flora, with reference to frugivory by lizards. *New Zealand Journal of Botany*, **39**, 567–76.

Loreto F, Pinelli P, Manes F and Kollist H 2004. Impact of ozone on monterpene emissions and evidence for an isoprene-like antioxidant action of monoterpenes emitted by *Quercus ilex* leaves. *Tree Physiology*, **24**, 361–7.

Luckenbach MW and Orth RJ 1999. Effects of a deposit-feeding invertebrate on the entrapment of *Zostera marina* L. seeds. *Aquatic Botany*, **62**, 235–47.

Lunau K 1992. A new interpretation of flower guide colouration: absorption of ultraviolet light enhances colour saturation. *Plant Systematics and Evolution*, **183**, 51–65.

Lunau K 2000. The ecology and evolution of visual pollen signals. *Plant Systematics and Evolution*, **222**, 89–111.

Lunau K, Wacht S and Chittka L 1996. Colour choices of naive bumble bees and their implications for colour perception. *Journal of Comparative Physiology A*, **178**, 477–89.

Lynn SK, Cnaani J and Papaj DR 2005. Peak shift discrimination learning as a mechanism of signal evolution. *Evolution*, **59**, 1300–305.

Mack AL 1990. Is frugivory limited by secondary compounds in fruits? *Oikos*, **57**, 135–8.

Mack AL 2000. Did fleshy fruit pulp evolve as a defence against seed loss rather than as a dispersal mechanism? *Journal of Biosciences*, **25**, 93–7.

Maier EJ and Bowmaker JK 1993. Colour vision in the passeriform bird, Leiothrix lutea: correlation of visual pigment absorbance and oil droplet transmission with spectral sensitivity. *Journal of Comparative Physiology A*, **172**, 295–301.

Majetic CJ, Raguso RA and Ashman T-L 2009. The sweet smell of success: floral scent affects pollinator attraction and seed fitness in *Hesperis matronalis*. *Functional Ecology*, **23**, 480–7.

Male LH and Smulders TV 2007. Hyperdispersed cache distributions reduce pilferage: a field study. *Animal Behaviour*, **73**, 717–26.

Malo JE and Suárez F 1995. Herbivorous mammals as seed dispersers in a Mediterranean Dehesa. *Oecologia*, **104**, 246–55.

Mancinelli AL 1983. The photoregulation of anthocyanin synthesis. In: Shropshire Jr W and Mohr H (eds) *Photomorphogenesis*. Springer Verlag, Berlin.

Manetas Y 2006. Why some leaves are anthocyanic and why most anthocyanic leaves are red? *Flora*, **201**, 163–77.

Manning K 1998. Isolation of a set of ripening-related genes from strawberry: their identification and possible relationsip to fruit quality traits. *Planta*, **205**, 622–31.

Mant J, Peakall R and Schiestl FP 2005. Does selection on floral odor promote differentiation among populations and species of the sexually deceptive orchid genus *Ophrys*? *Evolution*, **59**, 1449–63.

Mantyla E, Alessio GA, Blande JD, Heijari J, Holopainen JK, Laaksonen T, Piirtola P and Klemola T 2008. From plants to birds: Higher avian predation rates in trees responding to insect herbivory. *PloS ONE*, **3**, e2832.

Mantyla E, Klemola T and Haukioja E 2004. Attraction of willow warblers to sawfly-damaged mountain birches: novel function of inducible plant defences? *Ecology Letters*, **7**, 915–8.

Manzano P and Malo JE 2006. Extreme long-distance seed dispersal via sheep. *Frontiers in Ecology and the Environment*, **4**, 244–8.

Marino P, Raguso R et al. 2009. The ecology and evolution of fly dispersed dung mosses (family Splachnaceae): manipulating insect behaviour through odour and visual cues. *Symbiosis*, **47**, 61–76.

Mark S and Olesen JM 1996. Importance of elaiosome size to removal of ant-dispersed seeds. *Oecologia*, **107**, 95–101.

Marler P 1977. The evolution of communication. In: Sebeok TA (ed) *Animals communicate*, Indiana University Press, Bloomington, IN.

Marshall J, Cronin TW and Kleinlogel S 2007. Stomatopod eye structure and function: a review. *Arthropod Structure and Development*, **36**, 420–48.

Martin DM, Toub O, Chiang A, Lo BC, Ohse S, Lund ST and Böhlmann J 2009. The bouquet of grapevine (*Vitis vinifera* L. cv. Cabernet Sauvignon) flowers arise from the biosynthesis of sequiterpene volatiles in pollen grains. *Proceedings National Academy of Sciences*, **106**, 7245–50.

Martin NH 2004. Flower size preferences of the honeybee (*Apis mellifera*) foraging on Mimulus guttatus (Scorphulariaceae). *Evolutionary Ecology Research*, **6**, 777–82.

Martinez del Rio C and Stevens BR 1989. Physiological constraint on feeding behavior: intestinal membrane disaccharidases of the starling. *Science*, **243**, 794–6.

Massei G, Cotterill JV, Coats JC, Bryning G and Cowan DP 2007. Can Batesian mimicry help plants to deter herbivores? *Pest Management Science*, **63**, 559–63.

Matsuura K, Yashiro T, Shimizu K, Tatsumi S and Tamura T 2009. Cuckoo fungus mimics termite eggs by producing the cellulose-digesting enzyme beta-glucosidase. *Current Biology*, **19**, 30–6.

Mauricio R 2000. Natural selection and the joint evolution of toleranceand resistance as plant defenses. *Evolutionary Ecology*, **14**, 491–507.

Mayer V, Schaber D and Hadacek F 2008. Volatiles of myrmecophytic Piper plants signal stem tissue damage to inhabiting Pheidole ant-partners. *Journal of Ecology*, **96**, 962–70.

Maynard-Smith J and Harper DGC 1995. Animal signals: models and terminology. *Journal of Theoretical Biology*, **177**, 305–11.

Maynard Smith J and Harper D 2003. *Animal Signals*. Oxford University Press, Oxford.

Mazeh S, Korine C, Pinshow B and Dudley R 2008. The influence of ethanol on feeding in the frugivorous yellow-vented bulbul (*Pycnonotus xanthopygos*). *Behavioural Processes*, **77**, 369–75.

McAdoo JK, Evans CC, Roundy BA, Young JA and Evans RA 1983. Influence of heteromyid rodents on *Oryzopsis hymenoides* germination. *Journal of Range Management*, **36**, 61–4.

McCall AC and Irwin RE 2006. Florivory: the intersection of pollination and herbivory. *Ecology Letters*, **9**, 1351–65.

McEwen J and Vamosi JC 2010. Floral colour versus phylogeny in structuring subalpine flowering communities. *Proceedings Royal Society B*, **277**, 2957–65.

McKey D 1975. The ecology of coevolved seed dispersal systems. In: Gilbert LE and Raven PH (eds) *Coevolution of Animals and Plants*. University of Texas Press, Austin.

Medel R, Botto-Mahan C and Kalin-Arroyo M 2003. Pollinator-mediated selection on the nectar guide phenotype in the andean monkey flower, *Mimulus luteus*. *Ecology*, **84**, 1721–32.

Medel R, Valiente A, Botto-Mahan C, Carvallo G, Pérez F, Pohl N and Navarro L 2007. The influence of insects and hummingbirds on the geographical variation of the flower phenotype in *Mimulus luteus*. *Ecography*, **30**, 812–8.

Medel R, Vergara E, Silva A and Kalin-Arroyo M 2004. Effects of vector behavior and host resistance on mistletoe aggregation. *Ecology*, **85**, 120–6.

Meehan TD, Lease HM and Wolf BO 2005. Negative indirect effects of an avian insectivore on the fruit set of an insect-pollinated herb. *Oikos*, **109**, 297–304.

Menzel R 1985. Neurobiology of learning and memory in honeybees. In: Goodman LJ and Fisher RC (eds) *The behaviour and physiology of bees*. CAB International, Wallingford, UK.

Menzel R 1999. Memory dynamics in the honeybee. *Journal of Comparative Physiology A*, **185**, 323–40.

Merbach MA, Merbach DJ, Maschwitz U, Booth WE, Fiala B and Zizka G 2002. Carnivorous plants–Mass march of termites into the deadly trap. *Nature*, **415**, 36–7.

Merbach MA, Zizka G, Fiala B, Maschwitz U and Booth WE 2001. Patterns of nectar secretion in five Nepenthes species from Brunei Darussalam, Northwest Borneo, and implications for ant–plant relationships. *Flora*, **196**, 153–60.

Merilaita S, Lyytinen A and Mappes J 2001. Selection for cryptic coloration in a visually heterogeneous habitat. *Proceedings of the Royal Society B*, **268**, 1925–9.

Merzlyak MN, Gitelson AA, Chivkunova OB and Rakitin VY 1999. Non-destructive optical detection of pigment changes during leaf senescence and fruit ripening. *Physiologia Plantarum*, **106**, 135–41.

Mesler MR and Lu KL 1983. Seed dispersal of Trillium ovatum (Lilliaceae) in second-growth redwood forest. *American Journal of Botany*, **70**, 1460–7.

Meyers DG and Strickler JR 1979. Capture enhancement in a carnivorous aquatic plant: function of antennae and bristles in Utricularia vulgaris. *Science*, **203**, 1022–5.

Midgley JJ 2004. Why are Acacia thorns white? *African Journal of Range and Forage Science*, **21**, 211–2.

Minguez-Mosquera MI and Garrido-Fernández J 1989. Chlorophyll and carotenoid presence in olive fruit (*Olea europaea*). *Journal of Agricultural and Food Chemistry*, **37**, 1–7.

Moegenburg SM and Levey DJ 2003. Do frugivores respond to fruit harvest? An experimental study of short-term responses. *Ecology*, **84**, 2600–12.

Moehs CP, Tian L, Osteryoung KW and DellaPenna D 2001. Analysis of carotenoid biosynthetic gene expression during marigold petal development. *Plant Molecular Biology*, **45**, 281–93.

Moermond TC and Denslow JS 1983. Fruit choice in Neotropical birds: effects of fruit type and accessibility on selectively. *Journal of Animal Ecology*, **52**, 407–20.

Mollon JD 1989. 'Tho' she kneel'd in that place where they grew': the uses and origin of primate colour vison. *Journal of Experimental Biology*, **146**, 21–38.

Montoya D, Zavala MA, Rodriguez MA and Purves DW 2008. Animal versus wind dispersal and the robustness of tree species to deforestation. *Science*, **320**, 1502–1504.

Moore PD 1991. Prey for the plundering. *Nature*, **350**, 192.

Moore PD 1995. Location is a sticky business. *Nature*, **387**, 442.

Morales JM and Carlo TS 2006. The effects of plant distribution and frugivore density on the scale and shape of dispersal kernels. *Ecology*, **87**, 1489–96.

Moran JA 1996. Pitcher dimorphism, prey composition and the mechanisms of prey attraction in the pitcher plant *Nepenthes rafflesiana* in Borneo. *Journal of Ecology*, **84**, 515–25.

Moran JA, Booth WE and Charles JK 1999. Aspects of pitcher morphology and spectral characteristics of six bornean Nepenthes pitcher plant species: implications for prey capture. *Annals of Botany*, **83**, 521–8.

Morden-Moore AL and Willson MF 1982. On the ecological significance of fruit color in *Prunus serotina* and *Rubus occidentalis*: field experiments. *Canadian Journal of Botany*, **60**, 1554–60.

Mothershead K and Marquis RJ 2000. Fitness impacts of herbivory through indirect effcts on plant–pollinator interactions in *Oenothera macrocarpa*. *Ecology*, **81**, 30–40.

Mount A and Pickering CM 2009. Testing the capacity of clothing to act as a vector for non-native seed in protected areas. *Journal of Environmental Management*, **91**, 168–79.

Mouquet N, Daufresne T, Gray SM and Miller TE 2008. Modelling the relationship between a pitcher plant (*Sarracenia purpurea*) and its phytotelma community: mutualism or parasitism? *Functional Ecology*, **22**, 728–37.

Moya S and Ackerman JD 1993. Variation in the floral fragrance of Epidendrum ciliare (Orchidaceae). *Nordic Journal of Botany*, **13**, 41–7.

Muchhala N 2007. Adaptive trade-off in floral morphology mediates specialization for flowers pollinated by bats and hummingbirds. *American Naturalist*, **169**, 494–504.

Mumm R and Hilker M 2005. The significance of background odour for an egg parasitoid to detect plants with host eggs. *Chemical Senses*, **30**, 337–43.

Muñoz AM and Arroyo MTK 2004. Negative impacts of a vertebrate predator on insect pollinator visitation and seed output in *Chuquiraga oppositifolia*, a high Andean shrub. *Oecologia*, **138**, 66–73.

Murza GL, Heaver JR and Davis AR 2006. Minor pollinator–prey conflict in the carnivorous plant, Drosera anglica. *Plant Ecology*, **184**, 43–52.

Musser RO, Hum-Musser SM, Eichenseer H, Peiffer M, Ervin G, Murphy JB and Felton GW 2002. Herbivory: caterpillar saliva beats plant defences – A new weapon emerges in the evolutionary arms race between plants and herbivores. *Nature*, **416**, 599–600.

Nagel T 1974. What is it like to be a bat? *The Philosophical Review*, **83**, 435–50.

Nakagawa T, Sakurai T, Nishioka T and Touhara K 2005. Insect sex-pheromone signals mediated by specific combinations of olfactory receptors. *Science*, **307**, 1638–42.

Nakanishi H 1994. Myrmecochorous adaptations of *Corydalis* species (Papaveraceae) in southern Japan. *Ecological Research*, **9**, 1–8.

Nathan R 2006. Long-distance dispersal of plants. *Science*, **313**, 786–8.

Naug D and Arathi HS 2007. Receiver bias for exaggerated signals in honeybees and its implications for the evolution of floral displays. *Biology Letters*, **3**, 635–7.

Ness JH and Bressmer K 2005. Abiotic influences on the behaviour of rodents, ants, and plants affect an ant–seed mutualism. *Ecoscience*, **12**, 76–81.

Neumeyer C 1998. Comparative colour constancy. In: Walsh V and Kulikowski J (eds) *Perceptual constancy: why things look as they do*. Cambridge University Press, Cambridge.

Newell SJ and Nastase AJ 1998. Efficiency of insect capture by *Sarracenia purpurea* (Sarraceniaceae), the northern pitcher plant. *American Journal of Botany*, **85**, 88–91.

Ngugi HK and Scherm H 2006. Biology of flower-infecting fungi. *Annual Review of Phytopathology*, **44**, 261–82.

Niinemets U and Tamm U 2005. Species differences in timing of leaf fall and foliage chemistry modify nutrient resorption efficiency in deciduous temperate forest stands. *Tree Physiology*, **25**, 1001–14.

Nikiforou C, Zeliou K, Kytridis VP, Kyzeridou A and Manetas Y 2010. Are red leaf phenotypes more or less fit? The case of winter leaf reddening in *Cistus creticus*. *Environmental and Experimental Botany*, **67**, 509–14.

Nilsson LA 1992. Orchid pollination biology. *Trends in Ecology and Evolution*, **7**, 255–9.

Nilsson LA, Jonsson L and Ralison L 1987. Angrecoid orchids and hawkmoths in central Madagascar: specialized pollination systems and generalist foragers. *Biotropica*, **19**, 310–18.

Nobel PS and delaBarrera E 2000. Carbon and water balances for young fruits of platyopuntias. *Physiologia Plantarum*, **109**, 160–6.

Nogales M, Padilla DP, Nieves C, Illera JC and Traveset A 2007. Secondary seed dispersal systems, frugivorous lizards and predatory birds in insular volcanic badlands. *Journal of Ecology*, **95**, 1394–403.

Novoplansky A 2009. Picking battles wisely: plant behaviour under competition. *Plant Cell and Environment*, **32**, 726–41.

Numata S, Kachi N, Okuda T and Manokaran N 2004. Delayed greening, leaf expansion, and damage to sympatric Shorea species in a lowland rain forest. *Journal of Plant Research*, **117**, 19–25.

Nuñez-Farfán J, Fornoni J and Valverde PL 2007. The evolution of resistance and tolerance to herbivores. *Annual Review of Ecology Evolution and Systematics*, **38**, 541–66.

Nystrand O and Granstrom A 1997. Post-dispersal predation on *Pinus sylvestris* seeds by Fringilla spp: Ground substrate affects selection for seed color. *Oecologia*, **110**, 353–9.

Olesen JM, Bascompte J, Dupont YL and Jordano P 2007b. The modularity of pollination networks. *Proceedings of the National Academy of Sciences of the United States of America*, **104**, 19891–6.

Olesen JM, Dupont YL, Ehlers BK and Hansen DM 2007a. The openness of a flower and its number of flower-visitor species. *Taxon*, **56**, 729–36.

Olesen JM, Rønsted N, Tolderlund U, Cornett C, Mølgaard P, Madsen J and Olsen CE 1998. Mauritian red nectar remains a mystery. *Nature*, **393**, 529.

Olesen JM and Valido A 2003. Lizards as pollinators and seed dispersers: an island phenomenon. *Trends in Ecology and Evolution*, **18**, 177–81.

Ono K, Terashima I and Watanabe A 1996. Interaction between nitrogen deficit of a plant and nitrogen content in the old leaves. *Plant and Cell Physiology*, **37**, 1083–9.

Ordano M, Fornoni J, Boege K and Domínguez CA 2008. The adaptive value of phenotypic floral integration. *New Phytologist*, **179**, 1183–92.

Osche G 1979. Zur Evolution optischer Signale bei Blütenpflanzen. *Biologie in unserer Zeit*, **9**, 161–70.

Osche G 1983. Optische Signale in der Coevolution von Pflanze und Tier. *Berichte Deutsche Botanische Gesellschaft*, **96**, 1–27.

Osorio D, Miklósi A and Gonda Z 1999. Visual ecology and perception of coloration patterns by domestic chicks. *Evolutionary Ecology*, **13**, 673–89.

Osorio D, Smith AC, Vorobyev M and Buchanan-Smith HM 2004. Detection of fruit and the selection of primate visual pigments for color vision. *American Naturalist*, **164**, 696–708.

Osorio D and Vorobyev M 2008. A review of the evolution of animal colour vision and visual communication signals. *Vision Research*, **48**, 2042–51.

Otte D 1974. Effects and functions in the evolution of signaling systems. *Annual Review of Ecology and Systematics*, **5**, 385–417.

Ougham HJ, Morris P and Thomas H 2005. The colors of autumn leaves as symptoms of cellular recycling and defenses against environmental stresses. *Current Topics in Developmental Biology*, **66**, 135–60.

Page JE and Towers GHN 2002. Anthocyanins protect light-sensitive thiarubrine phototoxins. *Planta*, **215**, 478–84.

Pakeman RJ, Digneffe G and Small JL 2002. Ecological correlates of endozoochory by herbivores. *Functional Ecology*, **16**, 296–304.

Parchman TL, Benkman CW and Britch SC 2006. Patterns of genetic variation in the adaptive radiation of New World crossbills (Aves: Loxia). *Molecular Ecology*, **15**, 1873–87.

Pasteur G 1982. A classificatory review of mimicry systems. *Annual Review of Ecology and Systematics*, **13**, 169–99.

Pavel EW and DeJong TM 1993. Estimating the photosynthetic contribution of developing peach (*Prunus persica*) fruits to their growth and maintenance carbohydrate requirements. *Physiologia Plantarum*, **88**, 331–8.

Pearce JM 2008. *Animal learning and cognition*, 3rd edn. Psychology Press, Hove, UK.

Pearson KM and Theimer TC 2004. Seed-caching responses to substrate and rock cover by two Peromyscus species: implications for pinyon pine establishment. *Oecologia*, **141**, 76–83.

Pellmyr O and Thien LB 1986. Insect reproduction and floral fragrances: keys to the evolution of the angiosperms? *Taxon*, **35**, 76–85.

Peres CA and vanRoosmalen MGM 1996. Avian dispersal of 'mimetic seeds' of *Ormosia lignivalvis* by terrestrial granivores: deception or mutualism? *Oikos*, **75**, 249–58.

Pérez F, Arroyo MTK and Medel R 2007. Phylogenetic analysis of floral integration in *Schizanthus* (Solanaceae):

does pollination truly integrate corolla traits? *Journal of Evolutionary Biology*, **20**, 1730–8.

Peter CI and Johnson SD 2008. Mimics and magnets: The importance of color and ecological facilitation in floral deception. *Ecology*, **89**, 1583–95.

Pfeiffer M, Huttenlocher H and Ayasse M 2010. Myrmecochorous plants use chemical mimicry to cheat seed-dispersing ants. *Functional Ecology*, **24**, 545–55.

Pichaud F, Briscoe AD and Desplan C 1999. Evolution of color vision. *Current Opinion in Neurobiology*, **9**, 622–7.

Pichersky E and Gershenzon J 2002. The formation and function of plant volatiles: perfumes for pollinator attraction and defense. *Current Opinion in Plant Biology*, **5**, 237–43.

Pichersky E, Noel JP and Dudareva N 2006. Biosynthesis of plant volatiles: nature's diversity and ingenuity. *Science*, **311**, 808–11.

Pierik R, Cuppens MLC, Voesenek L and Visser EJW 2004. Interactions between ethylene and gibberellins in phytochrome-mediated shade avoidance responses in tobacco. *Plant Physiology*, **136**, 2928–36.

Pierik R, Visser EJW, De Kroon H and Voesenek L 2003. Ethylene is required in tobacco to successfully compete with proximate neighbours. *Plant Cell and Environment*, **26**, 1229–34.

Pietrini F, Iannelli MA and Massacci A 2002. Anthocyanin accumulation in the illuminated surface of maize leaves enhances protection from photo-inhibitory risks at low temperature, without further limitation to photosynthesis. *Plant Cell and Environment*, **25**, 1251–9.

Pigliucci M 2001. *Phenotypic plasticity: Beyond nature and nurture*. Johns Hopkins University Press, Baltimore, MD.

Pigliucci M, Paoletti C, Fineschi S and Malvolti ME 1991. Phenotypic integration in chestnut (*Castanea sativa* Mill.): leaves versus fruits. *Botanical Gazette*, **152**, 514–21.

Plachno BJ, Kozieradzka-Kiszkurno M and Swiatek P 2007. Functional utrastructure of Genlisea (Lentibulariaceae) digestive hairs. *Annals of Botany*, **100**, 195–203.

Podos J 2001. Correlated evolution of morphology and vocal signal structure in Darwin's finches. *Nature*, **409**, 185–8.

Pollux BJA, Ouborg NJ, Van Groenendael JM and Klaassen M 2007. Consequences of intraspecific seed-size variation in Sparganium emersum for dispersal by fish. *Functional Ecology*, **21**, 1084–91.

Polyak S 1957. *The vertebrate visual system*. University of Chicago Press, Chicago.

Postma E and Noordwijk AJv 2005. Gene flow maintains a large genetic difference in clutch size at small spatial scale. *Nature*, **433**, 65–8.

Potthoff M, Johst K, Gutt J and Wissel C 2006. Clumped dispersal and species coexistence. *Ecological Modelling*, **198**, 247–54.

Poulsen JR, Clark CJ and Smith TB 2001. Seed dispersal by a diurnal primate community in the Dja Reserve, Cameroon. *Journal of Tropical Ecology*, **17**, 787–808.

Prabhu C and Cheng K 2008. One day is all it takes: circadian modulation of the retrieval of colour memories in honeybees. *Behavioral Ecology and Sociobiology*, **63**, 11–22.

Pratt TK and Stiles EW 1983. How long fruit-eating birds stay in the plants where they feed: implications for seed dispersal. *American Naturalist*, **122**, 797–805.

Price PW, Bouton CE, Gross P, McPheron BA, Thompson JN and Weis AE 1980. Interactions among three trophic levels: influence of plants on interactions between insects herbivores and natural enemies. *Annual Review of Ecology and Systematics*, **11**, 41–65.

Provenza FD 1995. Postingestive feedback as an elementary determinant of food preference and intake in ruminants. *Journal of Range Management*, **48**, 2–17.

Puckey HL, Lill A and O'Dowd DJ 1996. Fruit color preferences of captive silvereyes (*Zosterops lateralis*). *Condor*, **98**, 217–26.

Puente M, Magori K, Kennedy GG and Gould F 2008a. Impact of herbivore-induced plant volatiles on parasitoid foraging success: a spatial simulation of the *Cotesia rubecula*, *Pieris rapae*, and *Brassica oleracea* system. *Journal of Chemical Ecology*, **34**, 959–70.

Puente ME, Kennedy GG and Gould F 2008b. The impact of herbivore-induced plant volatiles on parasitoid foraging success: a general deterministic model. *Journal of Chemical Ecology*, **34**, 945–58.

Raguso RA 2004. Why are some floral nectars scented? *Ecology*, **85**, 1486–94.

Raguso RA 2008. Wake up and smell the roses: the ecology and evolution of floral scent. *Annual Review of Ecology and Systematics*, **39**, 549–69.

Raguso RA 2009. Floral scent in a whole-plant context: moving beyond pollinator attraction. *Functional Ecology*, **23**, 837–40.

Raguso RA, Levin RA, Foose SE, Holmberg MW and McDade LA 2003. Fragrance chemistry, nocturnal rhyths and pollination 'syndromes' in *Nicotiana*. *Phytochemistry*, **63**, 265–84.

Raine NE and Chittka L 2007. The adaptive significance of sensory bias in a foraging context: floral colour preferences in the Bumblebee *Bombus terrestris*. *PLoS ONE*, **2**, e556.

Ramirez CC, Lavandero B and Archetti M 2008. Coevolution and the adaptive value of autumn tree colours: colour preference and growth rates of a southern beech aphid. *Journal of Evolutionary Biology*, **21**, 49–56.

Rapparini F, Baraldi R and Facini O (2001) Seasonal variation of monoterpene emisson from *Malus domestica* and *Prunus avium*. *Phytochemistry*, **57**, 681–7.

Rasmann S and Turlings TCJ 2007. Simultaneous feeding by aboveground and belowground herbivores attenuates plant-mediated attraction of their respective natural enemies. *Ecology Letters*, **10**, 926–36.

Ratzka A, Vogel H, Kliebenstein DJ, Mitchell-Olds T and Kroymann J 2002. Disarming the mustard oil bomb. *Proceedings National Academy of Sciences*, **99**, 11223–8.

Rausher MD 2006. The evolution of flavonoids and their genes. In: Grotewold E (ed) *The science of flavonoids*, Springer, Berlin.

Rausher MD 2008. Evolutionary transitions in floral color. *International Journal of Plant Sciences*, **169**, 7–21.

Regan BC, Julliot C, Simmen B, Vienot F, Charles-Dominique P and Mollon JD 2001. Fruits, foliage and the evolution of primate colour vision. *Philosophical Transactions Royal Society London B*, **356**, 229–83.

Reisenman CE, Riffell JA, Bernays EA and Hildebrand JG 2010. Antagonistic effects of floral scent in an insect–plant interaction. *Proceedings Royal Society B*, **277**, 2371–9.

Remis MJ and Kerr ME 2002. Taste responses to fructose and tannic acid among gorillas (*Gorilla gorilla gorilla*). *International Journal of Primatology*, **23**, 251–61.

Rendall D, Owren MJ and Ryan MJ 2009. What do animal signals mean? *Animal Behaviour*, **78**, 233–40.

Renner SS 2005. Relaxed molecular clocks for dating historical plant dispersal events. *Trends in Plant Science*, **10**, 550–8.

Retana J, Pico FX and Rodrigo A 2004. Dual role of harvesting ants as seed predators and dispersers of a non-myrmechorous Mediterranean perennial herb. *Oikos*, **105**, 377–85.

Riba-Hernández P, Stoner KE and Osorio D 2004. Effect of polymorphic colour vision for fruit detection in the spider monkey *Ateles geoffroyi*, and its implications for the maintenance of polymorphic colour vision in platyrrhine monkeys. *Journal of Experimental Biology*, **207**, 2465–70.

Ribeiro da Luz B 2006. Attenuated total reflectance spectroscopy of plant leaves: a tool for ecological and botanical studies. *New Phytologist*, **172**, 305–18.

Richard S, Lapointe G, Rutledge RG and Seguin A 2000. Induction of chalcone synthase expression in white spruce by wounding and jasmonate. *Plant and Cell Physiology*, **41**, 982–7.

Richardson AD, Berlyn GP and Duigan SP 2003. Reflectance of Alaskan black spruce and white spruce foliage in relation to elevation and latitude. *Tree Physiology*, **23**, 537–44.

Riffell JA, Lei H, Christensen TA and Hildebrand JG 2009. Characterization and coding of behaviorally significant odor mixtures. *Current Biology*, **19**, 335–40.

Rissing SW 1986. Indirect effects of granivory by harvester ants: plant species composition and reproductive increase near ant nests. *Oecologia*, **68**, 231–4.

Roberts JA, Elliott KA and Gonzalez-Carranza ZH 2002. Abscission, dehiscence, and other cell separation processes. *Annual Review of Plant Biology*, **53**, 131–58.

Rodd FH, Hughes KA, Grether GF and Baril CT 2002. A possible non-sexual origin of mate preference: are male guppies mimicking fruit? *Proceedings Royal Society B*, **269**, 475–81.

Rodríguez-Gironés MA and Santamaría L 2004. Why are so many bird flowers red? *PLoS Biology*, **2**, e350.

Rodriguez A, Nogales M, Rumeu B and Rodriguez B 2008. Temporal and spatial variation in the diet of the endemic lizard *Gallotia galloti* in an insular Mediterranean scrubland. *Journal of Herpetology*, **42**, 213–22.

Rodriguez E and Levin DA 1976. Biochemical parallelism of repellants and attractants in higher plants and arthropods. In: Wallace JM and Mansell RL (eds) *Biochemical interactions between plants and insects*. Plenum Press, New York.

Rogers SM and Simpson SJ 1997. Experience-dependent changes in the number of chemosensitive sensilla on the mouthparts of *Locusta migratoria*. *Journal of Experimental Biology*, **200**, 2313–21.

Rolshausen G and Schaefer HM 2007. Do aphids paint the tree red (or yellow) – can herbivore resistance or photoprotection explain colourful leaves in autumn? *Plant Ecology*, **191**, 77–84.

Romero GQ and Izzo TJ 2004. Leaf damage induces ant recruitment in the Amazonian ant–plant Hirtella myrmecophila. *Journal of Tropical Ecology*, **20**, 675–82.

Rose USR, Lewis J and Tumlinson JH 2006. Extrafloral nectar from cotton (Gossypium hirsutum) as a food source for parasitic wasps. *Functional Ecology*, **20**, 67–74.

Rosenquist JK and Morrison JC 1989. Some factors affecting cuticle and wax accumulation on grape berries. *American Journal of Enology and Viticulture*, **40**, 241–4.

Roxburgh L and Nicolson SW 2005. Patterns of host use in two African mistletoes: the importance of mistletoe–host compatibility and avian disperser behaviour. *Functional Ecology*, **19**, 865–73.

Roxburgh L and Nicolson SW 2008. Differential dispersal and survival of an African mistletoe: does host size matter? *Plant Ecology*, **195**, 21–31.

Roy BA 1993. Floral mimicry by a plant pathogen. *Nature*, **362**, 56–8.

Roy BA 1994. The effects of pathogen-induced pseudo-flowers and buttercups on each other's insect visitation. *Ecology*, **75**, 352–8.

Roy BA and Raguso RA 1997. Olfactory versus visual cues in a floral mimicry system. *Oecologia*, **109**, 414–26.

Roy BA and Widmer A 1999. Floral mimicry: a fascinating yet poorly understood phenomenon. *Trends in Plant Science*, **4**, 325–30.

Rudall P and Bateman RM 2003. Evolutionary change in flowers and inflorescences: evidence from naturally occurring terata. *Trends in Plant Science*, **8**, 76–82.

Ruhren S and Dudash MR 1996. Consequences of the timing of seed release of *Erythronium americanum* (Liliaceae), a deciduous forest myrmecochore. *American Journal of Botany*, **83**, 633–40.

Runyon JB, Mescher MC and De Moraes CM 2006. Volatile chemical cues guide host location and host selection by parasitic plants. *Science*, **313**, 1964–7.

Russo SE and Augspurger CK 2004. Aggregated seed dispersal by spider monkeys limits recruitment to clumped patterns in *Virola calophylla*. *Ecology Letters*, **7**, 1058–67.

Ruxton GD 2009. Non-visual crypsis: a review of the empirical evidence for camouflage to senses other than vision. *Philosophical Transactions of the Royal Society B, Biological Sciences*, **364**, 549–57.

Ruxton GD, Sherratt TN and Speed MP 2004. *Avoiding attack: the evolutionary ecology of crypsis, warning signals and mimicry*. Oxford University Press, Oxford.

Ryan MJ and Rand A 1993. Sexual selection and signal evolution: the ghost of biases past. *Philosophical Transactions Royal Society London B*, **340**, 187–95.

Sabelis MW and Dejong MCM 1988. Should all plants recruit bodyguard: conditions for a polymorphic ess of synomone production in plants. *Oikos*, **53**, 247–52.

Sadof CS, Neal JJ and Cloyd RA 2003. Effect of variegation on stem exudates of coleus and life history characteristics of citrus mealybug (Hemiptera: Pseudococcidae). *Environmental Entomology*, **32**, 463–9.

Sadof CS and Raupp MJ 1992. Effect of leaf variegation in *Euonymus japonica* on *Tetranychus urticae* (Acari, Tetranychidae). *Environmental Entomology*, **21**, 827–31.

Saleh N, Scott AG, Bryning GP and Chittka L 2007. Distinguishing signals and cues: bumblebees use general footprints to generate adaptive behaviour at flowers and nest. *Arthropod – Plant Interactions*, **1**, 119–27.

Sallabanks R 1993. Hierarchical mechanisms of fruit selection by an avian frugivore. *Ecology*, **74**, 1326–36.

Sánchez F, Korine C, Pinshow B and Dudley R 2004. The possible roles of ethanol in the relationship between plants and frugivores: first experiments with Egyptian fruit bats. *Integrative and Comparative Biology*, **44**, 290–4.

Sánchez F, Kotler BP, Korine C and Pinshow B 2008. Sugars are complementary resources to ethanol in foods consumed by Egyptian fruit bats. *Journal of Experimental Biology*, **211**, 1475–81.

Santamaría M and Franco AM 2000. Frugivory of salvin's curassow in a rainforest of the Colombian Amazon. *Wilson Bulletin*, **112**, 473–81.

Sapir Y, Shmida A and Ne'eman G 2006. Morning floral heat as a reward to the pollinators of the *Oncocyclus* irises. *Oecologia*, **147**, 53–9.

Saracino A, D'Alessandro CM and Borghetti M 2004. Seed colour and post-fire bird predation in a Mediterranean pine forest. *Acta Oecologica-International Journal of Ecology*, **26**, 191–6.

Saracino A, Pacella R, Leone V and Borghetti M 1997. Seed dispersal and changing seed characteristics in a *Pinus halepensis* Mill. forest after fire. *Plant Ecology*, **130**, 13–19.

Sargent RD 2004. Floral symmetry affects speciation rates in angiosperms. *Proceedings Royal Society B*, **271**, 603–8.

Scarantino, A 2010. Animal communication between information and influence. *Anim. Behav.* **79**, e1–e5.

Schaberg PG, Van den Berg AK, Murakami PF, Shane JB and Donnelly JR 2003. Factors influencing red expression in autumn foliage of sugar maple trees. *Tree Physiology*, **23**, 325–33.

Schaefer HM and Braun J 2009. Reliable cues and signals of fruit quality are contingent on the habitat in black elder (*Sambucus nigra*). *Ecology*, **90**, 1564–73.

Schaefer HM and Gould KS 2007. Modelling the evolution of leaf colouration with binary assumptions is barking up the wrong tree. *Journal of Theoretical Biology*, **249**, 638–9.

Schaefer HM, Levey DJ, Schaefer V and Avery ML 2006. The role of chromatic and achromatic signals for fruit detection by birds. *Behavioral Ecology*, **17**, 784–9.

Schaefer HM, McGraw K and Catoni C 2008b. Birds use fruit colour as honest signal of dietary antioxidant rewards. *Functional Ecology*, **22**, 303–10.

Schaefer HM, Rentzsch M and Breuer M 2008a. Anthocyanins reduce fungal growth in fruits. *Natural Product Communications*, **3**, 1267–72.

Schaefer HM and Rolshausen G 2006. Plants on red alert: do insects pay attention? *Bioessays*, **28**, 65–71.

Schaefer HM and Rolshausen G 2007. Aphids do not attend to leaf colour as visual signal, but to the handicap of reproductive investment. *Biology Letters*, **3**, 1–4.

Schaefer HM and Ruxton GD 2008. Fatal attraction: carnivorous plants roll out the red carpet to lure insects. *Biology Letters*, **4**, 153–5.

Schaefer HM and Ruxton GD 2009. Deception in plants: mimicry or perceptual exploitation? *Trends in Ecology and Evolution*, **24**, 676–85.

Schaefer HM and Schaefer V 2006. The fruits of selectivity: how birds forage on *Goupia glabra* fruits of different ripeness. *Journal of Ornithology*, **147**, 638–43.

Schaefer HM and Schaefer V 2007. The evolution of visual fruit signals: concepts and constraints. In: Dennis AJ, Schupp EW, Green RJ and Wescott DA (eds) *Seed dispersal: Theory and its application in a changing world*. CABI, Wallingford, UK.

Schaefer HM, Schaefer V and Levey DJ 2004. How plant–animal interactions signal new insights in communication. *Trends in Ecology and Evolution*, **19**, 577–84.

Schaefer HM, Schaefer V and Vorobyev M 2007. Are fruit colors adapted to consumer vision and birds equally efficient in detecting colorful signals? *American Naturalist*, **169**, S159–69.

Schaefer HM and Schmidt V 2004. Detectability and content as opposing signal characteristics in fruits. *Proceedings Royal Society B*, **271**, S370–3.

Schaefer HM, Schmidt V and Bairlein F 2003a. Discrimination abilities for nutrients: which difference matters for choosy birds and why? *Animal Behaviour*, **65**, 531–41.

Schaefer HM, Schmidt V and Bairlein F 2003b. Discrimination abilities for nutrients: which difference matters for choosy birds and why? *Animal Behaviour*, **65**, 531–41.

Schaefer HM, Spitzer K and Bairlein F 2008c. Long-term effects of previous experience determine nutrient discrimination abilities in birds. *Frontiers in Zoology*, **5**, 4.

Schaefer HM and Wilkinson DM 2004. Red leaves, insects and coevolution: a red herring? *Trends in Ecology and Evolution*, **19**, 616–18.

Schemske DW and Bierzychudek P 2007. Spatial differentiation for flower color in the desert annual *Linanthus parryae*: was Wright right? *Evolution*, **61**, 2528–43.

Schiestl FP 2005. On the success of a swindle: pollination by deception in orchids. *Naturwissenschaften*, **92**, 255–64.

Schiestl FP 2010. The evolution of floral scent and insect communication. *Ecology Letters*, **13**, 643–56.

Schiestl FP, Peakall R, Mant JG, Ibarra F, Schulz C, Franke S and Franke W 2003. The chemistry of sexual deception in an orchid-wasp pollination system. *Science*, **302**, 437–8.

Schluter D and Price T 1993. Honesty, perception and population divergence in sexually selected traits. *Proceedings Royal Society London B*, **253**, 117–22.

Schmidt V and Schaefer HM 2004. Unlearned preference for red may facilitate recognition of palatable food in young omnivorous birds. *Evolutionary Ecology Research*, **6**, 919–25.

Schmidt V, Schaefer HM and Winkler H 2004. Conspicuousness, not colour as foraging cue in plant–animal interactions. *Oikos*, **106**, 551–7.

Schnell DE 1976. *Carnivorous plants of the United States and Canada*. Lebanon Valley Offset Company, Annville.

Schoonhoven LM, van Loom JAJ and Dicke M 2005. *Insect – plant biology*. Oxford University Press, Oxford.

Schorkopf DLP, Jarau S, Francke W, Twele R, Zucchi R, Hrncir M, Schmidt VM, Ayasse M and Barth FG 2007. Spitting out information: *Trigona* bees deposit saliva to signal resource locations. *Proceedings Royal Society B*, **274**, 895–9.

Scott-Phillips T 2010. Animal communication: insights from linguistic pragmatics. *Animal Behaviour*, **79**, e1–e4.

Searcy WA and Nowicki S 2005. *The evolution of animal communication*. Princeton University Press, Princeton.

Seehausen, O., Terai, Y., Magalhaes, IS, Carleton, KL, Mrosso, HDJ, Miyagi, R., van der Sluijs, I., Schneider, MV, Maan, ME, Tachida, H., Imai, H. and Okada, N. 2008. Speciation through sensory drive in cichlid fish. *Nature* **455**, 620–626.

Seidel JL, Epstein WW and Davidson DW 1990. Neotropical ant gardens.1. Chemical constituents. *Journal of Chemical Ecology*, **16**, 1791–816.

Seidler TG and Plotkin JB 2006. Seed dispersal and spatial pattern in tropical trees. *PLoS Biology*, **4**, 2132–7.

Seyfarth, RM, Cheney, DL, Bergman, T, Fischer, J, Zuberbühler, K. and Hammerschmidt, K. 2010. The central importance of information in studies of animal communication. *Anim. Behav.* **80**, 3–8.

Seymour RS, Gibernan M and Ito K 2003. Thermogenesis and respiration of inflorescences of the dead horse arum *Helicodiceros muscivorus*, a pseudo-termoregulatory aroid associated with fly pollination. *Functional Ecology*, **17**, 886–94.

Shannon CE and Weaver W 1949. *The mathematical theory of communication*. University of Illinos Press, Urbana.

Sherratt TN, Wilkinson DM and Bain RS 2005. Explaining dioscorides' 'double difference': why are some mushrooms poisonous, and do they signal their unprofitability? *American Naturalist*, **166**, 767–75.

Shiojiri K, Karban R and Ishizaki S 2009. Volatile communication among sagebrush branches affects herbivory: timing of active cues. *Arthropod – Plant Interactions*, **3**, 99–104.

Shiojiri K, Takabayashi J, Yano S and Takafuji A 2001. Infochemically mediated tritrophic interaction webs on cabbage plants. *Population Ecology*, **43**, 23–9.

Shiojiri K, Takabayashi J, Yano S and Takafuji A 2002. Oviposition preferences of herbivores are affected by tritrophic interaction webs. *Ecology Letters*, **5**, 186–92.

Shuttleworth A and Johnson SD 2010. The missing stink: sulphur compounds can mediate a shift between fly and wasp pollination systems. *Proceedings Royal Society B*, **277**, 2811–19.

Siddiqi A, Cronin TW, Loew ER, Vorobyev M and Summers K 2004. Interspecific and intraspecific views of color signals in the strawberry poison frog Dendrobates pumilio. *Journal of Experimental Biology*, **207**, 2471–85.

Siepielski AM and Benkman CW 2008. A seed predator drives the evolution of a seed dispersal mutualism. *Proceedings Royal Society B*, **275**, 1917–25.

Siitari H, Honkavaara J and Viitala J 1999. Ultraviolet reflection of berries attracts foraging birds: a laboratory study with redwings (*Turdus iliacus*) and bilberries (*Vaccinium myrtillus*). *Proceedings Royal Society B*, **266**, 2125–9.

Simpson SJ and Raubenheimer D 1999. Geometric models of macronutrient selection. In: Berthoud HR and Seeley RJ (eds) *Metabolic control of macronutient intake*. CRC Press, Boca Raton.

Simpson SJ and Raubenheimer D 2000a. Geometric models of macronutrient selection. In: Berthoud H-R and Seeley RJ (eds) *Neural and metabolic control of macronutrient intake,*. CRC Press, Danvers, MA.

Simpson SJ and Raubenheimer D 2000b. The hungry locust. *Advances in the Study of Behavior*, **29**, 1–44.

Sims DA and Gamon JA 2002. Relationship between leaf pigment content and spectral reflectance across a wide range of species, leaf structures and developmental stages. *Remote Sensing of Environment*, **81**, 337–54.

Sinkkonen A 2006a. Do autumn leaf colours serve as a reproductive insurance against sucking herbivores? *Oikos*, **113**, 557–62.

Sinkkonen A 2006b. Sexual reproduction advances autumn leaf colours in mountain birch (*Betula pubescens* ssp. *czerepanovii*). *Journal of Evolutionary Biology*, **19**, 1722–4.

Sinkkonen A 2009. Ultraviolet leaf pigments as components of autumn colours: a constructive comment on Archetti et al. *Trends in Ecology and Evolution*, **24**, 236–7.

Skelhorn J, Rowland HM and Ruxton GD 2010b. The evolution and ecology of masquerade. *Biological Journal of the Linnean Society*, **99**, 1–8.

Skelhorn J, Rowland HM, Speed MP and Ruxton GD 2010a. Masquerade: camouflage without crypsis. *Science*, **327**, 51.

Skorupski P and Chittka L 2010. Differences in photoreceptor processing speed for chromatic and achromatic vision in the bumblebee, *Bombus terrestris*. *Journal of Neuroscience*, **30**, 3896–3903.

Skyrms B 2009. Evolution of signalling systems with multiple senders and receivers. *Philosophical Transactions Royal Society B*, **364**, 771–9.

Smith AC, Buchanan-Smith HM, Surridge AK, Osorio D and Mundy NI 2003. The effect of ocolour vision status on the detection and selection of fruits by tamarins (*Saguinus* spp.). *Journal of Experimental Biology*, **206**, 3159–65.

Smith AP 1986. Ecology of a leaf color polymorphism in a tropical forest species: habitat segregation and herbivory. *Oecologia*, **69**, 283–7.

Smith C, Barber I, Wooton RJ and Chittka L 2004. A receiver bias in the origin of three-spined stickleback mate choice. *Proceedings Royal Society London B*, **271**, 949–55.

Smith RA and Rausher MD 2008. Selection for character displacement is constrained by the genetic architecture of floral traits in the ivyleaf morning glory. *Evolution*, **62**, 2829–41.

Snow DW 1971. Evolutionary aspects of fruit-eating by birds. *Ibis*, **113**, 194–202.

Solfanelli C, Poggi A, Loreti E, Alpi A and Perata P 2006. Sucrose-specific induction of the anthocyanin biosynthetic pathway in Arabidopsis. *Plant Physiology*, **140**, 637–46.

Soltau U, Dotterl S and Liede-Schumann S 2009. Leaf variegation in *Caladium steudneriifolium* (Araceae): a case of mimicry? *Evolutionary Ecology*, **23**, 503–12.

Solvia M and Widmer A 2003. Gene flow across species boundaries in sympatric, sexually deceptive *Ophrys* (Orchidaceae) species. *Evolution*, **57**, 2252–61.

Somanathan H, Borges RM, Warrant EJ and Kelber A 2008. Nocturnal bees learn landmark colours in starlight. *Current Biology*, **18**, R996–7.

Sorensen AE 1986. Seed dispesal by adhesion. *Annual Review of Ecology and Systematics*, **17**, 443–63.

Sosinsky A, Glusman G and Lancet D 2000. The genomic structure of human olfactory receptor genes. *Genomics*, **70**, 49–61.

Spaethe J, Tautz J and Chittka L 2001. Visual constraints in foraging bumblebees: flower size and color affect search time and flight behavior. *Proceedings National Academy of Sciences*, **98**, 3898–903.

Sprengel CK 1793. *Das entdecke Geheimnis in der Natur im Bau und in der Befruchtung der Blumen*. Repr. 1972, Weldon and Wesley, New York.

Stafford HA 1994. Anthocyanins and betalains: evolution of the mutually exclusive pathways. *Plant Science*, **101**, 91–8.

Steele MA, Smallwood PD, Spunar A and Nelsen E 2001. The proximate basis of the oak dispersal syndrome: detection of seed dormancy by rodents. *American Zoologist*, **41**, 852–64.

Steidle JLM, Fischer A and Gantert C 2005. Do grains whisper for help? Evidence for herbivore-induced synomones in wheat grains. *Entomologia Experimentalis Et Applicata*, **115**, 239–45.

Steiger S, Schmitt T and Schaefer HM 2011. On the origin and dynamic evolution of chemical information transfer. *Proceedings Royal Society B*, **278**, (in press).

Steinebrunner F, Twele R, Franke W, Leuchtmann A and Schiestl FP 2008. Role of odour compounds in the attraction of gamete vectors in endophytic *Epichloë* fungi. *New Phytologist*, **178**, 401–11.

Stensmyr MC, Urru I, Collu I, Celander M, Hansson BS and Angioy AM 2002. Rotting smell of dead-horse arum florets: these blooms chemically fool flies into pollinating them. *Nature*, **420**, 625–6.

Stevens M and Cuthill IC 2007. Hidden messages: are ultraviolet signals a special channel in avian communication? *BioScience*, **57**, 501–507.

Stevens M and Merilaita S 2009. Animal camouflage: current issues and new perspectives. *Philosophical Transactions Royal Society*, **364**, 423–557.

Steyn WJ, Wand SJE, Holcroft DM and Jacobs G 2002. Anthocyanins in vegetative tissues: a proposed unified function in photoprotection. *New Phytologist*, **155**, 349–61.

Steyn WJ, Wand SJE, Jacobs G, Rosecrance RC and Roberts SC 2009. Evidence for a photoprotective function of low-temperature-induced anthocyanin accumulation in apple and pear peel. *Physiologia Plantarum*, **136**, 461–72.

Stiles EW 1982. Fruit flags: two hypotheses. *American Naturalist*, **120**, 500–509.

Stinchcombe JR and Rausher MD 2001. Diffuse selection on resistance to deer herbivory in the ivyleaf morning glory, *Ipomoea hederacea*. *American Naturalist*, **158**, 376–88.

Stinzig FC and Carle R 2004. Functional properties of anthocyanins and betalains in plants, food, and in human nutrition. *Trends in Food Science and Technology*, **15**, 19–35.

Stökl J, Schlüter PM, Stuessy TF, Paulus HF, Assum G and Ayasse M 2008. Scent variation and hybridization cause the displacement of a sexually deceptive orchid species. *American Journal of Botany*, **95**, 472–81.

Stökl J, Schlüter PM, Stuessy TF, Paulus HF, Fraberger R, Erdmann D, Schulz C, Franke W, Assum G and Ayasse M 2009. Speciation in sexually deceptive orchids: pollinator-driven selection maintains discrete odour phenotypes in hybridizing species. *Biological Journal of the Linnean Society*, **98**, 439–51.

Stone BC 1979. Protective coloration of young leaves in certain Malaysian palms. *Biotropica*, **11**, 126.

Stranden M, Røstelien T, Liblikas I, Almaas TJ, Borg-Karlson AK and Mustaparta H 2003. Receptor neurones in three heliothine moths responding to floral and inducible plant volatiles. *Chemoecology*, **13**, 143–54.

Strauss SY 1997. Floral characters link herbivores, pollinators, and plant fitness. *Ecology*, **78**, 1640–5.

Strauss SY and Irwin RE 2004. Ecological and evolutionary consequences of multispecies plant–animal interactions. *Annual Review of Ecology Evolution and Systematics*, **35**, 433–66.

Strauss SY, Irwin RE and Lambrix VM 2004. Optimal defence theory and flower petal colour predict variation in the secondary chemistry of wild radish. *Journal of Ecology*, **92**, 132–41.

Strauss SY and Whittall JB 2006. Non-pollinator agents of selection on floral traits. In: Harder LD and Barrett SCH (eds) *Ecology and evolution of flowers*. Oxford University Press, Oxford.

Strong JN and Fragoso JMV 2006. Seed dispersal by *Geochelone carbonaria* and *Geochelone denticulata* in northwestern Brazil. *Biotropica*, **38**, 683–6.

Struempf HM, Schondube JE and Martinez del Rio C 1999. The cyanogenic glycoside amygdalin does not deter consumption of ripe fruit by Cedar Waxwings. *Auk*, **116**, 749–58.

Su J-F, Gong Y-B, Renner SS and Huang S-Q 2008. Multifunctional bracts in the dove tree *Davidia involucrata* (Nyssaceae:Cornales): rain protection and pollinator attraction. *American Naturalist*, **171**, 119–24.

Suárez LH, Gonzáles WL and Gianoli E 2009. Foliar damage modifies floral attractiveness to pollinators in *Alstroemeria exerens*. *Evolutionary Ecology*, **23**, 545–55.

Sumner P and Mollon JD 2000a. Chromaticity as a signal of ripeness in fruits taken by primates. *Journal of Experimental Biology*, **203**, 1987–2000.

Sumner P and Mollon JD 2000b. Catarrhine photopigments are optimised for detecting targets against a foliage background. *Journal of Experimental Biology*, **203**, 1963–86.

Sun C, Ives AR, Kraeuter HJ and Moermond TC 1997. Effectiveness of three turacos as seed dispersers in a tropical montane forest. *Oecologia*, **112**, 94–103.

Surridge AK, Osorio D and Mundy NI 2003. Evolution and selection of trichromatic vision in primates. *Trends in Ecology and Evolution*, **18**, 198–205.

Takabayashi J, Sabelis MW, Janssen A, Shiojiri K and van Wijk M 2006. Can plants betray the presence of multiple herbivore species to predators and parasitoids? The role of learning in phytochemical information networks. *Ecological Research*, **21**, 3–8.

Takabayashi J, Takahashi S, Dicke M and Posthumus MA 1995. Developmental stage of herbivore *Pseudaletia separata* affects produciton of herbivore-induced synomone by corn plants. *Journal of Chemical Ecology*, **21**, 273–87.

Takeuchi A, Matsumoto S and Hayatsu M 1994. Chalcone synthase from *Camellia sinensis*: isolation of the cDNAs and the organ-specific and sugar-responsive expression of the genes. *Plant Cell Physiology*, **35**, 1011–8.

Takhtajan A 1991. *Evolutionary trends in flowering plants*. Columbia University Press, New York.

Tamura N and Hayashi F 2008. Geographic variation in walnut seed size correlates with hoarding behaviour of two rodent species. *Ecological Research*, **23**, 607–14.

Tanaka T, Yagi S and Nakamatu Y 1992. Regulation of parasitoid sex allocation and host growth by *Cotesia* (*apan-*

teles) kariyai (Hymenoptera, Braconidae). *Annals of the Entomological Society of America*, **85**, 310–6.

Tang AMC, Corlett RT and Hyde KD 2005. The persistence of ripe fleshy fruits in the presence and absence of frugivores. *Oecologia*, **142**, 232–7.

Tellería JL, Ramirez A and Pérez-Tris J 2008. Fruit tracking between sites and years by birds in Mediterranean wintering grounds. *Ecography*, **31**, 381–8.

Temeles EJ and Kress WJ 2003. Adaptation in a plant–hummingbird association. *Science*, **300**, 630–3.

Tentelier C and Fauvergue X 2007. Herbivore-induced plant volatiles as cues for habitat assessment by a foraging parasitoid. *Journal of Animal Ecology*, **76**, 1–8.

Tewksbury J 2002. Fruit, frugivores and the evolutionary arms race. *New Phytologist*, **156**, 137–9.

Tewksbury JJ, Levey DJ, Huizinga M, Haak DC and Traveset A 2008b. Costs and benefits of capsaicin-mediated control of gut retention in dispersers of wild chillies. *Ecology*, **89**, 107–17.

Tewksbury JJ, Reagan KM, Machnicki NJ, Carlo TA, Haak DC, Calderón Peñaloza AJ and Levey DJ 2008a. Evolutionary ecology of pungency in wild chilies. *Proceedings National Academy of Sciences*, **105**, 11808–11.

Tewksbury JT and Nabhan GP 2001. Why are chillies hot? Directed deterrence of capsaicin in wild chillies. *Nature*, **412**, 403–404.

Thaler JS 1999. Jasmonate-inducible plant defences cause increased parasitism of herbivores. *Nature*, **399**, 686–8.

Theis N and Lerdau M 2003. The evolution of function in plant secondary metabolites. *International Journal of Plant Sciences*, **164**, S93–102.

Theis N, Lerdau M and Raguso RA 2007. The challenge of attracting pollinators while evading floral herbivores: Patterns of fragrance emission in *Cirsium arvense* and *Cirsium repandum* (Asteraceae). *International Journal of Plant Sciences*, **168**, 587–601.

Thiery D and Visser JH 1986. Masking of host plant odor in the olfactory orientation of the colorado potato beetle. *Entomologia Experimentalis Et Applicata*, **41**, 165–72.

Thies W, Kalko EKV and Schnitzler H-U 1998. The roles of echolocation and olfaction in two Neotropical fruit-eating bats, *Carollia perspicillata* and *C. castanea*, feeding on Piper. *Behavioral Ecology and Sociobiology*, **42**, 397–409.

Thompson JN 2005b. Coevolution: the geographic mosaic of coevolutionary arms races. *Current Biology*, **15**, R992–4.

Thompson JN 2005a. *The geographic mosaic of coevolution*. University of Chicago Press, Chicago.

Thompson JN and Fernandez CC 2006. Temporal dynamics of antagonism and mutualism in a geographically variable plant–insect interaction. *Ecology*, **87**, 103–12.

Thompson JN and Willson MF 1979. Evolution of temperate fruit/bird interactions: phenological strategies. *Evolution*, **33**, 973–82.

Thomson JD and Thomson BA 1989. Dispersal of *Erythronium grandiflorum* pollen by bumblebees: implications for gene flow and reproductive success. *Evolution*, **43**, 657–61.

Thouhara K and Vosshall LB 2009. Sensing odorants and pheromones with chemosensory receptors. *Annual Reviews of Physiology*, **71**, 307–32.

Thum M 1986. Segregation of habitat and prey in two sympatric carnivorous plant species, *Drosera rotundifolia* and *Drosera intermedia*. *Oecologia*, **70**, 601–605.

Thum M 1988. The significance of carnivory for the fitness of Drosera in its natural habitat. 1. The reactions of *Drosera intermedia* and *Drosera rotundifolia* to supplementary feeding. *Oecologia*, **75**, 472–80.

Tibbetts EA and Dale J 2004. A socially enforced signal of quality in a paper wasp. *Nature*, **432**, 218–22.

Tielborger K and Prasse R 2009. Do seeds sense each other? Testing for density-dependent germination in desert perennial plants. *Oikos*, **118**, 792–800.

Tiffney BH 2004. Vertebrate dispersal of seed plants through time. *Annual Review of Ecology Evolution and Systematics*, **35**, 1–29.

Tinbergen N 1952. 'Derived' activities; their causation, biological significance, origin, and emancipation during evolution. *Quarterly Review of Biology*, **27**, 1–32.

Touhara K and Vosshall LB 2009. Sensing odorants and pheromones with chemosensory receptors. *Annual Review of Physiology*, **71**, 307–32.

Traveset A, Riera N and Mas RE 2001. Ecology of fruit-colour polymorphism in *Myrtus communis* and differential effects of birds and mammals on seed germination and seedling growth. *Journal of Ecology*, **89**, 749–60.

Traveset A and Willson MF 1998. Ecology of the fruit-colour polymorphism in Rubus spectabilis. *Evolutionary Ecology*, **12**, 331–45.

Traveset A, Willson MF and Verdú M 2004. Characteristics of fleshy fruits in southeast Alaska: phylogenetic comparison with fruits from Illinois. *Ecography*, **27**, 41–8.

Treisman AM and Gelade G 1980. A feature-integration theory of attention. *Cognitive Psychology*, **12**, 97–136.

Türke M, Heinze E, Andreas K, Svendsen SM, Gossner MM and Weisser WW 2010. Seed consumption and dispersal of ant-dispersed plants by slugs. *Oecologia*, **163**, 681–93.

Turček F 1961. *Oekologische Beziehung der Vögel und Gehölze*. Slovak. Akad. Wiss., Bratislava.

Turlings TCJ, Loughrin JH, McCall PJ, Rose USR, Lewis WJ and Tumlinson JH 1995. How caterpillar-damaged plants protect themselves by attracting parasitic wasps.

Proceedings of the National Academy of Sciences of the United States of America, **92**, 4169–74.

Uemura S and Sugiura Y 2007. Density-dependent hoarding by rodents contributes to large variation in seed mass of the woodland herb *Symplocarpus renifolius*. *Canadian Journal of Forest Research – Revue Canadienne De Recherche Forestiere*, **37**, 1675–80.

Ushimaru A and Hyodo F 2005. Why do bilaterally symmetrical flowers orient vertically? Flower orientation influences pollinator landing behaviour. *Evolutionary Ecology Research*, **7**, 151–60.

Valburg LK 1992. Eating infested fruits: interactions in a plant–disperser–pest triad. *Oikos*, **65**, 25–8.

Valdivia CE and Niemeyer HM 2006. Do pollinators simultaneously select for inflorescence size and amount of floral scents? An experimental assessment of *Escallonia myrtoidea*. *Austral Ecology*, **31**, 897–903.

Valido A and Olesen JM 2007. The importance of lizards as frugiores and seed dispersers. In: Dennis AJ, Schupp EW, Green RJ and Wescott DA (eds) *Seed dispersal: theory and its application in a changing world*. CABI, Wallingford, UK.

Valido, A, Schaefer, HM and Jordano, P. 2011. Color, design, and reward: phenotypic integration of fleshy fruit displays. Journal of Evolutionary Biology, 24, in press.

Van Bael SA, Brawn JD and Robinson SK 2003. Birds defend trees from herbivores in a Neotropical forest canopy. *Proceedings of the National Academy of Sciences of the United States of America*, **100**, 8304–307.

van den Berg AK and Perkins TD 2007. Contribution of anthocyanins to the antioxidant capacity of juvenile and senescing sugar maple (*Acer saccharum*) leaves. *Functional Plant Biology*, **34**, 714–9.

van der Meijden E and Klinkhamer PGL 2000. Conflicting interests of plants and the natural enemies of herbivores. *Oikos*, **89**, 202–208.

Van der Pijl L 1960. Ecological aspects of flower evolution. I. Phyletic evolution. *Evolution*, **14**, 403–16.

van der Pijl L 1982. *Principles of dispersal in higher plants*. Springer Verlag, New York.

van Loon JJA, de Boer JG and Dicke M 2000. Parasitoid – plant mutualism: parasitoid attack of herbivore increases plant reproduction. *Entomologia Experimentalis Et Applicata*, **97**, 219–27.

van Ommeren RJ and Whitham TG 2002. Changes in interactions between juniper and mistletoe mediated by shared avian frugivores: parasitism to potential mutualism. *Oecologia*, **130**, 281–8.

van Rheede van Outshoorn K and van Rooyen MW 1999. *Dispersal biology of desert plants*. Springer Verlag, Berlin.

van Tol R, van der Sommen ATC, Boff MIC, van Bezooijen J, Sabelis MW and Smits PH 2001. Plants protect their roots by alerting the enemies of grubs. *Ecology Letters*, **4**, 292–4.

Vander Wall SB 1990. *Food hoarding in animals*. University of Chicago Press, Chicago.

Vander Wall SB 2001. The evolutionary ecology of nut dispersal. *Botanical Review*, **67**, 74–117.

Vander Wall SB, Beck MJ, Briggs JS, Roth JK, Thayer TC, Hollander JL and Armstrong JM 2003. Interspecific variation in the olfactory abilities of granivorous rodents. *Journal of Mammalogy*, **84**, 487–96.

Vander Wall SB, Kuhn KM and Gworek JR 2005. Two-phase seed dispersal: linking the effects of frugivorous birds and seed-caching rodents. *Oecologia*, **145**, 282–7.

Vander Wall SB and Longland WS 2004. Diplochory: are two seed dispersers better than one? *Trends in Ecology and Evolution*, **19**, 155–61.

Vanderwall SB 1992. The role of animals in dispersing a wind-dispersed pine. *Ecology*, **73**, 614–21.

Vanderwall SB 1993. Cache site selection by chipmunks (*Tamias spp*) and its influence on the effectiveness of seed dispersal in Jeffrey pine (*Pinus jeffreyi*). *Oecologia*, **96**, 246–52.

Vane-Wright RI 1976. A unified classification of mimetic resemblances. *Biological Journal of the Linnean Society*, **8**, 25–56.

Vázquez DP and Simberloff D 2004. Indirect effects of an introduced ungulate on pollination and plant reproduction. *Ecological Monographs*, **74**, 281–308.

Venail J, Dell'Olivo A and Kuhlemeier C 2010. Speciation genes in the genus *Petunia*. *Philosophical Transactions of the Royal Society B*, **365**, 461–8.

Vereecken NJ, Mant J and Schiestl FP 2007. Population differentiation in female sex pheromone and male preferences in a solitary bee. *Behavioral Ecology and Sociobiology*, **61**, 811–21.

Vereecken NJ and Schiestl FP 2008. The evolution of imperfect floral mimicry. *Proceedings of the National Academy of Sciences of the United States of America*, **105**, 7484–8.

Verhoeven AS, Adams III WW, Demming-Adams B, Croce R and Bassi R 1999. Xanthophyll cycle pigment localization and dynamics during exposure to low temperatures and light stress in *Vinca major*. *Plant Physiology*, **120**, 727–37.

Vibrans H 1999. Epianthropochory in Mexican weed communities. *American Journal of Botany*, **86**, 476–81.

Visser JH 1986. Host odor perception in phytophagous insects. *Annual Review of Entomology*, **31**, 121–44.

Vogel ER, Neitz M and Dominy NJ 2007. Effect of color vision phenotype ono the foraging of wild white-faced capuchins, *Cebus capucinus*. *Behavioral Ecology*, **18**, 292–7.

Vogel S 1978. Evolutionary shifts from reward to deception in pollen flowers. In: Richards AJ (ed) *The pollination of flowers by insects*. Academic Press for the Linnean Society of London, London.

Vogelmann TC 1993. Plant tissue optics. *Annual Review of Plant Physiology and Plant Molecular Biology*, **44**, 231–51.

Voigt D and Gorb S 2008. An insect trap as habitat: cohesion-failure mechanism prevents adhesion of *Pameridea roridulae* bugs to the sticky surface of the plant *Roridula gorgonias*. *Journal of Experimental Biology*, **211**, 2647–57.

Voigt FA, Bleher B, Fietz J, Ganzhorn JU, Schwab D and Böhning-Gaese K 2004. A comparison of morphological and chemical fruit traits between two sites with different frugivore assemblages. *Oecologia*, **141**, 94–104.

Volkov AG, Adesina T, Markin VS and Jovanov E 2008. Kinetics and mechanism of *Dionaea muscipula* trap closing. *Plant Physiology*, **146**, 694–702.

von Dahl CC and Baldwin IT 2007. Deciphering the role of ethylene in plant–herbivore interactions. *Journal of Plant Growth Regulation*, **26**, 201–209.

von Helversen D, Holderied MW and von Helversen O 2003. Echoes of bat-pollinated bell-shaped flowers: conspicuous for nectar-feeding bats? *Journal of Experimental Biology*, **206**, 1025–34.

von Helversen D and von Helversen O 1999. Acoustic guide in bat-pollinated flower. *Nature*, **398**, 759–60.

von Helversen D and von Helversen O 2003. Object recognition by echolocation: a nectar-feeding bat exploiting the flowers of a rain forest vine. *Journal of Comparative Physiology A*, **189**, 327–36.

von Helversen O and Winter Y 2003. Glossophagine bats and their flowers: costs and benefits for plants and pollinators. In: Kunz TH and Fenton MB (eds) *Bat ecology*. University of Chicago Press, Chicago.

Vorobyev M and Osorio D 1998. Receptor noise as a determinant of colour thresholds. *Proceedings Royal Society B*, **265**, 351–8.

Waelti MO, Muhlemann JK, Widmer A and Schiestl FP 2008. Floral odour and reproductive isolation in two species of *Silene*. *Journal of Evolutionary Biology*, **21**, 111–21.

Wagner GP, Kennedy-Hunt JP, Pavlicev M, Peck JR, Waxman D and Cheverud JM 2008. Pleiotropic scaling of gene effects and the 'cost of complexity'. *Nature*, **452**, 470–3.

Wahaj SA, Levey DJ, Sanders AK and Cipollini ML 1998. Control of gut retention time by secondary metabolites in ripe Solanum fruits. *Ecology*, **79**, 2309–19.

Waite TA 1988. A field-test of density-dependent survival of simulated gray jay caches. *Condor*, **90**, 247–9.

Wallace AR 1879. Colour in nature. *Nature*, **19**, 580–1.

Wallace HM and Trueman SJ 1995. Dispersal of Eucalyptus torelliana seeds by the resin-collecting stingless bee *Trigona carbonaria*. *Oecologia*, **104**, 12–16.

Wannes WA, Mhamdi B and Marzouk B 2009. Variations in essential oil and fatty acid composition during *Myrtus communis* var. italica fruit maturation. *Food Chemistry*, **112**, 621–6.

Ward MJ and Paton DC 2007. Predicting mistletoe seed shadow and patterns of seed rain from movements of the mistletoebird, *Dicaeum hirundinaceum*. *Austral Ecology*, **32**, 113–21.

Waser NM and Campbell DR 2004. Ecological speciation in flowering plants. In: Dieckmann U, Doebeli M, Metz JAJ and Tautz D (eds) *Adaptive speciation* Cambridge University Press, Cambridge.

Waser NM, Chittka L, Price MV, Williams NM and Ollerton J 1996. Generalization in pollination systems, and why it matters. *Ecology*, **77**, 1043–60.

Watson MA and Caspar BB 1984. Morphogenetic contraints on patterns of carbon distribution in plants. *Annual Review of Ecology and Systematics*, **15**, 233–58.

Weberling F 1989. *Morphology of flowers and inflorescences*. Cambridge University Press, Cambridge.

Weckerly FW, Nicholson KE and Semlitsch RD 1989. Experimental test of discrimination by squirrels for insect-infested and noninfested acorns. *American Midland Naturalist*, **122**, 412–15.

Weins D 1978. Mimicry in plants. *Evolutionary Biology*, **11**, 365–403.

Weiss MR 1991. Floral colour changes as cues for pollinators. *Nature*, **354**, 227–9.

Weiss MR 1995. Floral color change: a widespread functional convergence. *American Journal of Botany*, **82**, 167–85.

Wenny DG 1999. Two-stage dispersal of *Guarea glabra* and *G-kunthiana* (Meliaceae) in Monteverde, Costa Rica. *Journal of Tropical Ecology*, **15**, 481–96.

Wenny DG 2000. Seed dispersal, seed predation, and seedling recruitment of a neotropical montane tree. *Ecological Monographs*, **70**, 331–51.

Wenny DG and Levey DJ 1998. Directed seed dispersal by bellbirds in a tropical cloud forest. *Proceedings of the National Academy of Sciences of the United States of America*, **95**, 6204–7.

West-Eberhard MJ 2003. *Developmental plasticity and evolution*. Oxford University Press, Oxford.

Westoby M, French K, Hughes L, Rice B and Rodgerson L 1991. Why do more plant species use ants for dispersal on infertile compared with fertiel soils. *Australian Journal of Ecology*, **16**, 445–55.

Westoby M, Rice B and Howell J 1990. Seed size and plant growth form as factors in dispersal spectra. *Ecology*, **71**, 1307–15.

Wheelwright NT and Janson CH 1985. Colors of fruit displays of bird dispersed plants in two tropical forests. *American Naturalist*, **126**, 777–99.

Whelan CJ, Schmidt KA, Steele BB, Quinn WJ and Dilger S 1998. Are bird-consumed fruits complementary resources? *Oikos*, **83**, 195–205.

White TCR 2009. Catching a red herring: autumn colours and aphids. *Oikos*, **118**, 1610–12.

Whitehead MR and Peakall R 2009. Integrating floral scent, pollination ecology and population genetics. *Functional Ecology*, **23**, 863–74.

Whitney HM, Kolle M, Andrew P, Chittka L, Steiner U and Glover BJ 2009. Floral irisdescence, produced by diffractive optics, acts as a cue for animal pollinators. *Science*, **323**, 130–3.

Whitney KD 2005. Linking frugivores to the dynamics of a fruit color polymorphism. *American Journal of Botany*, **92**, 859–67.

Whitney KD 2009. Comparative evolution of flower and fruit morphology. *Proceedings Royal Society B*, **276**, 2941–7.

Whitney KD and Lister CE 2004. Fruit colour polymorphism in *Acacia ligulata*: seed and seedling performance, clinal patterns, and chemical variation. *Evolutionary Ecology*, **18**, 165–86.

Whitney KD and Stanton ML 2004. Insect seed predators as novel agents of selection on fruit color. *Ecology*, **2004**, 2153–60.

Wichmann MC, Alexander MJ, Soons MB, Galsworthy S, Dunne L, Gould R, Fairfax C, Niggemann M, Hails RS and Bullock JM 2009. Human-mediated dispersal of seeds over long distances. *Proceedings of the Royal Society B*, **276**, 523–32.

Wickler W 1968. *Mimicry in plants and animals*. Weidenfeld and Nicholson, London.

Wiens D 1978. Mimicry in plants. *Evolutionary Biology*, **11**, 365–403.

Wiens F, Zitzmann A, Lachance M-A, Yegles M, Pragst F, Wurst FM, von Holst D, Guan SL and Spanagel R 2008. Chronic intake of fermented floral nectar by wild treeshrews. *Proceedings National Academy of Sciences of the United States of America*, **105**, 10426–31.

Wilkinson DM, Sherratt TN, Phillip DM, Wratten SD, Dixon AFG and Young AJ 2002. The adaptive significance of autumn leaf colours. *Oikos*, **99**, 402–407.

Williams SE 1976. Comparative sensory physiology of the Droseraceae – the evolution of a plant sensory system. *Proceedings of the American Philosophical Society*, **120**, 187–204.

Williamson GB 1982. Plant mimicry: evolutionary constraints. *Biological Journal of the Linnean Society*, **18**, 49–58.

Willson MF 1994. Fruit choices by captive American robins. *Condor*, **96**, 494–502.

Willson MF and Comet TA 1993. Food choices by Northwestern Crows: experiments with captive, free-ranging, and hand-raised birds. *Condor*, **95**, 596–615.

Willson MF, Graff DA and Whelan CJ 1990a. Color preferences of frugivorous birds in relation to the colors of fleshy fruits. *Condor*, **92**, 545–55.

Willson MF and Hoppes WG 1986. Foliar 'flags' for avian frugivore: signal or serendipity? In: Estrada A and Fleming TH (eds) *Frugivore and seed dispersal*. Junk Publisher, Boston.

Willson MF and Melampy MN 1983. The effect of bicolored fruit displays on fruit removal by avian frugivores. *Oikos*, **41**, 27–31.

Willson MF and O'Dowd DJ 1989. Fruit colour polymorphism in a bird-dispersed shrub (*Rhagodia parabolica*) in Australia. *Evolutionary Ecology*, **3**, 40–50.

Willson MF, Rice BL and Westoby M 1990b. Seed dispersal spectra: a comparison of temperate plant communities. *Journal of Vegetation Science*, **1**, 547–62.

Willson MF and Thompson JN 1982. Phenology and ecology of color in bird-dispersed fruits or why some fruits are red when they are 'green'. *Canadian Journal of Botany*, **60**, 701–13.

Willson MF and Whelan CJ 1989. Ultraviolet reflectance of fruits of vertebrate-dispersed plants. *Oikos*, **55**, 341–8.

Willson MF and Whelan CJ 1990. The evolution of fruit color in fleshy-fruited plants. *American Naturalist*, **136**, 790–809.

Wilms J and Eltz T 2008. Foraging scent marks of bumblebees: footprint cues rather than pheromone signals. *Naturwissenschaften*, **95**, 149–53.

Wingler A, Purdy S, MacLean JA and Pourtau N 2006. The role of sugars in integrating environmental signals during the regulation of leaf senescence. *Journal of Experimental Botany*, **57**, 391–9.

Winkel-Shirley B 2002. Biosynthesis of flavanoids and effects of stress. *Current Opinion in Plant Biology*, **5**, 218–23.

Winterrowd MF and Weigl PD 2006. Mechanisms of cache retrieval in the group nesting southern flying squirrel (*Glaucomys volans*). *Ethology*, **112**, 1136–44.

Witjes S and Eltz T 2009. Hydrocarbon footprints as a record of bumblebee flower visitation. *Journal of Chemical Ecology*, **35**, 1320–5.

Witmer MC 2001. Nutritional interactions and fruit removal: springtime consumption of Viburnum opulus fruits by Cedar Waxwings. *Ecology*, **82**, 3120–30.

Witmer MC and Cheke AS 1991. The dodo and the tambalacoque tree: an obligate mutualism reconsidered. *Oikos*, **61**, 133–7.

Wittstock U, Agerbirk N, Stauber EJ, Olsen CE, Hippler M, Mitchell-Olds T, Gershenzon J and Vogel H 2004. Successful herbivore attack due to metabolic diversion of a plant chemical defense. *Proceedings National Academy of Sciences of the United States of America*, **101**, 4859–64.

Wolf BO, Martinez del Rio C and Babson J 2002. Stable isotope reveal that saguaro fruit provides different resources to two desert dove species. *Ecology*, **83**, 1286–93.

Wood WF and Wood BJ 2004. Chemical released from host Acacia by feeding herbivores is detected by symbiotic Acacia-ants. *Caribbean Journal of Science*, **40**, 396–9.

Wright G and Schiestl FP 2009. The evolution of floral scent: the influence of olfactory learning by insect pollinators on the honest signalling of floral rewards. *Functional Ecology*, **23**, 841–51.

Wright S 1988. Surfaces of selective value revisited. *American Naturalist*, **131**, 115–23.

Xiao ZS, Chang G and Zhang ZB 2008. Testing the high-tannin hypothesis with scatter-hoarding rodents: experimental and field evidence. *Animal Behaviour*, **75**, 1235–41.

Yamazaki K 2008. Autumn leaf colouration: a new hypothesis involving plant–ant mutualism via aphids. *Naturwissenschaften*, **95**, 671–6.

Yamazaki K 2010. Leaf mines as visual defensive signals to herbivores. *Oikos*, **119**, 796–801.

Yang JS and Sadof CS 1995. Variegation in *Coleus blumei* and the life history of citrus mealybug (Homoptera: Pseudococcidae). *Environmental Entomology*, **24**, 1650–5.

Yanoviak SP, Kaspari M, Dudley R and Poinar G 2008. Parasite-induced fruit mimicry in a tropical canopy ant. *American Naturalist*, **171**, 536–44.

Yarmolinsky DA, Zuker CS and Ryba NJP 2009. Common sense about taste: from mammals to insects. *Cell*, **139**, 234–44.

Yearsley JM, Villaba JJ, Gordon IJ, Kyriazakis I, Speakman JR, Tolkamp BJ, Illius AW and Duncan AJ 2006. A theory of associating food types with their postingestive consequences. *American Naturalist*, **167**, 705–16.

Youngsteadt E, Nojima S, Haberlein C, Schulz S and Schal C 2008. Seed odor mediates an obligate ant–plant mutualism in Amazonian rainforests. *Proceedings of the National Academy of Sciences of the United States of America*, **105**, 4571–5.

Yu DW and Davidson DW 1997. Experimental studies of species-specificity in Cecropia–ant relationships. *Ecological Monographs*, **67**, 273–94.

Yu F and Utsumi R 2009. Diversity, regulation, and genetic manipulation of plant mono- and sesquiterpenoid biosynthesis. *Cellular and Molecular Life Sciences*, **66**, 3043–52.

Zahavi A 1975. Mate-selection – a selection for a handicap. *Journal of Theoretical Biology*, **53**, 204–14.

Zamora R 1990. The feeding ecology of a carnivorous plant (*Pinguicula nevadense*) – prey analysis and capture constraints. *Oecologia*, **84**, 376–9.

Zamora R 1995. The trapping success of a carnivorous plant, Pinguicula vallisnerifolia – the cumulative effects of availability, attraction, retention and robbery of prey. *Oikos*, **73**, 309–22.

Zamora R 1999. Conditional outcomes of interactions: The pollinator–prey conflict of an insectivorous plant. *Ecology*, **80**, 786–95.

Zhang HM, Cheng JR, Xiao ZS and Zhang ZB 2008. Effects of seed abundance on seed scatter-hoarding of Edward's rat (*Leopoldamys edwardsi* Muridae) at the individual level. *Oecologia*, **158**, 57–63.

Zhang PJ, Zheng SJ, van Loon JJA, Boland W, David A, Mumm R and Dicke M 2009. Whiteflies interfere with indirect plant defense against spider mites in Lima bean. *Proceedings of the National Academy of Sciences of the United States of America*, **106**, 21202–207.

Zhang S, Schwarz S, Pahl M, Hong Z and Tautz J 2006. A honey bee knows what to do and when. *Journal of Experimental Biology*, **209**, 4420–8.

Zimmermann Y, Ramírez SR and Eltz T 2009. Chemical niche differentiation among sympatric species of orchid bees. *Ecology*, **90**, 2994–3008.

Zufall RA and Rausher MD 2004. Genetic changes associated with floral adaptation restrict future evolutionary potential. *Nature*, **428**, 847–50.

Index

shape 122
size 75, 81, 96
softness 106
temporarily bicoloured fruit
  display 87
texture 105–6
water content 106
fruit contrast hypothesis 78
fruit-flagging theory 157
fruit predators 84–5, 93; *see also*
  frugivores
functional information 227
fungi 173; *see also* pseudoflowers
Fusarium semitectum 102, 104

G
galls 173
garden warblers *see Sylvia borin*
gas chromatography 33
generalization 9–10, 128, 173, 222,
  224, 230
generalized sensory bias 218; *see also*
  sensory bias
geographical pollinator mosaic
  hypothesis 122
geographic mosaic of co-
  evolution 14, 18, 104, 112, 209,
  219, 226, 229
grey squirrel *see Sciurus carolinensis*
grooming 57
guelder rose *see Viburnum opulus*
gulpers (versus mashers) 106–7
gustatory response 43–4

H
habitat fragmentation 63
Hakea trifurcata 157
Hamilton, W. 158
handicaps 7, 82, 118, 125, 132–3, 225
headspace analysis 38
heat reward 117
Heliamphora heterodoxa 214
Helicodiceros 180
Helicoverpa zea 199
Helleborus foetidus 14
herbivore damage toleration 185; *see
  also* tolerance
herbivores 193–4, 199
Hesperis matronalis 139
hoarding 54–6, 65
honest signal 7, 82, 118, 146, 189,
  225, 234
  between plants 192, 195, 198
  in leaf colouration 155–7, 160–1, 163
  in nectar 132
  in olfactory communication 142
honey bee

odour perception 34
hormones 94, 108, 114
host marking behaviour 115; *see also*
  chemical footprints
host morphogen hypothesis 182
hue 23–4
hummingbirds 128
husk 73, 75, 93
hybrids 133
hybrid zones 121
Hydrophyllum virginatum 165

I
ichthocory 61
index *4*, 130, 132, 142, 155, 174,
  227–8, 234
indirect defence 187, 199–202, 204
indirect defence hypothesis 163, 195
induced plant defence 144, 187–8,
  192, 198–9, 228, 234
induced reaction 192
infauna 207, 217, 232; *see also*
  inquilines
information, 1–2
innate aversion 213
innate biases 75–6, 175–6, 178, 186,
  232
innate preferences 123, 177
inquilines 207, 221, 223; *see also*
  infauna
insectivorous plants 206
inselbergs 212
intrafloral integration 129
intra-guild predators 197, 201
Ipomea viscosum 129
Ipomopsis aggregata 121, 133
Ipomopsis tenuituba 121, 133
iridiscence 30
isoprene 41
ivy-leaf morning glory *see Ipomea
  viscosum*

J
Janzen-Connell hypothesis 48
Janzen, D. 104
Japanese maple *see Acer palmatum*
jasmonate 94
  methyl jasmonate 205

K
kin grouping 193
kin-selected benefits 190
Kölreuter, J. 8

L
larder hoarding 55, 65
leaf miners 165–6

leaf senescence 149, 153, 164
leaf signal theory 158, 167
learned aversion 213
learned biases 186
learning 222
Leptinotarsa decemlineata 202
life-dinner principle 225, 230
light habitat 26
lilac aldehyde 9
linalool 33, 146
Linanthus parryae 17, 143
lipids 95, 98–9, 106
locusts 42–4
lodgepole pine *see Pinus contorta*
Loxia curvirostra 112–13
Loxonites spp. *see* elephant
luminance 23
Lycopersicon esculentum 200
Lycopersicon hirsutum 202

M
magnet effect 177–9
Maieta guianensis 203
Manduca sexta 33, 39, 136, 144, 231
market hypothesis 102, 118–20, 122,
  129
marking 119
mashers (versus gulpers) 106–7
masquerade 184
masting 66
megafauna 64
Melanerpes formicivorus 111
memory 23, 120
methyl jasmonate 205
microbes 104–5
mimicry 169, 173, 174, 181–2, 231–2,
  234
  Batesian mimicry 169, 173, 175–8,
    210–11
  brood-site mimicry 180
  floral mimicry 140, *174–180*, 210,
    217
  fruit mimicry 181, 185, 217
  imperfect mimicry 185
  Müllerian mimicry 169, 173,
    177–8, 184
  non-floral mimicry in plants 180
  of attack 204
  of carrion 123, 142, 180
  of crops by weeds 183–4
  of dung 180
  of female insects *see* orchids
  of flowers *see* pitcher plants;
    pseudoflowers
  of mates *see* orchids
  of pollen 130–1, 232
  of seeds 184